MICROWAVE COMPONENTS AND SYSTEMS

ELECTRONIC SYSTEMS ENGINEERING SERIES

Consulting Editor **E L Dagless**
University of Bristol

OTHER TITLES IN THE SERIES

Advanced Microprocessor Architectures *L Ciminiera and A Valenzano*

Modern Logic Design *D Green*

Data Communications and Computer Networks *F Halsall*

MICROWAVE COMPONENTS AND SYSTEMS

K F Sander
Emeritus Professor of Electronic Engineering,
University of Bristol

Addison-Wesley Publishing Company
Wokingham, England • Reading, Massachusetts • Menlo Park, California
New York • Don Mills, Ontario • Amsterdam • Bonn • Sydney
Singapore • Tokyo • Madrid • Bogota • Santiago • San Juan

The programs in this book have been included for their instructional value. They have been tested with care but are not guaranteed for any particular purpose. The publisher does not offer any warranties or representations, nor does it accept any liabilities with respect to the programs.

Cover design by Sampson/Tyrrell Limited.
Typeset by Advanced Filmsetters (Glasgow) Limited.
Printed and bound in Great Britain by R J Acford.

First printed 1987.

British Library Cataloguing in Publication Data

Sander, K.F.
Microwave components and systems. —
(Electronic systems engineering series).
1. Microwave devices
I. Title II. Series
621.381′3 TK7876

ISBN 0–201–14544–8

Library of Congress Cataloging in Publication Data

Sander, K. F. (Kenneth Frederick), 1921–
Microwave components and systems/K.F. Sander.
p. cm.—(Electronic systems engineering series)
Bibliography: p.
Includes index.
ISBN 0–201–14544–8
1. Microwave devices. I. Title. II. Series.
TK7876.S26 1987
621.381′3—dc1987–22876

PREFACE

This book, as one of a series on systems, was conceived as an introduction to the design of microwave systems, suitable for final year undergraduate or graduate courses on microwaves. It should also provide a useful introduction for a graduate or professional engineer who has been trained in a different discipline. Microwave engineering is a mature discipline: much of present day practice stems directly from the intensive work done during World War II, work still of great practical importance. The range of relevant subjects is wide, a fact demonstrated by the 27 volume series prepared by the staff of the MIT Radiation Laboratory summarizing wartime work. The chief problem for an author writing an introductory text is therefore one of selection. The selection for this book has been made with an eye to relevance to present day systems, rather than 'state of the art' technology which will take decades to come into widespread use.

In some contexts microwaves as a subject can seem rather an offshoot of electromagnetic theory, with its frequently forbidding mathematical apparatus. Electromagnetic theory must indeed come into the detailed design of equipment directly related to propagation or transmission, but as far as use in a system is concerned, such equipment is characterized by relatively few engineering parameters: the emphasis in this book is on the latter. An acquaintance with the elements of electromagnetic theory will be an advantage, providing a background against which transmission and propagation can be discussed. Wave propagation in transmission lines and waveguides has been formulated largely in terms of scalar transmission line theory, although for completeness outline derivations from Maxwell's equations have been included in the appendix: the text confines itself to quoting such results where relevant.

The knowledge assumed of the reader is that which might be expected of a final year undergraduate in electrical engineering: a working knowledge is assumed of circuit theory, as applied at frequencies below the microwave spectrum; mathematical demands are mainly limited to algebraic manipulation, although acquaintance with the concepts of Fourier analysis is desirable.

In certain instances wider knowledge may be presumed (as in Chapter 4 for example, where electron ballistics and semiconductor theory are relevant): it will be found, however, that such applications go no further than the immediate section in which they appear. More advanced mathematical techniques have been reserved for the Appendix.

The book is effectively in two sections: the first, Chapters 1–7, deals with the 'building blocks' of microwave systems; the second, comprising Chapters 8–10, describes and discusses radar and communication systems, giving (where practicable) descriptions of typical systems in operation. Chapter 1 is introductory; Chapter 2 may be considered as revision on transmission lines and waveguides, although it has its own point of view; Chapter 3 deals with propagation in free space and with the properties of antennas; Chapter 4 considers the most common sources of microwave power, both electron tubes and semiconductor devices; Chapter 5 covers a variety of components commonly encountered in transmission line and waveguide technology; Chapter 6 is conveniently entitled 'Microwave transistor amplifiers', although it includes much material applicable to any linear amplifier; Chapter 7 deals specifically with the superheterodyne receiver including discussion of the mixing process and of RF front ends. Chapter 8 is concerned with radar, Chapter 9 with terrestrial communications and Chapter 10 with satellite communications. In these chapters principles are described and the relevance of previous considerations to the complete system shown. Apart from Chapter 1, all these chapters finish with a selection of problems of varying difficulty, on occasion extending the subject matter of the chapter. Notes on solutions to these problems have been placed at the end of the book, sufficiently detailed to assist the student on unfamiliar ground. No system can be designed, commissioned or operated successfully without quantitative specification, backed up by precise measurement, so it was thought appropriate to include a review of methods for measuring the various quantities of interest, such as power or network parameters (Chapter 11).

The greater number of books on microwaves in publication are devoted to only one of the aspects of the subject range included here. It is hoped that this book will provide a comprehensive introduction of practical relevance which an undergraduate may be expected to assimilate.

The breadth of the subject matter covered has correspondingly brought the author into contact with many people, specialists in their own areas, who have willingly given of their time and expertise, and in some cases read all or part of the manuscript. To these and to all who have assisted in any way the author offers grateful thanks, conscious of a genuine debt of gratitude. Specifically, Messrs M. Esterson of EEV, J. D. Mason of Marconi Radar and G. Temple of Marconi Instruments, as well as their respective companies, must be thanked for advice and material; Professor J. E. Carroll of Cambridge University Engineering Department, Professor A. L. Cullen of University College London and Messrs L. Davies, A. Fairhead and R. Garnham of

RSRE undertook the labour of reading the entire manuscript, making many invaluable suggestions. What credit the book may receive will owe much to their efforts. It is a pleasure to have once again occasion to thank Mrs A. R. Tyler for her unfailing help in the preparation of the manuscript. Finally, I must thank my wife for her continuous support throughout the project.

Acknowledgements

The Publisher would like to thank the following for permission to reproduce material within this book:

Cambridge University Press for Figures 2.9, 2.10, 2.13–2.15, 5.3–5.6, 5.11, 5.17, 5.23 and 11.7(a) taken from Sander, K. F. and Reed, G. A. L., *Transmission and Propagation of Electromagnetic Waves* (1978).

Methuen & Co. for Figure 8.12 taken from Eastwood, E., *Radar Ornithology* (1967).

McGraw-Hill Book Company for Figure 8.28 from Skolnik, M., *Introduction to Radar Systems* (1962).

MacMillan Publishers for Figures 9.14 and 9.15 from Picquenard, A., *Radio Wave Propagation* (1974).

Edward Arnold for Figures 4.17, 4.18 and 4.24 from Carroll, J. E., *Hot Electron Microwave Generators* (1970)

English Electric Valve Company for Figures 4.6, 4.9(a), 4.15 and 8.21, and accompanying information.

British Telecom for Plates 1 and 3.

Figures 8.29–8.32 have been drawn from information supplied by Marconi Radar Ltd who also supplied the photographs of Plate 2.

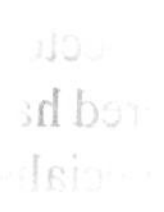

CONTENTS

LIST OF ABBREVIATIONS

The table below lists the more important abbreviations and acronyms used in the text, with their expanded form.

ADT	Automatic detection and tracking
AM/PM conversion	Amplitude modulation to phase modulation conversion
ATR	Anti transmit-receive cells
BWO	Backward wave oscillator
CFA	Crossed-fields amplifier
EIRP	Effective isotropic radiated power
FDM	Frequency division multiplex
FET	Field-effect transistor
FM	Frequency modulation
IF	Intermediate frequency
IMPATT	Impact avalanche and transit time (diode)
Maser	Microwave amplification by stimulated emission of radiation
MTI	Moving target indicator
PPI	Plan position indicator
prf	Pulse repetition frequency
PSD	Phase sensitive detector
PSK	Phase shift keying
QPSK	Quadrature phase shift keying
RF	Radio frequency
SAR	Synthetic aperture radar
SAW	Surface acoustic wave
STC	Sensitivity time control
TDM	Time division multiplex
TE	Transverse electric (wave)
TED	Transferred electron device
TEM	Transverse electric and magnetic (wave)
TM	Transverse magnetic (wave)

TR	Transmit receive (cell)
TWT	Travelling wave tube
TWTA	Travelling wave tube amplifier
VCO	Voltage controlled oscillator
VSWR	Voltage standing wave ratio
YIG	Yttrium iron garnet (ferrite)

LIST OF SYMBOLS

Lower case:

*	conjugate complex
$\boldsymbol{a}_x, \boldsymbol{a}_y$ etc.	unit vector in specified coordinate direction
a, b	complex wave amplitudes lengths
b	normalized susceptance
c	in general, velocity of plane electromagnetic waves in medium of constants μ, ε, $(\mu\varepsilon)^{-\frac{1}{2}}$; in particular, velocity of light in free space, $(\mu_0\varepsilon_0)^{-\frac{1}{2}}$
d	distance
$\boldsymbol{e}$	modal pattern of electric field in transmission line and waveguide propagation
e	electronic charge, 1.6×10^{-19} C
f	frequency
g	normalized conductance
h	Planck's constant, 6.63×10^{-34} J s
h	length
$h(t)$	impulse response
i	normalized current
j	imaginary
i, j	integer indices
k	Boltzmann's constant, 1.38×10^{-23} J K^{-1} propagation constant for waves in free space, ω/c
l	length integer index
m	integer index electronic mass, 9.1×10^{-31} kg

n	integer index refractive index electron number density
p	power flux density (Wm^{-2})
$\mathbf{r}, r$	radius vector
r	normalized resistance
s	a variable
$s(t)$	pulse shape, Fourier transform pair $S(f)$
s_{mn}	zero of Bessel function derivative $J'_m(u)$
t	time, a variable
t_{mn}	zero of Bessel function $J_m(u)$
u, v	variables real and imaginary parts of reflection coefficient velocities
v	normalized voltage
w	length
$w(f)$	power spectral density
x, y, z	Cartesian coordinates
x	normalized reactance
y	normalized admittance
z	normalized impedance

Upper case:

A	constant effective receiving area of antenna
B	constant bandwidth
$\mathbf{B}, B$	magnetic flux density
C	capacitance capacitance per unit length
$\mathscr{C}$	Pierce's constant in travelling wave tube theory
$\mathbf{E}, E$	electric field strength
E	energy
$\mathbf{E}_t$	component of electric field strength transverse to direction of propagation
F	noise figure
$H(f)$	transfer function Fourier transform pair of $h(t)$

$\boldsymbol{H}, H$	magnetic field strength
G	amplifier gain antenna power gain
I	current
J	current density
L	inductance inductance per unit length
M	noise measure
P	power
R	resistance
R_s	skin resistance (ohms per square)
R_a	antenna radiation resistance
S	voltage standing wave ratio, defined > 1
$S(f)$	Fourier transform pair of $s(t)$
S_{ij}	components of scattering matrix
T	time interval
V	potential function voltage
W	stored energy
X	reactance parameter in klystron theory
X_a	antenna radiation reactance
Y	admittance
Z	impedance
Z_0	characteristic impedance of transmission line
Z_g	wave impedance

Greek letters:

α	attenuation constant ionization rate coefficient
β	phase constant
α, β	angles
δ	loss angle of dielectric
ε	permittivity small quantity
ε_r	relative permittivity (dielectric constant)
ε_0	primary electric constant, $8.854 \times 10^{-12}\,\mathrm{C\,m^{-1}}$

γ	propagation constant ($= \alpha + j\beta$)
κ	constant
λ	free space wavelength, c/f
λ_g	guide wavelength, $2\pi/\beta$
ξ, η	coordinate variables
η	wave impedance, $(\mu/\varepsilon)^{\frac{1}{2}}$
μ	permeability
μ_0	primary magnetic constant, $4\pi \times 10^{-7}\,\mathrm{H\,m^{-1}}$
ω	angular frequency
ρ	radius in cylindrical polar coordinates charge density
σ	conductivity
τ	time time interval
θ	polar angle in spherical polar coordinates angle
ϕ	azimuthal angle in spherical polar coordinates angle
ψ	angle potential function
Γ	reflection coefficient ($= u + jv$)
Ω	solid angle

CHAPTER 1

THE BASICS OF MICROWAVE SYSTEMS

OBJECTIVES

This chapter considers the possible overall configuration of a microwave system. Three main segments, transmission, propagation and reception are identified to form the matter for detailed treatment in succeeding chapters.

1.1 Microwaves

The term **microwave** is used to denote that part of the electromagnetic spectrum for which the **free space wavelength** is less than about 0.5 m, extending into the region of millimetric wavelengths. In terms of frequency the coverage is about 0.5 GHz to 100 GHz, and over. One characteristic of microwaves is clearly that the wavelength is, at most, of the order of (and often much less than) the dimensions of the circuits ordinarily used at lower frequencies: this is a factor that will clearly influence circuit design. Another characteristic is that it becomes possible to consider radiation at such wavelengths in terms of quasi-optical behaviour. The analogy is useful: structures of many wavelengths in dimension become possible, although it must not be forgotten that a wavelength of 0.1 m is some 10^5 wavelengths of visible light. It is the ability to form well defined beams of radiation that has made the use of such frequencies attractive for a range of purposes, coupled with the wide bandwidths available as communication channels on modulation of such high frequency carriers.

For convenience various regions of the microwave spectrum have been given internationally recognized alphabetical designation. This is shown in Section A.1.

Although diverse applications of microwaves exist they have certain aspects in common: there will usually, although not always, be a transmitter at a suitable power level feeding a transmitting antenna; the radiated waves will traverse a medium before falling on a receiving antenna; and finally the (usually) low-level signal received will require amplification and processing for display. The 'receive only' part of the system would be relevant to the measurement of radiation coming from natural sources as in radio astronomy, for example. These three aspects – power sources, propagation and reception – will be briefly considered here, before more detailed treatment in succeeding chapters.

1.2 Power sources

What may be termed conventional oscillators of the type used at lower frequencies can be used in the microwave spectrum provided suitable transistors and circuit configuration are used. Bipolar silicon transistors are usable up to about 4 GHz; gallium arsenide field effect transistors up to about 10 GHz. (These limits are continually being extended by improvements in materials and manufacturing techniques.)

Transistors such as these are a comparatively recent development. Formerly conventional devices, such as vacuum triodes, would not work at these high frequencies, and a number of devices using different methods of electron-wave interaction were developed, such as the klystron, magnetron and travelling wave tube. These devices still have their place in microwave systems, performing tasks not possible to solid-state devices. In particular the need for high power can only be met using these devices (the peak powers required in radar systems for example are of the order of megawatts). Despite the potential failure problems in thermionic vacuum devices, many of the power amplifiers flown in satellites use travelling wave tubes.

It is important to realize that as the frequency increases it becomes more difficult to generate a given power. In general terms, Pf^2 is constant for a device.

1.3 Propagation

Power from the transmitter is fed to an antenna, which is designed to have directional properties appropriate to the application. The distances involved in connecting the transmitter and its antenna (and also receiver to its antenna) are likely to be many wavelengths, so transmission lines or waveguides must be used. This applies to most interconnections between and within circuits, so an understanding of wave propagation on transmission lines and waveguides is necessary.

The half-wave dipole used as a primary radiator at lower frequencies becomes of limited use as its size decreases. Most antennas will use a reflector irradiated from a primary feed antenna. Although a dipole might be used for this purpose, waveguide feeds are usually more suitable. Apart from the power levels involved, the receiving antenna may be identical, and indeed the same antenna may be used for both transmission and reception.

The medium in which propagation between antennas most often takes place is the atmosphere. To a first approximation this may be considered as free space, and the characteristics of an antenna radiating into free space will apply. In detail however the small but finite refractive index of the atmosphere affects propagation between earth stations, and it is necessary to take these effects into account. For 'point to point' transmission the ray bending caused is of the same order of magnitude as the apparent ray bending resulting from the finite radius of the earth. In addition to ray bending, which does not depend greatly on frequency, attenuation due to atmospheric gases becomes important above 10 GHz. Precipitation in the form of rain or snow can cause severe attenuation.

1.4 Reception

The energy incident on the receiving antenna will be fed via a transmission line or waveguide to a receiver. Sometimes the effects of losses in the connecting line are avoided by mounting part of the receiver very close to, or on, the antenna. This part is then referred to as a head amplifier. The receiver itself must amplify the received signal and extract the modulation carrying the information. As at lower radio frequencies, receivers are usually of the superheterodyne type, in which the modulation on the carrier is transferred to a lower, intermediate, frequency by the process of frequency changing. This takes place by mixing the signal with a local oscillator in a non-linear device, the mixer or frequency changer, from which is extracted the difference frequency. The non-linear device at microwave frequencies is a semiconductor diode: formerly a 'crystal' of silicon with a tungsten 'cat's whisker', now more usually a Schottky barrier diode. The local oscillator is required to produce perhaps 10 mW at a frequency differing from the carrier by the intermediate frequency, often 70 MHz. The gain–frequency characteristic of the receiver is then determined by suitable IF filters. Finally the modulation is extracted from the intermediate frequency signal in a detector, or demodulator. Depending on the performance required and complexity warranted, some amplification at carrier frequency may be included. In other cases the signal from the antenna is taken straight to the mixer.

1.5 Noise

Even at this early stage in the discussion of the transmission of microwaves it is desirable to introduce the concept of noise: that background level of random signals which can mask a low level wanted signal. An antenna at any frequency connected to a sensitive receiver will show such a background. Some will arise from natural sources, such as the sun (where it forms the low frequency part of a spectrum peaking in the visible) or interstellar gas clouds; some will be man made, such as is caused by sparking in ignition systems. With any electrical signal there will be associated electrical noise, and it is the ratio of signal to noise which determines the accuracy with which information in the signal can be extracted. Noise initially present with a signal will be increased by noise sources inherent in amplifiers, so that the goodness and utility of an amplifier must be judged by its effect on the signal-to-noise ratio as well as by gain. To give an order of magnitude to noise signals it is convenient to quote the formula for the noise power delivered by a resistor into an equal resistor:

$$P = kTB \tag{1.1}$$

in which T is the resistor temperature (Kelvin), B the bandwidth of the measuring circuit (Hz) and k is Boltzmann's constant, $1.38 \times 10^{-23}\,\mathrm{J\,K^{-1}}$. P is given in watts. Thus a resistor at room temperature (290 K) delivers 4×10^{-15} W into a bandwidth of 1 MHz. This may seem small but calculation will verify that for a 50 Ω resistor the corresponding rms open circuit voltage is about 1 μV, which does not seem so insignificant.

Thus in our treatment we must also be concerned with signal-to-noise ratios.

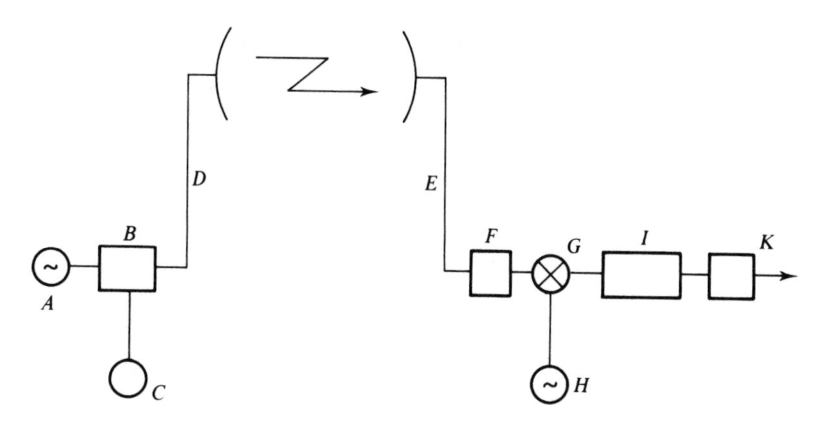

Figure 1.1 A microwave communication system: A, source of RF power; B, amplifier; C, source of modulation; D, transmitting antenna; E, receiving antenna; F, RF amplifier; G, mixer; H, local oscillator; I, intermediate frequency amplifier; K, demodulator or detector.

Summary

This chapter has briefly reviewed the important elements in a microwave system and described the basic system configuration shown in Figure 1.1. Transmitting and receiving elements have been identified, and propagation between antennas considered. The importance of electrical noise was emphasized. The various elements of the system will be considered more fully in succeeding chapters.

CHAPTER 2

GUIDED TRANSMISSION OF MICROWAVES

OBJECTIVES

This chapter considers the properties of transmission lines and waveguides used as connectors between the various parts of a microwave system. It deals with characterization in general terms, applicable to any linear guiding system, and highlights the importance of description by means of complex wave amplitudes. The results of applying electromagnetic theory to some important lines and waveguides are given in summary form. The Smith Chart is described and its use in transmission line problems illustrated. Finally, the use of scattering parameters to characterize microwave networks is described.

As discussed in Chapter 1, the connections between the various parts of a microwave system have to be transmission lines or waveguides. The fact that the wavelength is short also means that it is difficult, or even impossible, to realize simple lumped components such as inductors or capacitors, and these components are frequently replaced by short lengths of transmission line. It is therefore important to consider the theoretical and practical aspects of using such guiding structures.

It is simpler and does not affect the utility of the results to consider these structures in the first instance as lossless: in practice all are low loss, and the effects of loss can be estimated as perturbations on the lossless situation. Losses arise in this context because of the resistivity in conductors and imperfect dielectrics.

2.1 Some guiding systems

The ideal guiding system is considered to be a configuration of conductors and dielectrics of constant cross-section with respect to a linear axis. Thus, all that is necessary for description is to give this cross-section.

A range of structures is shown in Figure 2.1: (a) represents the familiar coaxial line, frequently made with a braided outer for flexibility and having the great advantage of containing the field within the outer; (b) and (c) are open wire structures; (d) is a triplate type; (e) is the widely used microstrip, made using printed circuit board techniques using a low loss substrate such as alumina; (f) is coplanar line, a variant on microstrip using only a single sided board; (g) is a completely shielded line with the inner mounted on a thin dielectric sheet; (h) and (i) are hollow metal tube waveguides; (j) is a dielectric rod, most often used as an optical waveguide. Although these structures are diverse, they have in common the ability to guide energy along the structure in the form of waves. The precise forms of the electric and magnetic fields are of great importance when it comes to considering junctions between different structures, or termination in components, for example, but for many purposes only power flow need be considered.

Before entering into further discussion about the field patterns, attention should be drawn to one vital difference between the 2-conductor structures and the waveguides. It is evident that 2-conductor structures will transmit energy at zero frequency, whereas the waveguides cannot. This transmission line wave in the case of a homogeneous dielectric medium has special properties:

- the field patterns are those applying to the static situations;
- there are no axial components of electric or magnetic field, so that voltage between lines and current in the line can unambiguously be defined;
- the wave velocity is that of electromagnetic waves in the homogeneous dielectric filling.

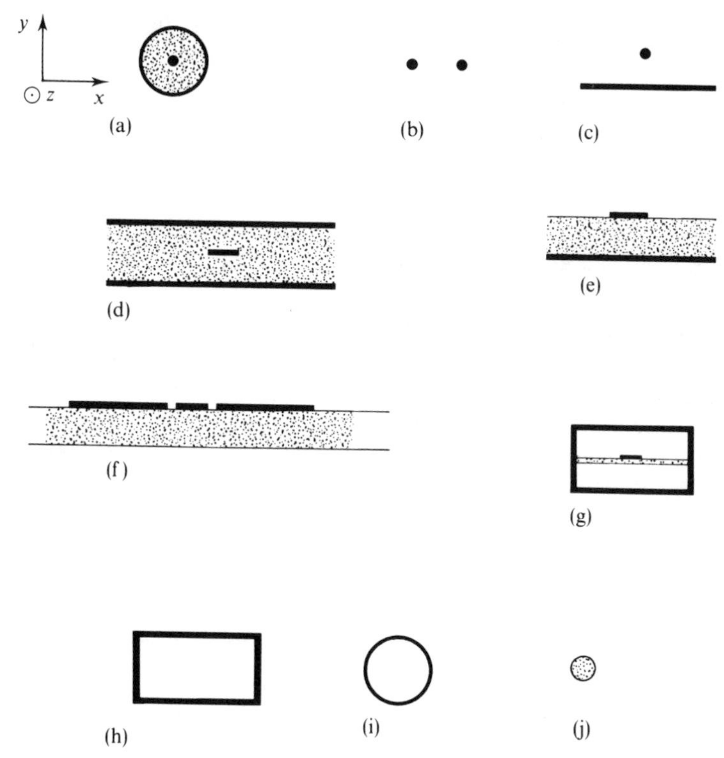

Figure 2.1 Linear guiding systems in cross-section. Heavy lines indicate conductors; stippled areas, dielectric: (a) coaxial transmission line, (b) balanced two wire pair, (c) conductor plus ground plane, (d) triplate type transmission line, (e) microstrip transmission line, (f) slot line, (g) shielded microstrip line, (h) rectangular waveguide, (i) circular waveguide, (j) dielectric waveguide or optical fibre.

It is termed the **principal wave**, and is designated **TEM (transverse electric and magnetic)**. The 2-conductor structures with inhomogeneous filling cannot strictly support a principal wave, but in practice it is usually adequate for the short lengths involved to assume a TEM wave with a suitable effective value for the dielectric constant.

It is clear that the hollow tubes and dielectric rod cannot propagate energy at low frequencies. Analysis shows that propagation commences at a frequency for which the wavelength is of the order of the transverse dimension, called the cut-off frequency for the guide.

2.2 Wave propagation

Theory shows that any structures of the type shown, considered to be of infinite extent in the axial direction, can support one or more waves for which the components of electric and magnetic field in a plane transverse to the axis (i.e. the x–y plane in Figure 2.1) are given by formulae of the form

$$\boldsymbol{E}_t = A\boldsymbol{e}(x, y)\exp(-\mathrm{j}\beta z) \tag{2.1}$$

$$\boldsymbol{H}_t = Z_g^{-1}\boldsymbol{a}_z \times \boldsymbol{E}_t \tag{2.2}$$

In these formulae $\boldsymbol{a}_z$ is the unit vector in the direction of the axis (direction of wave propagation); $\beta(\omega)$ is the **propagation constant**, assumed real, which is a function of the angular frequency ω; Z_g is a constant which has dimensions of impedance, and which is usually called the **wave impedance**. The vector function $\boldsymbol{e}(x, y)$ is real for lossless guides and specific to a particular mode of wave propagation. Cartesian coordinates are used solely to emphasize that $\boldsymbol{e}$ is independent of the axial coordinate z. (They will not in general be the correct choice for analytical purposes.) A is an arbitrary constant. In addition to these transverse components there may be axial components, $\boldsymbol{E}_z$ and $\boldsymbol{H}_z$, of the total electric and magnetic field. The actual analytical form for $\boldsymbol{e}(x, y)$ must be determined by solution of Maxwell's equations subject to the appropriate boundary conditions at surfaces of discontinuity. Some particular solutions will be given in later sections.

It should be noted that in parallel with the forward wave described by equations (2.1) and (2.2) a backward wave can propagate described by

$$\boldsymbol{E}_t = B\boldsymbol{e}(x, y)\exp(\mathrm{j}\beta z) \tag{2.3}$$

$$\boldsymbol{H}_t = -Z_g^{-1}\boldsymbol{a}_z \times \boldsymbol{E}_t \tag{2.4}$$

2.2.1 Power flow and complex amplitude

Attention has been concentrated on the transverse components of electric and magnetic fields since these determine axial power flow. Using Poynting's theorem, for a forward wave the total power flow is given by

$$\begin{aligned} P &= \tfrac{1}{2} Re \int \boldsymbol{E}_t \times \boldsymbol{H}_t^* \cdot \boldsymbol{a}_z \, \mathrm{d}S \\ &= \frac{1}{2Z_g} \int \boldsymbol{E}_t \cdot \boldsymbol{E}_t^* \, \mathrm{d}S \\ &= \frac{1}{2Z_g} |A|^2 \int \boldsymbol{e} \cdot \boldsymbol{e} \, \mathrm{d}S \end{aligned} \tag{2.5}$$

in which the integral is taken over the entire cross-section. Since in many instances it is only power flow that is of interest, a forward wave may conveniently be characterized by a **complex wave amplitude**, a, such that

$$P = \tfrac{1}{2}|a|^2 \tag{2.6}$$

We may formally identify

$$a = [\langle e^2 \rangle / Z_g]^{\frac{1}{2}} A \exp(-j\beta z)$$

where

$$\langle e^2 \rangle = \iint e^2(x, y)\, dx\, dy$$

Similarly a complex amplitude, b, can be used to describe the backward propagation wave of equations (2.3) and (2.4). The power flow will then be given by

$$P = \tfrac{1}{2}[|a|^2 - |b|^2] \tag{2.7}$$

2.2.2 Guide wavelength, phase and group velocity

For a single frequency wave travelling in one direction along a waveguide filled with uniform dielectric, the spatial periodicity or **guide wavelength**, is given by

$$\lambda_g = \frac{2\pi}{\beta} \tag{2.8}$$

In general, this is not equal to the wavelength of a plane wave of the same frequency in an unbounded medium with the same permittivity and permeability as in the waveguide. This is given by

$$\lambda = \frac{c}{f} \quad \text{where} \quad c = (\mu\varepsilon)^{-\frac{1}{2}}$$

The symbol c will be used to denote this quantity, and will not be reserved solely for the case when the dielectric medium is free space.

Since we have to envisage the phasor expressions of equations (2.1) and (2.2) multiplied by $\exp(j\omega t)$, a surface of constant phase is defined by a constant value of $(\omega t - \beta z)$. Such a surface moves with the phase velocity

$$c_p = \frac{\omega}{\beta} \tag{2.9}$$

If an amplitude modulated wave of the form $(1 + m\cos pt)\exp(j\omega t)$, where $p \ll \omega$, passes along the guide, it can simply be shown that the maxima travel with the group velocity

$$c_g = \frac{d\omega}{d\beta} \tag{2.10}$$

For a TEM wave $\beta = \omega/c$, so that if c (i.e. μ and ε) is independent of frequency $c_p = c_g = c$. If a structure does not support a TEM wave, the phase and group velocities will not be equal and will vary with frequency. The structure is then said to be **dispersive**. Dispersion will also arise when the material properties μ and ε vary with frequency.

2.3 Transmission lines

In a 2-conductor system a wave exists which is TEM, with

$$\left.\begin{aligned} \beta &= \omega/c \\ Z_g &= (\mu/\varepsilon)^{\frac{1}{2}} = \eta \end{aligned}\right\} \tag{2.11}$$

in which μ and ε refer to the dielectric medium. (From here, the symbol η will be used to denote $(\mu/\varepsilon)^{\frac{1}{2}}$ in which μ and ε refer to the dielectric medium.) Further, it is proper to speak of a voltage between conductors in any given plane z constant. The voltage is given for a forward wave by

$$V_+(z) = -\int_1^2 \boldsymbol{E}_t \cdot d\boldsymbol{l} = -A\exp(-j\beta z)\int_1^2 \boldsymbol{e} \cdot d\boldsymbol{l}$$

the integral being taken along any path between the conductors in the plane. Using this equation to eliminate A from equation (2.5) we have

$$P = \tfrac{1}{2}|V_+|^2 Z_0^{-1} \tag{2.12}$$

where

$$Z_0 = \eta\left(\int_1^2 \boldsymbol{e} \cdot d\boldsymbol{l}\right)^2 \Big/ \langle e^2 \rangle \tag{2.13}$$

Equation (2.12) expresses the power in the forward wave in terms of voltage across a resistance Z_0, for which an expression is given by equation (2.13). This resistance is termed the **characteristic impedance** of the line. Comparing further with equation (2.6) we can identify

$$V_+ = aZ_0^{\frac{1}{2}}$$

The wave will be associated with a forward current

$$I_+ = V_+ Z_0^{-1} = aZ_0^{-\frac{1}{2}}$$

For a backward wave these two relations become

$$V_- = bZ_0^{\frac{1}{2}}$$
$$I_- = -bZ_0^{-\frac{1}{2}}$$

In the presence of both forward and backward waves

$$\left.\begin{aligned} V &= V_+ + V_- = (a+b)Z_0^{\frac{1}{2}} \\ I &= I_+ + I_- = (a-b)Z_0^{-\frac{1}{2}} \end{aligned}\right\} \tag{2.14}$$

It is often convenient to use normalized voltage and current defined by

$$\left.\begin{aligned} v &= VZ_0^{-\frac{1}{2}} = a+b \\ i &= IZ_0^{\frac{1}{2}} = a-b \end{aligned}\right\} \tag{2.15}$$

Both normalized voltage and current, as well as complex wave amplitudes have dimensions of $(\text{power})^{\frac{1}{2}}$. It will be observed that

$$P = \tfrac{1}{2}(|a|^2 - |b|^2) = \tfrac{1}{2}Re(vi^*)$$

2.3.1 Reflection coefficients

If a source is connected to an infinite transmission line of characteristic impedance Z_0, it will be as though it were connected to an impedance Z_0. Similarly, at any point along the line the finite length between that point and the source will see a termination equal to Z_0, as indicated in Figure 2.2. A line

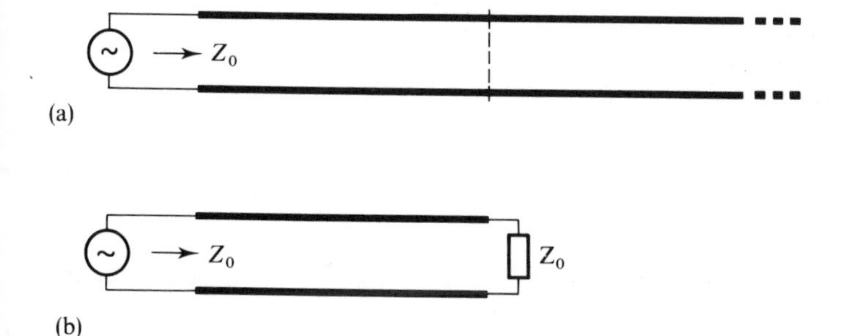

Figure 2.2 Illustrating the concept of characteristic impedance: the infinite line in (a) loads the source in exactly the same way as the terminated line in (b).

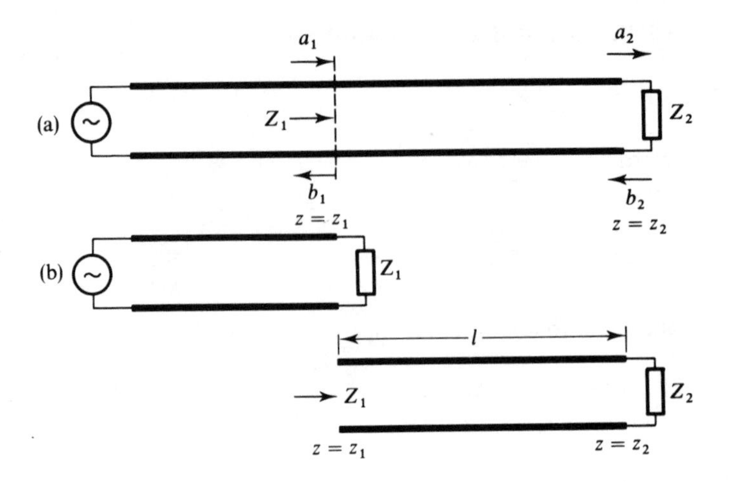

Figure 2.3 Impedance transformation through a length of transmission line: the impedance Z_2 is transformed through the line of length l to Z_1. Conditions between source and z_1 are the same the same in both (a) and (b).

terminated in Z_0 is said to be **matched**. More generally any load that does not produce a reflection is said to be a **matched load**. Any other load will produce a reflection. Consider the load to be at $z = z_2$, where the complex wave amplitudes will be a_2, b_2. This is shown in Figure 2.3. The reflection coefficient at an arbitrary point z is defined by the ratio

$$\Gamma = b/a \tag{2.16}$$

The complex amplitudes at a point z_1 between the source and z_2 will be related to a_2, b_2 by

$$a_2 = a_1 \exp\{-j\beta(z_2 - z_1)\}$$
$$b_1 = b_2\{\exp -j\beta(z_2 - z_1)\}$$

Thus a reflection coefficient, Γ_1, may be defined at z_1, related to Γ_2 at z_2, by

$$\Gamma_1 = \frac{b_1}{a_1} = \Gamma_2 \exp\{-2j\beta(z_2 - z_1)\} \tag{2.17}$$

Through equations (2.15) the reflection coefficient may be associated with an impedance Z_2 defined by

$$\frac{Z_2}{Z_0} = \frac{v_2}{i_2} = \frac{a_2 + b_2}{a_2 - b_2} = \frac{1 + \Gamma_2}{1 - \Gamma_2} \tag{2.18}$$

An impedance Z_2 connected at z_2 will give rise to the reflection coefficient Γ_2. Similarly the reflection coefficient Γ_1 at z_1 would be caused by an impedance Z_1 given by

$$\frac{Z_1}{Z_0} = \frac{1 + \Gamma_1}{1 - \Gamma_1}$$

The length of transmission line between z_1 and z_2 may be said to have transformed the impedance Z_2 to Z_1. Eliminating Γ_1 and Γ_2 between these equations yields the explicit formula

$$Z_1 = Z_0 \frac{Z_2 + jZ_0 \tan \beta l}{Z_0 + jZ_2 \tan \beta l} \tag{2.19}$$

where $l = z_2 - z_1$.

2.3.2 Open and short circuits

If we put $Z_2 = 0$ in equation (2.19) we find

$$Z_1 = jZ_0 \tan \beta l$$

Using the expression of equation (2.11) for β this becomes

$$Z_1 = jZ_0 \tan (\omega l/c) \tag{2.20}$$

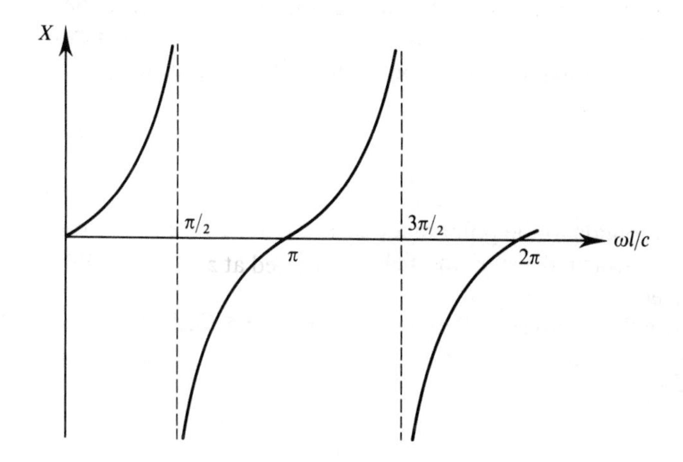

Figure 2.4 Variation of input reactance of a length l of line short-circuited at the far end. The abscissa is $\omega l/c = 2\pi l/\lambda$. Parallel resonance occurs when $l = \lambda/4$, series resonance when $l = \lambda/2$.

For $\omega l \ll c$

$$Z_1 \simeq jZ_0\omega l/c$$

Thus a loop comprising a short length of transmission line appears to have an inductance L per unit length where

$$L = Z_0/c \tag{2.21}$$

In terms of wavelength, the restriction on l is

$$l \ll \lambda/(2\pi)$$

The variation of reactance exhibited by equation (2.20) with frequency is indicated in Figure 2.4.

For an open circuit at z_2, $Z_2 \rightarrow \infty$, in which case

$$Z_1 = -jZ_0 \cot \beta l = -jZ_0 \cot(\omega l/c) \tag{2.22}$$

If $\omega l \ll c$

$$Z_1 \simeq cZ_0/(j\omega l)$$

which represents a capacitance per unit length given by

$$C = (cZ_0)^{-1} \tag{2.23}$$

It is frequently convenient either to measure or to calculate the inductance and capacity per unit length (which can be done at low frequency or DC) and then to derive values for velocity and characteristic impedance in the forms

$$\left.\begin{aligned} c &= (LC)^{-\frac{1}{2}} \\ Z_0 &= (L/C)^{\frac{1}{2}} \end{aligned}\right\} \tag{2.24}$$

This corresponds to the point made earlier that in a transmission line propagating the principal wave, the field configurations are the same as for the static cases.

It will be observed that the reactance specified by equation (2.22) is given by the graph of Figure 2.4, provided that the origin of the abscissa is displaced by $\pi/2$.

2.4 Waveguides

Whereas it is possible to relate the propagating field in a transmission line to the static field, no such comparison is possible with a waveguide: a static field

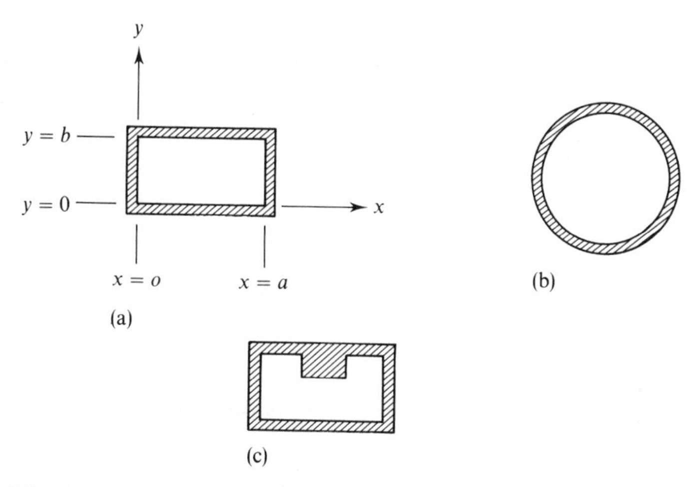

Figure 2.5 Some waveguide cross-sections: (a) rectangular, (b) circular, (c) ridged guide.

near the mouth of a hollow tube will merely decay into the tube. The waveguides that immediately spring to mind are the tubes of rectangular and circular cross-section as indicated in Figure 2.5(a) and (b). However, propagation can take place with any section, such as the ridged waveguide of Figure 2.5(c). Waveguide modes can exist inside a coaxial line as well as the principal wave, a factor which limits the frequency at which a coaxial line of given size may be used. Application of Maxwell's equations (Section A.2) shows that a waveguide mode in a hollow tube necessarily has an axial component of either electric or magnetic field. Fields are therefore divided into two classes: **transverse electric** (TE), for which the axial electric field is zero, and **transverse magnetic** (TM) for which the axial magnetic field is zero. With a dielectric rod waveguide each mode requires both electric and magnetic axial fields, and the modes are then termed **hybrid**.

2.4.1 Modes

Modes are characterized by two integers, so that a mode is classed in the form TE_{mn} or TM_{mn}. The indices relate to the pattern $e_{mn}(x, y)$ of equation (2.1) as well as the corresponding expression for β_{mn}. Each mode has a cut-off frequency, f_{mn}, which in general becomes higher as the field pattern becomes more complicated. The propagation constant is related to the cut-off frequency by the equation

$$\beta_{mn} = \frac{2\pi}{c}(f^2 - f_{mn}^2)^{\frac{1}{2}} \quad c = (\mu\varepsilon)^{-\frac{1}{2}} \tag{2.25}$$

in which f is frequency and μ and ε refer to (homogeneous) dielectric filling. Frequencies above f_{mn} propagate with flow of energy. For frequencies below

f_{mn}, β becomes imaginary and the wave is attenuated with amplitude proportional to $\exp(-\alpha z)$ where $\alpha = |\beta|$. Such a wave is said to be **evanescent**. For TE modes the wave impedance is given by

$$Z_g = \omega\mu/\beta = \eta\lambda_g/\lambda \tag{2.26}$$

For TM modes

$$Z_g = \beta/\omega\varepsilon = \eta\lambda/\lambda_g \tag{2.27}$$

(λ_g and λ were defined in Section 2.2.2.)

It is frequently convenient to express β and λ_g in terms of the 'free space' wavelength, λ_{mn}, corresponding to the cut-off frequency. Thus

$$\left.\begin{aligned} \lambda_{mn} &= c/f_{mn} \\ \beta_{mn} &= 2\pi\left(\frac{1}{\lambda^2} - \frac{1}{\lambda_{mn}^2}\right)^{\frac{1}{2}} \\ \frac{1}{\lambda_g^2} &= \frac{1}{\lambda^2} - \frac{1}{\lambda_{mn}^2} \end{aligned}\right\} \tag{2.28}$$

Using equations (2.9) and (2.10) it can be shown that the phase velocity exceeds c and the group velocity is less than c. Phase and group velocities are often displayed graphically on an ω–β diagram, as shown in Figure 2.6.

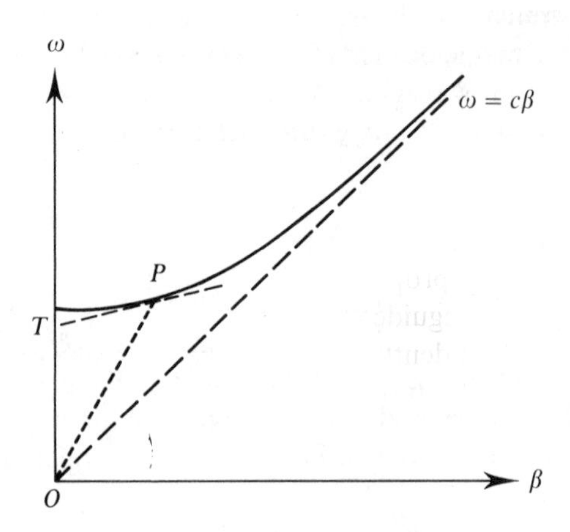

Figure 2.6 The ω–β diagram for waveguide transmission. The point P represents a particular frequency: the phase velocity, c_p, is the slope of the line OP; the group velocity, c_g, is the slope of the curve at P, given by the tangent TP. At very high frequencies propagation becomes non-dispersive with the free space velocity.

2.4.2 Degeneracy and the dominant mode

The cut-off frequencies for the various modes of a particular waveguide can be arranged in ascending order of magnitude: it may so happen that two or more modes have the same cut-off frequency. Such modes are referred to as **degenerate**. If the lowest mode is not degenerate, excitation of the waveguide at a frequency above the lowest cut-off but less than the second highest can only produce one propagating mode: other modes which may be initially produced will be attenuated along the waveguide, leaving only the lowest mode propagating. The lowest mode is then said to be **dominant**. Propagation in the dominant mode can be compared with the principal wave on a transmission line, in that only this mode can propagate at the frequency of use.

Although propagation in waveguides is clearly different from the principal wave in a transmission line, it equally accords with equations (2.1) and (2.2) and thus can be expressed in terms of complex wave amplitudes. A discontinuity in a waveguide, or a termination will produce reflections composed of many modes, but if the frequency is such that only the dominant mode propagates, then all the other modes evanesce rapidly away from the discontinuity, leaving the same mode as the input. A reflection coefficient can then be defined as in equation (2.16). Clearly such a reflection cannot be attributed to load impedance in the transmission line sense, but the right-hand side of equation (2.18) nevertheless can be evaluated. The value thus obtained allows us to ascribe to the load in association with the specific waveguide a normalized impedance (for which Z_0 is not defined). Further, although voltage and current do not have a direct significance in a waveguide, the normalized voltage and current can be defined. Because of these equivalences, we frequently use for waveguide circuits the same methods and language used for analysing transmission line circuits. The use of complex wave amplitudes and power flow covers both situations.

2.4.3 Mode coupling

It is sometimes necessary to operate a waveguide at a frequency for which more than one mode can propagate: the power in each mode will depend on the way in which the waveguide is excited. In an ideal waveguide such modes would propagate independently, but in a real waveguide irregularities in the walls cause **mode coupling** so that power initially confined to one mode may be shared among all possible modes after a distance along the waveguide. It is necessary to avoid this situation or to incorporate mode filters.

2.5 Losses

The above considerations have concerned lossless systems. The effect of losses is to dissipate as heat some of the electromagnetic energy flowing along

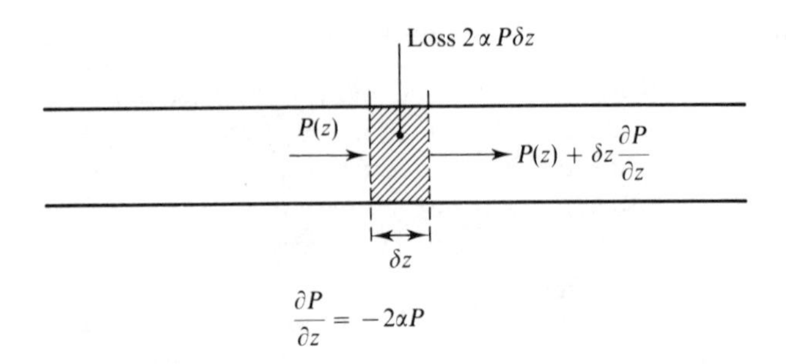

Figure 2.7 Illustrating the calculation of attenuation constant by considering power flow.

the transmission line or waveguide. Consider a short length, δz, of waveguide carrying power P, as in Figure 2.7. The lost power in the length δz will be proportional to the power, and may be expressed in the form $2\alpha P\delta z$. Balancing the loss against decrease in power flow we have

$$\frac{\mathrm{d}P}{\mathrm{d}z} = -2\alpha P$$

$$P = P_0 \exp(-2\alpha z) \tag{2.29}$$

The amplitude of the wave carrying power P will be proportional to $P^{\frac{1}{2}}$ and hence $\exp(-\alpha z)$. The quantity α is the **attenuation constant** and in the form given will be measured in nepers m^{-1}. The attenuation constant is usually combined with the propagation constant β to give a complex propagation constant denoted by

$$\gamma = \alpha + \mathrm{j}\beta$$

Practically, α is expressed in $\mathrm{dB\,m^{-1}}$, given by

$$\alpha_{\mathrm{dB}} = 10\log_{10}[\exp(2\alpha)] = 8.686\,\alpha\,\mathrm{dB\,m^{-1}}$$

Losses arise from two sources: imperfect dielectric and finite wall conductivity. The former is specified by the loss angle δ, the latter by the conductivity σ.

2.5.1 Dielectric loss

The advantage of hollow tube waveguides is that no dielectric filling is needed to provide mechanical support, so that we are not concerned with dielectric loss in this case. In a transmission line the effect of a lossy dielectric is simple to estimate: the mean loss per unit volume in a dielectric of permittivity ε and loss angle δ in a phasor electric field E is $\frac{1}{2}\omega\varepsilon \tan\delta\,|E|^2$. Applying this to a length δz of transmission line we find the lost power is given by

$$-\delta P = \tfrac{1}{2}\omega\varepsilon \tan\delta \int \boldsymbol{E}_t \cdot \boldsymbol{E}_t \,\mathrm{d}S\,\delta z$$

in which the integration is over the cross-section.

Using equation (2.5) with $Z_g = \eta$, this can be rewritten as

$$-\delta P = \omega\varepsilon\eta \tan\delta\, P\,\delta z$$

Hence comparing with equation (2.29)

$$\alpha_d = \tfrac{1}{2}\omega\varepsilon\eta \tan\delta = \frac{\pi f}{c}\tan\delta \text{ nepers m}^{-1} \tag{2.30}$$

This result shows the important fact that, since $\tan\delta$ is for many dielectrics approximately independent of frequency, the attenuation constant arising from dielectric loss is proportional to frequency. In deriving this result it has been assumed that it is permissible to replace Z_g by its value for a lossless dielectric. Exact analysis shows that Z_g will have a small phase angle, resulting only in a second order correction.

2.5.2 Wall loss

With perfectly conducting walls, wall currents flow in the surface without dissipation. When the wall material is resistive the wall current is distributed, as indicated in Figure 2.8. This shows the current distribution in a plane wall

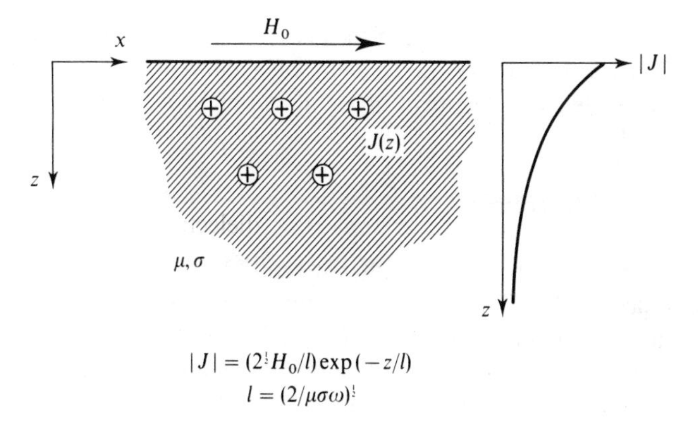

Figure 2.8 The distribution of current near the surface of a good conductor in an alternating magnetic field. The current density decays exponentially into the conductor at a rate determined by the skin depth, l. In terms of wall permeability, μ, conductivity, σ, and angular frequency, ω, $l = (2/\omega\mu\sigma)^{\frac{1}{2}}$.

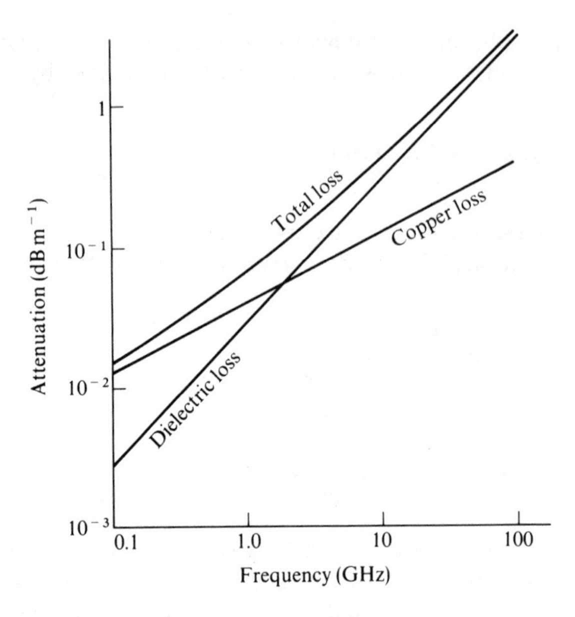

Figure 2.9 Attenuation constant for a dielectric filled coaxial line: $b = 12.7\,\text{mm}$, $b/a = 3.59$; dielectric $\varepsilon_r = 2.3$, $\mu = \mu_0$, $\tan\delta = 2 \times 10^{-4}$; copper wall, $\sigma = 5.9 \times 10^7\,\text{S m}^{-1}$.

at which a tangential magnetic field H is maintained. The parameter l is known as the **skin depth** and is given in terms of the permeability, μ, and conductivity, σ, of the wall material and angular frequency, ω, by the equation

$$l = (2/\omega\mu\sigma)^{\frac{1}{2}}$$

For copper ($\mu = \mu_0$, $\sigma = 5.9 \times 10^7\,\text{S m}^{-1}$) at 1 GHz, $l = 2.1\,\mu\text{m}$. Such a small value of penetration allows an element of a curved wall, over which $\boldsymbol{H}$ varies, to be treated as an element of a uniform plane wall, with a lost power density given by

$$p = (1/2\sigma l)\boldsymbol{H}\cdot\boldsymbol{H}^* = \tfrac{1}{2}R_s|H|^2$$

it being assumed that $\boldsymbol{H}$ is a peak phasor representation. R_s is known as the **skin resistance**, and has dimensions of ohms per square. It increases with frequency as $f^{\frac{1}{2}}$. For copper at 1 GHz, $R_s = 8.2\,\text{m}\Omega$.

Wall losses in a transmission line or waveguide can then be calculated as follows: the power lost within a short length δz is equal to

$$\tfrac{1}{2}R_s \oint_C |H|^2\,\mathrm{d}l\,\delta z$$

in which C is the contour of the conducting walls in the cross-section, and H is assumed to be the same field as would be present in the lossless case. Evaluation requires a knowledge of field distribution, but because H appears as a quadratic term, the result will be proportional to P, allowing determination of a value for α in equation (2.29).

We have

$$\alpha_c = \frac{1}{4}\frac{R_s}{P}\oint |H|^2 \, \mathrm{d}l \tag{2.31}$$

The attenuation constant will depend on frequency in two ways: the one through change of field configuration as frequency changes in a given waveguide, the other because R_s is proportional to $f^{\frac{1}{2}}$.

The above theory depends on the correctness of the assumption that the field pattern in the presence of wall resistivity is only slightly different from the perfectly conducting case. When two or more modes have the same cut-off this assumption may not be true: the existence of small but finite wall resistivity may require both modes to be present. In such cases a more detailed analysis is necessary.

A transmission line is difficult to make, certainly in long lengths, without dielectric supports, so that the attenuation coefficient inevitably contains a term for which the increase is proportional to frequency as the frequency becomes high. This is not the case with hollow tube waveguides, which are mechanically self supporting, and for long runs they may well give lower attenuation than a transmission line. This is illustrated by specific examples in Figures 2.9 and 2.10.

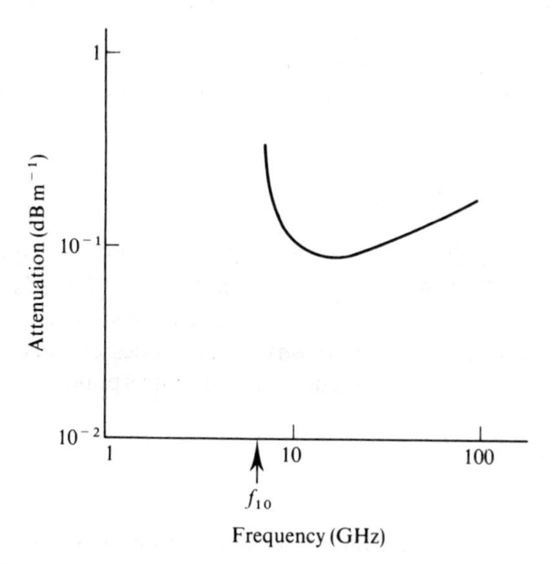

Figure 2.10 Attenuation constant for WG 16 rectangular copper waveguide: 22.9 mm × 10.2 mm, $\sigma = 5.9 \times 10^7 \mathrm{S\,m^{-1}}$.

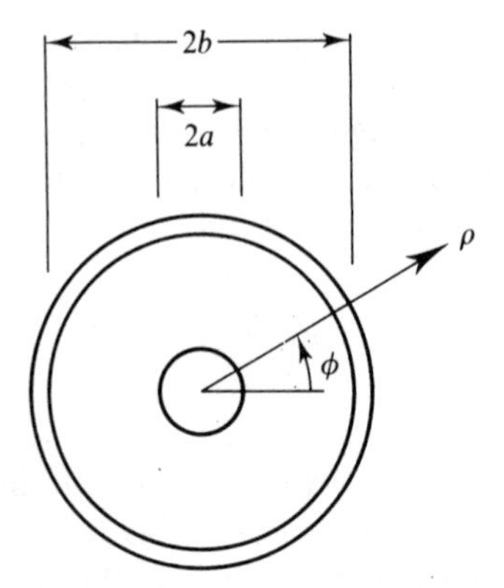

Figure 2.11 Coordinates for coaxial transmission line.

The above sections have dealt in general terms with wave propagation along transmission lines and waveguides. Particular results for some types of line and waveguide will now be given.

2.6 Coaxial transmission line

As mentioned earlier, the field configuration for the principal wave is the same as in static situations. Thus the vector function $\boldsymbol{e}$ may be conveniently taken as radial in direction and of magnitude ρ^{-1}, Figure 2.11. The propagation constant and wave impedance are given by equation (2.11). Making the appropriate substitutions the characteristic impedance and attenuation constant may be calculated. The principal formulae are derived in Section A.3, and are gathered in Table 2.1.

At fixed frequency and constant outer diameter the expression for α_c can be regarded as a function of the ratio (b/a). This function has a minimum when

$$\ln(b/a) = 1 + a/b$$

This gives $b/a \simeq 3.6$. The corresponding value of Z_0 for an air spaced line with this ratio is about 77 Ω. With polythene dielectric, $\varepsilon_r = 2.25$, $Z_0 \simeq 51\ \Omega$. Some numerical values for attenuation constant have been shown in Figure 2.9.

It has been mentioned that waveguide modes can also propagate in coaxial line. The lowest frequency cut-off corresponds to a free space wavelength

$$\lambda_c = 2.95\,(a + b)$$

The need to avoid this condition places a maximum on the values of a and b for any given frequency.

Table 2.1 Principal waves in coaxial line.

$$\boldsymbol{E}_t = A\boldsymbol{e}\exp(-\mathrm{j}\beta z)$$
$$\mathbf{H}_t = \eta^{-1}\boldsymbol{a}_z \times \boldsymbol{E}_t$$
$$e_\rho = \rho^{-1}$$
$$\langle e^2 \rangle = 2\pi \ln (b/a)$$
$$\beta = \omega/c$$
$$P = \frac{\pi}{\eta}\ln (b/a)|A|^2$$
$$Z_0 = \frac{\eta}{2\pi}\ln (b/a)$$
$$|E_\rho| = \frac{1}{\rho}\left(\frac{\eta P}{\pi \ln (b/a)}\right)^{\frac{1}{2}} = \frac{\eta}{\pi\rho}\left(\frac{P}{2Z_0}\right)^{\frac{1}{2}}$$
$$\alpha_c = \frac{1}{2}\frac{R_s}{\eta b}\left(1 + \frac{b}{a}\right)\Big/\ln (b/a) \quad R_s = (\omega\mu/2\sigma)^{\frac{1}{2}}$$
$$\alpha_d = \frac{\pi f}{c}\tan\delta$$

2.7 Microstrip

The convenience of printed circuit board techniques for transmission lines in microwave circuits has been mentioned, and Figure 2.1(d), (e), (f) and (g) show some configurations. Although the triplate type of Figure 2.1(d) is a true transmission line, there are practical difficulties of fabrication in that two regions of dielectric have to be bonded. The other types cannot support a true TEM wave, and propagation modes are hybrid, involving both TE and TM. However, theory predicts and experiment confirms that with the lengths and dimensions used propagation is effectively TEM up to frequencies of a few gigahertz. In choosing materials for constructing lines such as microstrip, many factors are involved. The dielectric substrate must have suitable mechanical properties, it must not warp during processing, must take a polished surface and must give good adherence to conductors. Electrically it must be low loss, uniform and temperature stable. Its thermal conductivity will control the rate at which heat can be removed from circuit devices. Likewise the conductors must be capable of close definition, usually by photolithography and etching, have good adherence, be smooth and of good conductivity.

Some typical substrate materials which have been used are shown in Table 2.2.

A discussion of these points is given in Sobol (1971). The general

Table 2.2 Substrate materials.

Material	Loss tangent at 10 GHz	Relative permittivity
Alumina	2.2×10^{-4}	9.6–9.9
Sapphire	10^{-4}	9.3–11.7
Teflon glass laminate	7×10^{-4}	2.2

principles by which the electrical parameters of these lines may be determined will be illustrated for the standard microstrip shown in Figure 2.12.

2.7.1 Microstrip line

It is convenient to use the definitions given in equations (2.24) for characteristic impedance and velocity of propagation in terms of inductance and capacity per unit length. These latter quantities may be determined by calculation or measurement. For non-magnetic dielectric substrates, the inductance is the same as for a completely air-dielectric line with the same conductors, which may be denoted by L_a. The capacity per unit length for an air line would be given by

$$C_a = 1/c^2 L_a \tag{2.32}$$

in which c is the velocity of light *in vacuo*.

The capacity per unit length, C, of the real line will be greater than C_a, and an effective dielectric constant ε_{rf} may be defined by

$$\varepsilon_{rf} = C/C_a \tag{2.33}$$

We also have the relationships

$$\left.\begin{aligned} Z_{0f} &= \varepsilon_{rf}^{-\frac{1}{2}} Z_a \\ c_f &= c\varepsilon_{rf}^{-\frac{1}{2}} \\ \lambda_f &= \varepsilon_{rf}^{-\frac{1}{2}} \lambda \end{aligned}\right\} \tag{2.34}$$

in which $Z_a = (L_a/C_a)^{\frac{1}{2}}$ and λ is the free space wavelength. Both Z_{0f} and ε_{rf} depend on the geometry as well as the permittivity of the substrate. Standard microstrip, Figure 2.12, has received much attention, yielding formulae such as those given in Schneider (1969), Wheeler (1977) or Edwards (1981) and examples of which are quoted below:

$$\left.\begin{aligned} Z_a &= 60 \ln (h/w + w/4h) \quad w/h < 1 \\ \varepsilon_{rf} &= \tfrac{1}{2}(\varepsilon_r + 1) + \tfrac{1}{2}(\varepsilon_r - 1)(1 + 10h/w)^{-\frac{1}{2}} \end{aligned}\right\} \tag{2.35}$$

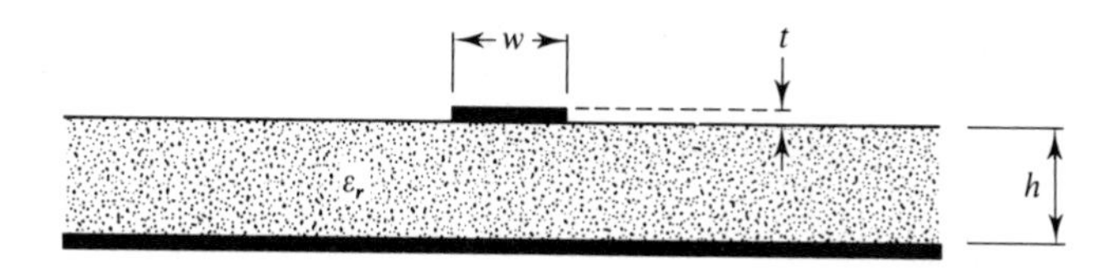

Figure 2.12 Notation for microstrip line.

It is necessary to appreciate that higher order modes may propagate at high frequencies. As a guide this may occur when the substrate is of the order of one-quarter wavelength in thickness.

Losses in microstrip arise from dielectric loss, which gives a contribution to the attenuation coefficient in the same way as in, for example, coaxial lines, and from conductor loss. A 'standard' conductor loss may be calculated as for other situations involving bulk conductors, but there are too many geometrical parameters to allow easy presentation of results, which are covered in earlier references. In addition, conductor roughness increases the 'standard' loss: the skin depth at 10 GHz in copper is of the order of 0.7 μm and in order to avoid significant extra loss a finish of rms roughness better than 0.1 μm is desirable, as noted by Schneider (1969). Losses of the order of 0.1 dB per wavelength occur at 30 GHz. Such figures apply to a smooth run of line: in a circuit junctions and bends are necessary, and such discontinuities give rise to radiation. This may effectively double the attenuation.

2.8 Rectangular waveguide

The results given in this section are derived in Section A.4. It is convenient to specify the cut-off frequencies f_{mn} introduced in Section 2.4 in terms of cut-off wavelength λ_{mn}, equations (2.28).

2.8.1 TE modes

Using axes and dimensions as specified in Figure 2.5(a), the mode vector $\boldsymbol{e}$ is given by the expressions

$$e_x = \frac{n\lambda_{mn}}{2b}\cos(m\pi x/a)\sin(n\pi y/b)$$

$$e_y = -\frac{m\lambda_{mn}}{2a}\sin(m\pi x/a)\cos(n\pi y/b)$$

in which

$$\lambda_{mn} = \left[\left(\frac{m}{2a}\right)^2 + \left(\frac{n}{2b}\right)^2\right]^{-\frac{1}{2}}$$

The axial magnetic field for this mode is given by

$$H_z = A\eta^{-1}(-j\lambda/\lambda_{mn})\cos(m\pi x/a)\cos(n\pi y/b)\exp(-j\beta z)$$

For the TE modes,

$$Z_g = \eta\lambda_g/\lambda$$

2.8.2 TM modes

Similar results apply for TM modes. The cut-off wavelengths are the same as for the TE modes.

$$e_x = -\frac{m\lambda_{mn}}{2a}\cos(m\pi x/a)\sin(n\pi y/b)$$

$$e_y = -\frac{n\lambda_{mn}}{2b}\sin(m\pi x/a)\cos(n\pi y/b)$$

$$E_z = A(-j\lambda_g/\lambda_{mn})\sin(m\pi x/a)\sin(n\pi y/b)\exp(-j\beta z)$$

In this case,

$$Z_g = \eta\lambda/\lambda_g$$

We find

$$\begin{aligned}\langle e^2\rangle &= ab/2 \quad m \text{ or } n = 0 \qquad \text{(TE only)}\\ &= ab/4 \quad m \neq 0 \quad n \neq 0 \quad \text{(TE and TM)}\end{aligned}$$

In Figure 2.13 the patterns of the transverse electric field $\boldsymbol{e}$ and transverse magnetic field $\boldsymbol{a}_z \times \boldsymbol{e}$ are shown. It is to be noted that these are not necessarily lines of force, for which the axial components must be taken into account.

2.8.3 The dominant mode

The cut-off wavelengths and frequencies for the first few values of m, n are given in Table 2.3.

Table 2.3 Cut-off wavelengths.

$\lambda_{10} = 2a$	$f_{10} = c/2a$
$\lambda_{20} = a$	$f_{20} = 2f_{10}$
$\lambda_{01} = 2b$	$f_{01} = f_{10}(a/b)$
$\lambda_{11} = 2ab/(a^2 + b^2)^{\frac{1}{2}}$	$f_{11} = f_{10}(1 + a^2/b^2)^{\frac{1}{2}}$

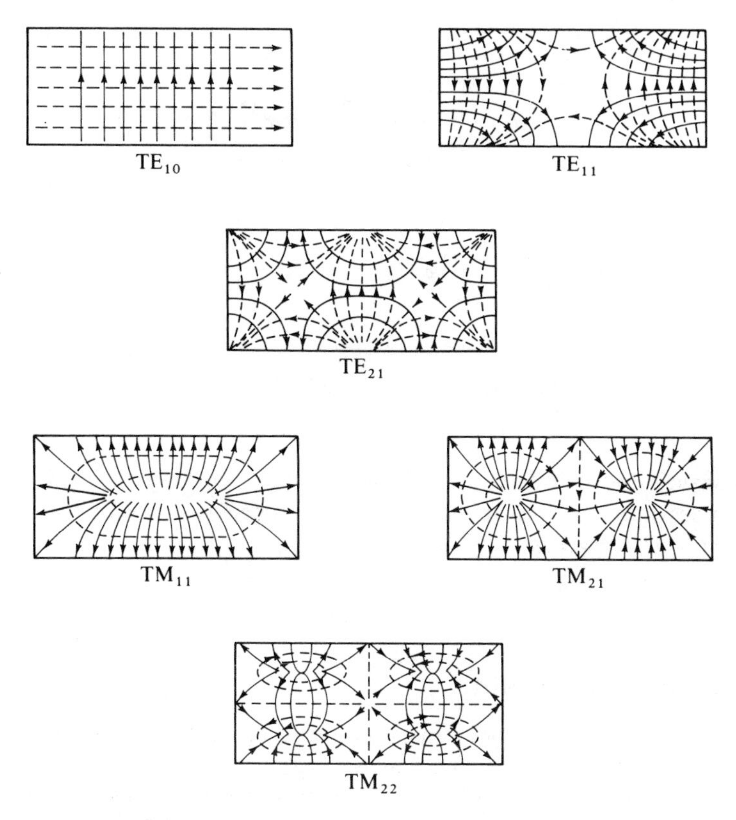

Figure 2.13 Modal patterns in rectangular waveguide (full lines, electric field; pecked lines, magnetic field).

From this it is seen that if $b < a$, of the values given f_{10} is the lowest and f_{11} the highest. It is common to choose $a = 2b$, when $f_{20} = f_{01}$. There is therefore a range of frequency in which only one mode can propagate, higher ones being evanescent. Furthermore, inspection of the equations of Section 2.8.2 shows that no TM mode exists for which either m or n is zero. The TE_{10} is therefore dominant in the range $f_{10} < f < f_{20}$.

As mentioned earlier, dominant mode propagation plays a part in power transmission equivalent to that of transmission lines. It is necessary to limit the frequency range for a given size of waveguide to ensure that only the dominant mode propagates, and hence a standard range of waveguide sizes has come into use. A selection of these is given in Section A.5. The range of frequency given is approximately $1.3f_{10} < f < 1.9f_{10}$, although due attention must be paid to the attenuation constant of the next modes at the high end of this range.

Formulae relating to the dominant mode are given in Table 2.4. The entry in Table 2.4 expressing E_y as a function of P indicates that, since a limit to the electric field exists at which breakdown will take place, the power any

Table 2.4 TE_{10} mode in rectangular waveguide.

$$e_{10} = -\boldsymbol{a}_y \sin(\pi x/a)$$
$$Z_g H_x = -E_y = A\sin(\pi x/a)\exp(-\mathrm{j}\beta z)$$
$$H_z = A(-\mathrm{j}\lambda/\lambda_{10}\eta)\cos(\pi x/a)\exp(-\mathrm{j}\beta z)$$
$$\lambda_{10} = 2a$$
$$\left(\frac{\beta}{2\pi}\right)^2 = \frac{1}{\lambda_g^2} = \frac{1}{\lambda^2} - \frac{1}{4a^2}$$
$$Z_g = \eta\lambda_g/\lambda$$
$$|E_y| = (4Z_gP/ab)^{\frac{1}{2}}\sin(\pi x/a)$$
$$\alpha_c = (R_s/b\eta)^{\frac{1}{2}}[1 + (2b/a)(\lambda/\lambda_{10})^2][1 - (\lambda/\lambda_{10})^2]^{-\frac{1}{2}}$$

given waveguide can handle is limited. When working at high power, waveguides may be pressurized with dry nitrogen to increase the field at which breakdown occurs.

2.9 Circular waveguide

The electric and magnetic fields in circular waveguide have to be expressed in terms of Bessel functions of the first kind, $J_m(u)$. To make the expressions more readily intelligible the first four of these functions are shown in the graphs of Figure 2.14. The zeros of these functions and of the first derivatives determine the cut-off frequencies. The zeros of $J_m(u)$ are denoted by t_{mn},

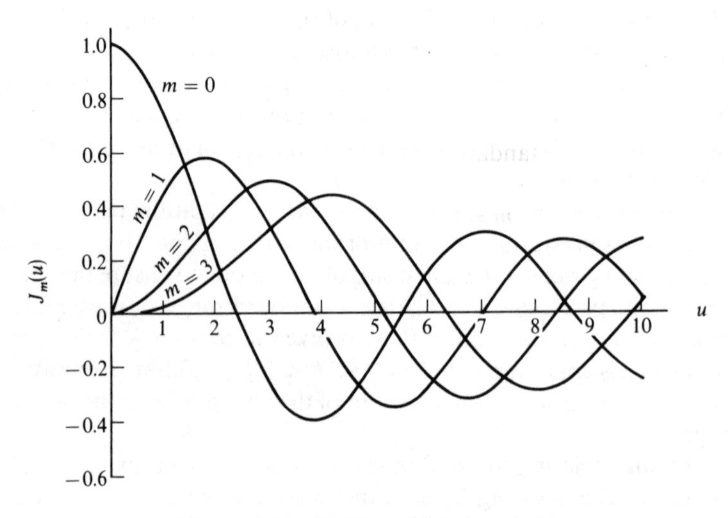

Figure 2.14 The Bessel functions $J_m(u)$.

Table 2.5 Zeros of $J_m(u)$ and $J'_m(u)$.

t_{mn}				s_{mn}			
n/m	0	1	2	n/m	0	1	2
1	2.405	3.832	5.136	1	3.832	1.841	3.054
2	5.520	7.016	8.417	2	7.016	5.331	6.706
3	8.654	10.173	11.620	3	10.173	8.536	9.969
	$J_m(t_{mn}) = 0$				$J'_m(s_{mn}) = 0$		

$n = 1, 2, \ldots$ and the zeros of the derivative $J'_m(u)$ by s_{mn}, $n = 1, 2, \ldots$ Some values are given in Table 2.5.

Derivation of the following results is given in Section A.6.

2.9.1 TE modes

The cut-off wavelengths are given in terms of the zeros s_{mn} of $J'_m(u)$, Table 2.5, by the expression

$$\lambda_{mn} = 2\pi a/s_{mn}$$

The components of the modal function $\boldsymbol{e}_{mn}$ are given by

$$e_\rho = \frac{ma}{\rho s_{mn}} J_m(s_{mn}\rho/a) \sin m\phi$$

$$e_\phi = J'_m(s_{mn}\rho/a) \cos m\phi$$

$$H_z = \eta^{-1} A(-\mathrm{j}\lambda/\lambda_{mn}) J_m(s_{mn}\rho/a) \cos m\phi \exp(-\mathrm{j}\beta z)$$

$$Z_g = \eta\lambda_g/\lambda$$

Evaluating $\langle e^2 \rangle$ involves standard Bessel integrals, and leads to

$$\langle e^2 \rangle_{\mathrm{TE}} = \pi a^2 J_0^2(s_{0n}) \quad m = 0$$

$$= \tfrac{1}{2}\pi a^2 \left(1 - \frac{m^2}{s_{mn}^2}\right) J_m^2(s_{mn}) \quad m \neq 0$$

2.9.2 TM modes

The cut-off wavelengths are given in terms of the zeros t_{mn} of $J_m(u)$, Table 2.5, by the expression

$$\lambda_{mn} = 2\pi a/t_{mn}$$

Table 2.6 Wavelength cut-offs.

Mode	$\lambda_{mn}/2\pi a$	af_{mn}(cm – GHz)
TE_{11}	0.543	8.79
TM_{01}	0.416	11.48
TE_{21}	0.327	14.58
TE_{01} } TM_{11} }	0.261	18.30
TE_{12}	0.188	25.45
TM_{02}	0.181	26.36

The components for the modal function $\boldsymbol{e}_{mn}$ are given by

$$e_\rho = -J'_m(t_{mn}\rho/a)\cos m\phi$$
$$e_\phi = (ma/\rho t_{mn})J_m(t_{mn}\rho/a)\sin m\phi$$
$$E_z = (-j\lambda_g/\lambda_{mn})AJ_m(t_{mn}\rho/a)\cos m\phi$$
$$Z_g = \eta\lambda/\lambda_g$$

Evaluation of $\langle e^2 \rangle$ gives

$$\langle e^2 \rangle_{\mathrm{TM}} = \pi a^2[J'_0(t_{0n})]^2 \quad m = 0$$
$$= \tfrac{1}{2}\pi a^2[J'_m(t_{mn})]^2 \quad m \neq 0$$

Some modal patterns for both TE and TM waves are shown in Figure 2.15.

2.9.3 The dominant mode

The cut-off frequencies for the various modes are best compared by tabulating $\lambda_{mn}/2\pi a$, the wavelength measured in terms of the circumference. This ratio is either s_{mn}^{-1} or t_{mn}^{-1}. Using Table 2.5 the cut-offs in descending order of wavelength are shown in Table 2.6 together with the product af_{mn} for an air filled waveguide.

This table shows that TE_{11} is a true dominant mode. Comparison of Figures 2.13 and 2.15 shows that there is a close similarity between the patterns of TE_{10} rectangular and TE_{11} circular. This is of use in designing transitions between waveguides of these sections.

Although TE_{11} is dominant, it is also degenerate in the sense that its orientation in the waveguide is quite arbitrary. Due to the imperfections in a real waveguide the orientation of a TE_{11} wave may change over a long run.

Formulae relating to the TE_{11} mode are given in Table 2.7 (p. 36).

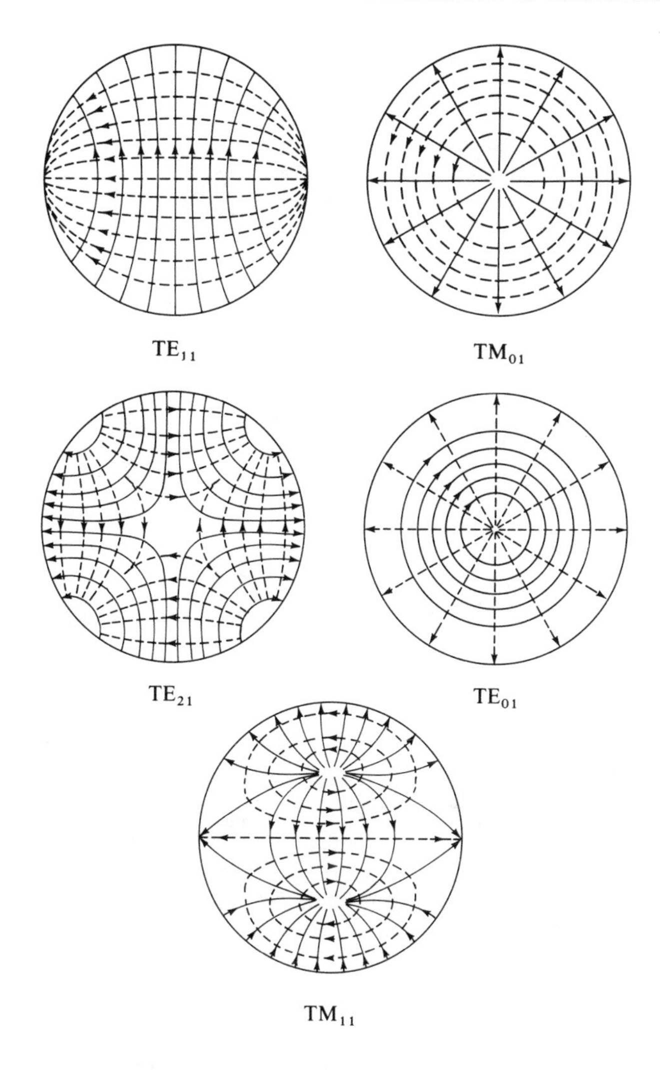

Figure 2.15 Modal patterns in circular waveguide (full lines, electric field; pecked lines, magnetic field).

2.10 Transmission line calculations: the Smith Chart

It has been shown that propagation along either transmission line or waveguide can be treated in terms of complex wave amplitudes for forward and backward waves. It has been shown further that it is possible to associate normalized voltage and current with the complex amplitudes by use of equations (2.15) and through them an arbitrary load may be specified as a normalized impedance. It has been assumed in putting forward power

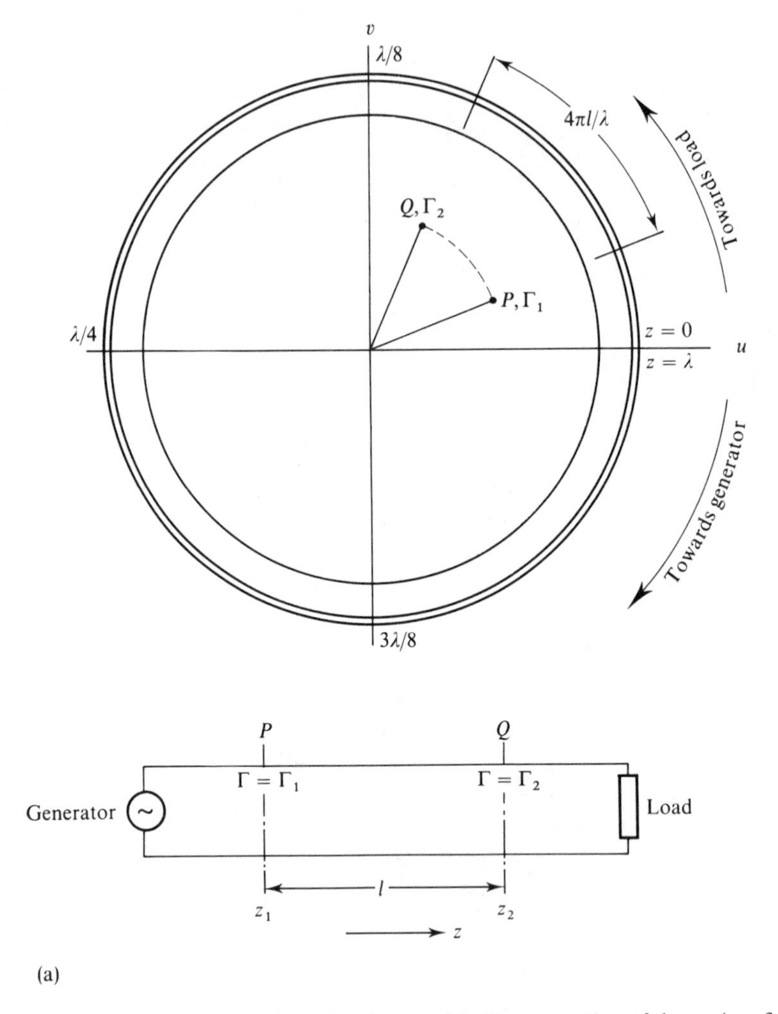

Figure 2.16 (a) The Smith Chart showing graphical interpretation of change in reflection coefficient ($\Gamma = u + \mathrm{j}v$) between points $P(z_1)$ and $Q(z_2)$ on a transmission line, $\Gamma_1 = \Gamma_2 \exp(-2\mathrm{j}\beta l) = \Gamma_2 \exp(-\mathrm{j}4\pi l/\lambda)$. (b) Impedance format of complete chart showing some loci of constant r and constant x. (c) Format for admittance chart: loci are those of the impedance chart inverted in the origin of Γ.

relationships as in equation (2.7) that the wave impedance and equivalent Z_0 are real. As long as this is true, small losses can if necessary be accommodated by use of a complex propagation constant. Transmission line or waveguide calculations are usually concerned with calculating power flow, and the effect on power flow of the various elements connected to the line. (The development of suitable components for use in waveguide is considered in Chapter 5.)

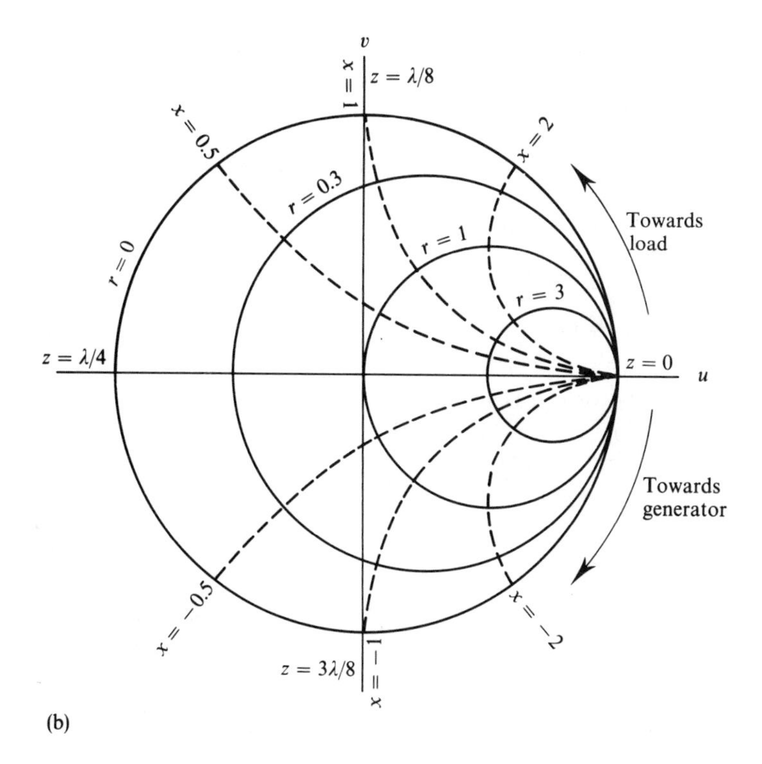
v
x = 1
z = λ/8
x = 0.5
x = 2
r = 0.3
r = 0
r = 1
r = 3
Towards load
z = λ/4
z = 0
u
Towards generator
x = −0.5
x = −2
z = 3λ/8
x = −1

(b)

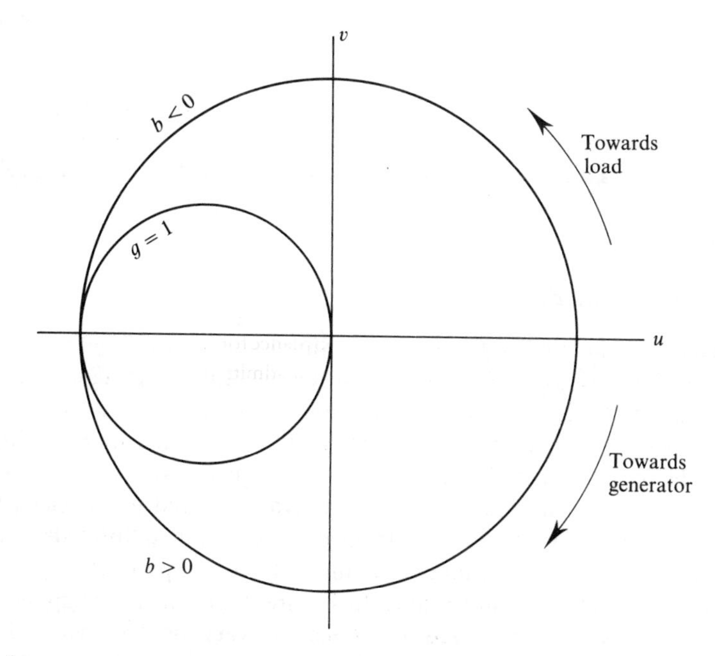
v
b < 0
g = 1
Towards load
u
Towards generator
b > 0

(c)

Table 2.7 TE_{11} mode in circular waveguide.

$$\boldsymbol{E}_t = A\boldsymbol{e}_{11}\exp(-\mathrm{j}\beta z)$$
$$\boldsymbol{H}_t = Z_g^{-1}\boldsymbol{a}_z \times \boldsymbol{E}_t$$
$$H_z = A\eta^{-1}(-\mathrm{j}\lambda/\lambda_{11})J_1(s_{11}\rho/a)\cos\phi\exp(-\mathrm{j}\beta z)$$
$$e_{11_\rho} = \frac{a}{\rho s_{11}}J_1(s_{11}\rho/a)\sin\phi$$
$$e_{11_\phi} = J_1'(s_{11}\rho/a)\cos\phi$$
$$s_{11} = 1.841$$
$$\lambda_{11} = 3.41a$$
$$\langle e_{11}\rangle^2 = 0.375a^2$$
$$\left(\frac{\beta}{2\pi}\right)^2 = \frac{1}{\lambda_g^2} = \frac{1}{\lambda^2} - \frac{1}{\lambda_{11}^2}$$
$$Z_g = \eta\lambda_g/\lambda$$
$$P = \frac{1}{2Z_g} \times 0.375\,a^2|A|^2$$
$$\rho = 0,\quad |E| = 1.15(PZ_g)^{\frac{1}{2}}a^{-1}$$
$$\rho = a,\quad |E| = 0.73(PZ_g)^{\frac{1}{2}}a^{-1}\sin\phi$$
$$\alpha_c = (R_s/a\eta)^{\frac{1}{2}}\left(\frac{1}{s_{11}^2-1} + \left(\frac{\lambda}{\lambda_{11}}\right)^2\right)\left(1-\left(\frac{\lambda}{\lambda_{11}}\right)^2\right)^{-\frac{1}{2}}$$

The Smith Chart is a graphical display which clarifies this type of problem: it can be used for calculation, or it can serve as a display for more accurate computations. This chart will first be described and then typical problems considered.

2.10.1 The Smith Chart

The basis of the Smith Chart is the simple graphical interpretation of equation (2.17) by use of Cartesian axes for the real and imaginary parts of the reflection coefficient $\Gamma = u + \mathrm{j}v$, as shown in Figure 2.16(a). On a lossless line the locus of the reflection coefficient at different points along the line is a circle. The angle between Γ_1 and Γ_2 at z_1, z_2 is equal to $4\pi(z_2 - z_1)/\lambda$, so that the circle can be calibrated directly in length measured in wavelengths. In addition the directions of rotation 'towards generator' and 'towards load' can be added. For passive impedances, having positive real parts, $|\Gamma| < 1$, so that the bounding circle has unit radius. In Figure 2.16(b) loci corresponding to loads with constant normalized resistance and varying reactance and those corresponding to constant normalized reactance and varying resistance have been added. These loci are derived from the relation

$$r + jx = \frac{1 + \Gamma}{1 - \Gamma}$$

The algebra is given in Section A.7. It is frequently convenient to use admittances rather than impedances, when

$$g + jb = \frac{1 - \Gamma}{1 + \Gamma}$$

This relation also gives loci of constant g and constant b, but they are in fact the r and x loci inverted with respect to the origin, as indicated in Figure 2.16(c). With this orientation of impedance and admittance charts, the reflection coefficient is consistently defined as the voltage reflection coefficient. A commercial chart is shown in Figure 2.17, on which the Cartesian axes are omitted.

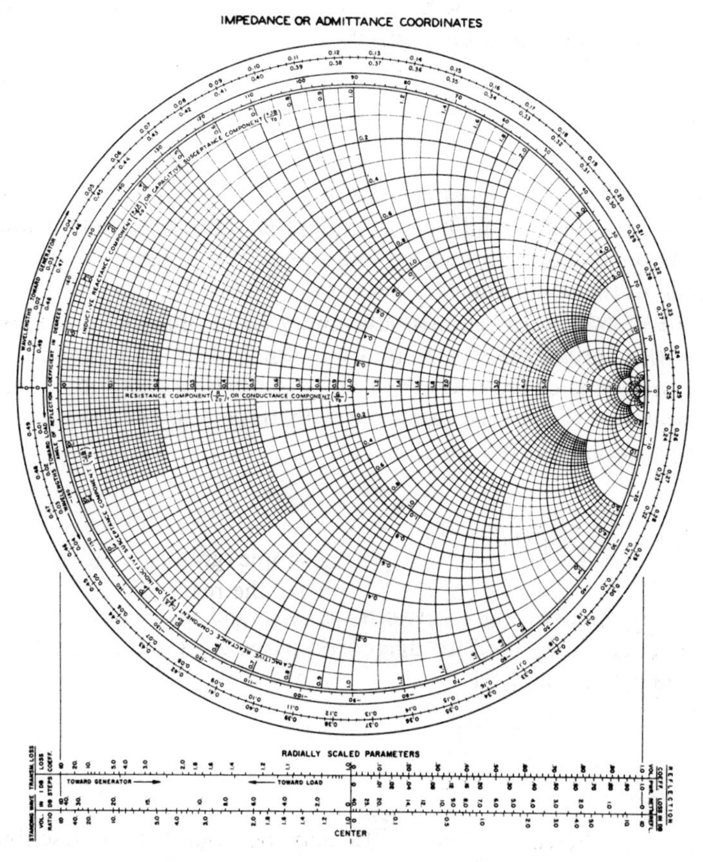

Figure 2.17 A commercial Smith Chart.

2.10.2 Voltage standing wave ratio (VSWR)

Figure 2.18(a) shows a line terminated at $z = z_2$ by a load of reflection coefficient Γ_2. The complex voltage at any point, equation (2.15), is given by the expression

$$\begin{aligned} v(z) &= a(z) + b(z) \\ &= a(z)(1 + \Gamma(z)) \end{aligned}$$

Thus

$$|v| = |a|\,|1 + \Gamma(z)|$$

The term $|a|$ is constant on a lossless line, and the term $|1 + \Gamma|$ is given by the length AP in Figure 2.18(c). It can be seen that as z alters $|v|$ shows maxima and minima as indicated in Figure 2.18(b). A maximum of $|v|$ occurs at B, when Γ is real and equal to $|\Gamma_2|$; a minimum occurs at C when Γ is real and negative, equal to $-|\Gamma_2|$. The ratio of maximum to minimum is termed the **voltage standing wave ratio** and is given by

$$S = \frac{1 + |\Gamma_2|}{1 - |\Gamma_2|}$$

Defined in this way S is greater than unity. (The VSWR could equally be defined as S^{-1}, which would be less than unity.) It can be seen that since Γ is real at the point B, the normalized impedance at that point is also real and equal to S. At the minimum, C, it is real and equal to S^{-1}.

The VSWR is an experimentally observable quantity in some important cases, when a probe can be inserted into the line or waveguide to give a measure proportional to the electric field strength. If, in addition, the position of the minimum, C, can be measured with respect to the load, Γ_2 can be found. By inspection of Figure 2.18(c)

$$\Gamma_2 = \frac{S - 1}{S + 1} \exp(\mathrm{j}(\beta l - \pi))$$

in which l is the line length between the load and the minimum. Thus the measurements can be used for diagnostic purposes. Practical problems are considered in Section 11.5.1.

2.10.3 Matching

The situation shown in Figure 2.18 may be regarded as describing an experimental situation with the value of Γ_2 represented in the diagram. It is

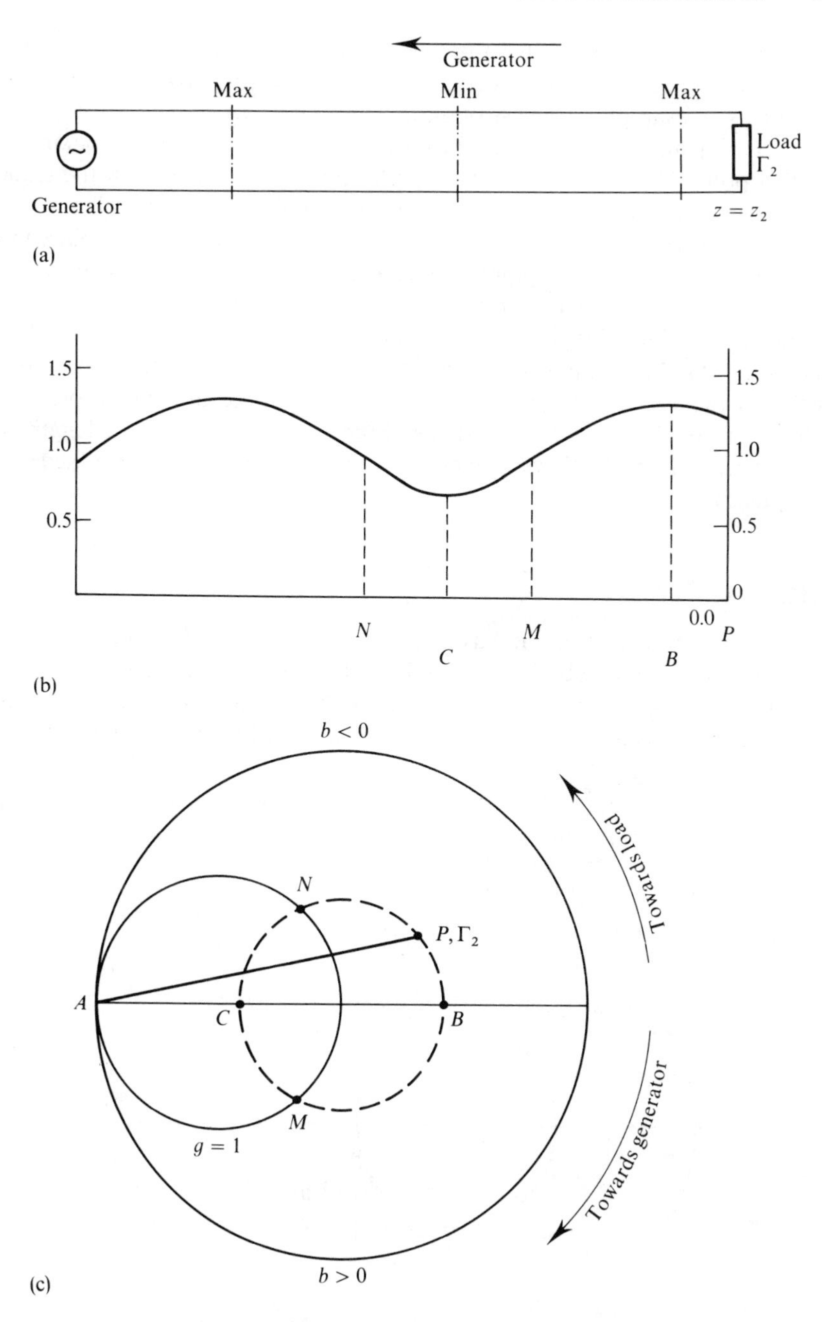

Figure 2.18 Demonstration of the Smith Chart: (a) a length of transmission line terminated in a load of reflection coefficient Γ_2, (b) the variation of voltage amplitude on the line, (c) the Smith Chart (admittance form) for the situation. P corresponds to Γ_2. Going towards the generator, at B a voltage maximum is reached, and at C a minimum. At the points M and N, added shunt susceptance can reduce the reflection coefficient to zero.

frequently necessary to match the load with reflection coefficient Γ_2 to the line by the use of a lossless network. One way of doing this is described in this section as an example of the use of the Smith Chart.

Going along the line from the termination, after passing the maximum at B, a point M is reached where the reflection coefficient lies on the circle $|g| = 1$, i.e. the normalized admittance is unit conductance in parallel with a capacitive susceptance. If an inductive susceptance of equal value is placed in shunt with the line at this point, the normalized admittance becomes unity, and the line to the generator is then matched. A similar process may be carried out at the point N, when an additional capacitive susceptance will be required. Matching only exists at the design frequency: change of frequency will alter the electrical lengths and also the value of the added susceptances. With transmission line the added susceptance will probably be a length of short-circuited line. With waveguides, components such as irises are used (see Chapter 5).

2.10.4 Active loads

In the preceding sections it has been tacitly assumed that the load was passive, corresponding to $0 < r < \infty$. If the load is active (e.g. a Gunn diode) then for some frequencies $r < 0$. In this case, the formulae for construction of the circles of constant g, b still apply, as indicated in Figure 2.19. The interior

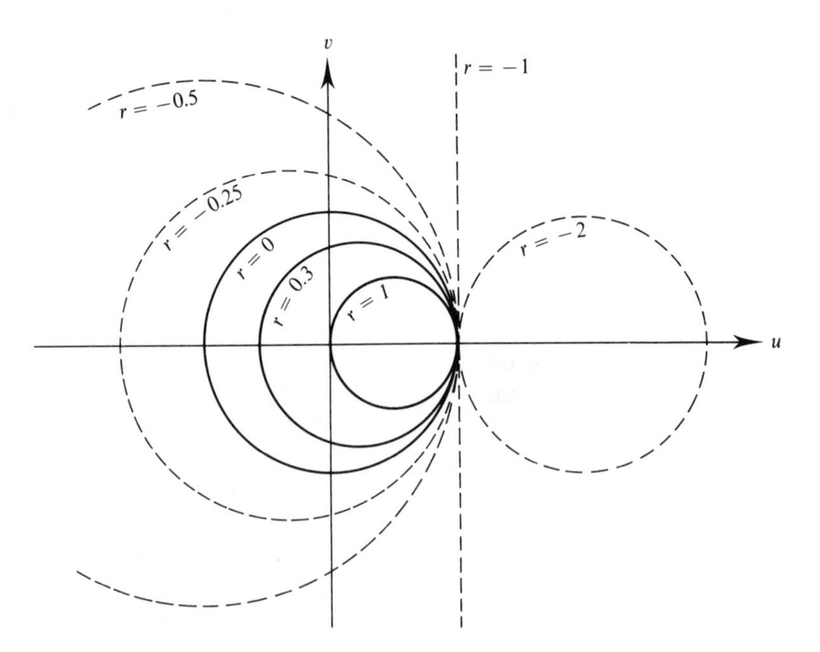

Figure 2.19 Extension of the Smith Chart to include negative resistances, $|\Gamma| > 1$.

of the circle $|\Gamma| = 1$ covers all impedances with $r > 0$; the region outside the circle corresponds to $r < 0$. It is precisely because all passive loads correspond to the interior that this form of chart is so useful.

2.11 Scattering parameters

As emphasized in previous sections, at microwave frequencies connections are made by transmission line or waveguide, and wave propagation in such lines can be described by the complex wave amplitudes of the forward and backward waves. Consider a 2-port device with lossless transmission lines of specified characteristic impedance on each port. A wave a_1 incident on port one will result in a reflected wave b_1 and a transmitted wave (which is leaving port two and therefore classed as a backward wave). If the line to port two is not matched, a reflected wave will be incident on port two. This is shown in Figure 2.20. For a linear 2-port there must be relations of the form

$$\begin{aligned} b_1 &= S_{11}a_1 + S_{12}a_2 \\ b_2 &= S_{21}a_1 + S_{22}a_2 \end{aligned} \tag{2.36}$$

To determine the complex wave amplitudes precisely, reference planes R_1, R_2 must be chosen. The four parameters S_{ij} are components of the **scattering matrix**

$$S = \begin{bmatrix} S_{11} & S_{12} \\ S_{21} & S_{22} \end{bmatrix}$$

The scattering parameters can be defined experimentally: in Figure 2.21(a) the junction with a matched load at end two is shown, which ensures no reflection at the termination. This means that in guide two, $a_2 \equiv 0$ and

$$\begin{aligned} b_1 &= S_{11}a_1 \\ b_2 &= S_{21}a_1 \end{aligned}$$

Thus S_{11} and S_{21} may be obtained by measuring reflected and transmitted waves. S_{11} can be recognized as the reflection coefficient defined in Section 2.3.1. Similarly S_{22} and S_{12} may be found from the configuration of Figure 2.21(b).

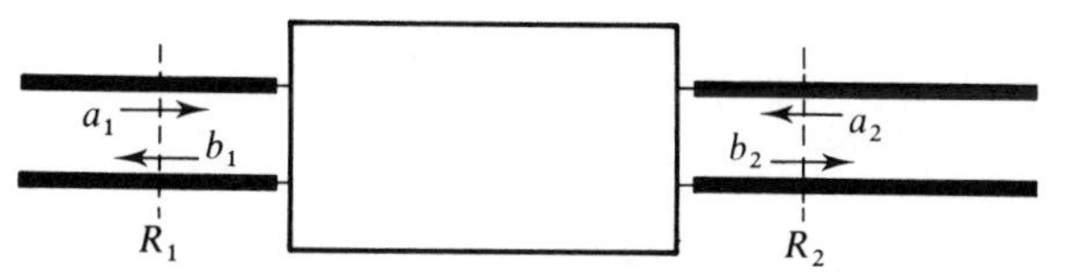

Figure 2.20 Complex amplitudes and reference planes for a 2-port network.

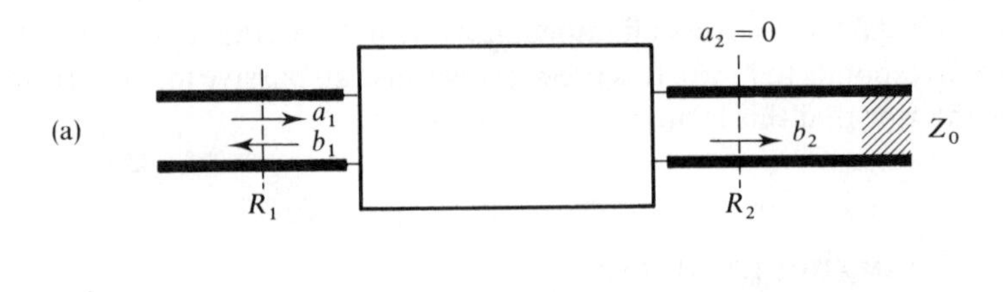

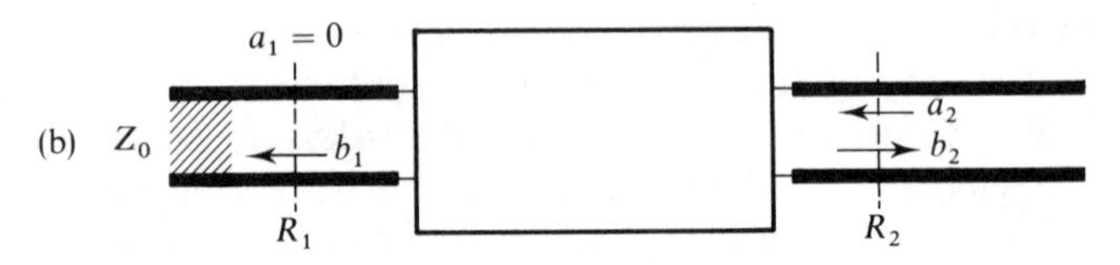

Figure 2.21 Configurations for measuring scattering parameters. (a) A matched load at end 2 ensures $a_2 = 0$, when $b_1 = S_{11}a_1$ and $b_2 = S_{21}a_1$; (b) with a matched load at end 1, $b_2 = S_{22}a_2$, $b_1 = S_{12}a_2$.

Scattering parameters are often used for the specification of microwave transistors: as just described, experimental determination involves only resistive loading rather than short or open circuits. It may not be possible to realize these latter conditions without oscillations developing. It must be emphasized that in the use of scattering matrices it is assumed that all connecting lines are lossless, and may be of very small length.

If the 2-port is passive and does not contain magnetic materials like ferrites, it will be reciprocal. In the context of Figure 2.21(a) and (b) this implies that interchanging ports will leave the ratio b_2/a_1 in Figure 2.21(a) the same as b_1/a_2 in Figure 2.21(b). Thus

$$S_{12} = S_{21}$$

Therefore, only three parameters are necessary to describe a reciprocal 2-port, in conformity with description by impedance or admittance matrices.

2.11.1 Multi-ports

Microwave components with several ports have found use, and can be described by scattering parameters. For an n-port component, equations (2.36) are extended to

$$b = S\,a \tag{2.37}$$

where a and b are the vectors $(a_1, a_2, \ldots, a_n)$ and $(b_1, b_2, \ldots, b_n)$ and S is an $n \times n$ matrix. There are various general relations among the coefficients.

Using the definition of reciprocity just given and considering all ports to be matched, excitation of port one followed by excitation of port two gives $S_{21} = S_{12}$. In general, for a junction not containing ferrite material.

$$S_{ij} = S_{ji} \tag{2.38}$$

Where lossless junctions are considered, the total power input must be zero. Hence

$$\sum_{i=1}^{n} (a_i a_i^* - b_i b_i^*) = 0 \tag{2.39}$$

Substituting from equation (2.37) for b this becomes

$$\sum_{i=1}^{n} \left[a_i a_i^* - \sum_{p=1}^{n} S_{ip} a_p \sum_{q=1}^{n} S_{iq}^* a_q^* \right] = 0 \tag{2.40}$$

Considering all ports except port one to have matched loads, ensuring $a_2 = a_3 = \ldots = a_n = 0$ we have

$$|a_1|^2 - \sum_{i=1}^{n} S_{i1} S_{i1}^* |a_1|^2 = 0$$

Hence

$$\sum_{i=1}^{n} |S_{i1}|^2 = 1$$

More generally

$$\sum_{i=1}^{n} |S_{ij}|^2 = 1 \quad i, j = 1, 2, \ldots, n \tag{2.41}$$

In equation (2.40) the identity has to be maintained for all values of $a_p a_q^*$ requiring that

$$\sum_{i=1}^{n} S_{ip} S_{iq}^* = 0 \quad p \neq q \tag{2.42}$$

In matrix terms these relations are summarized by

$$\tilde{S} S^* = I \tag{2.43}$$

where I is the $n \times n$ identity matrix and $\tilde{S}$ denotes the transpose of S. Hence

$$S^{-1} = \tilde{S}^*$$

The matrix S is said to be **unitary**.

As an example of the application of these relations, consider a 2-port lossless junction. Equations (2.41) and (2.42) give

$$|S_{11}|^2 + |S_{21}|^2 = 1$$
$$|S_{12}|^2 + |S_{22}|^2 = 1$$
$$S_{11}S_{12}^* + S_{21}S_{22}^* = 0$$

From the last equation

$$S_{11}/S_{21} = -(S_{22}/S_{12})^*$$

Hence

$$|S_{11}/S_{21}| = |S_{22}/S_{12}|$$

From the first and second equations

$$|S_{11}/S_{21}|^2 = |S_{21}|^{-2} - 1$$

and

$$|S_{22}/S_{12}|^2 = |S_{12}|^{-2} - 1$$

Hence

$$|S_{12}| = |S_{21}|$$
$$|S_{11}| = |S_{22}|$$

We may choose reference planes so that the reflection coefficients S_{11}, S_{22} become real. Suppose this to be done, giving

$$S_{11} = S_{22}$$

Hence

$$S_{21} = -S_{12}^*$$

If the junction is reciprocal $S_{21} = S_{12}$, requiring

$$S_{12} = S_{21} = \mathrm{j}K$$

and

$$S_{11} = S_{22} = (1 - K^2)^{\frac{1}{2}}$$

Further examples will be encountered which make use of the unitary relation combined with symmetry requirements imposed by the hardware geometry.

2.11.2 Admittance and impedance matrices

Since normalized voltage and current have been defined in terms of complex wave amplitudes, matrix relationships can be set up in terms of voltage and current. Equations (2.15) lead to

$$v - i = S(v + i)$$

in which v and i are now vectors of order equal to the number of ports. This equation may be rearranged to give

$$v = Zi$$

where

$$Z = (I - S)(I + S)^{-1}$$

in which I is the identity matrix. Alternatively

$$i = Yv$$

where

$$Y = (I + S)(I - S)^{-1}$$

If all lines have the same characteristic impedance, these are normalized impedance and admittance matrices for the n-port. If the values of the characteristic impedances are different, more complicated expressions result.

2.11.3 Source load configuration

A frequently encountered situation concerns a loaded 2-port fed from a source, as indicated in Figure 2.22. Reference planes are most conveniently taken at the ports of the source and the load, R_1 and R_2 in the figure. The load is characterized by its reflection coefficient, Γ_2, giving the relationship

$$b_2' = \Gamma_2 a_2'$$

In terms of the waves at the output port of the 2-port $b_2' = a_2$, $a_2' = b_2$. Thus

$$a_2 = \Gamma_2 b_2. \tag{2.44}$$

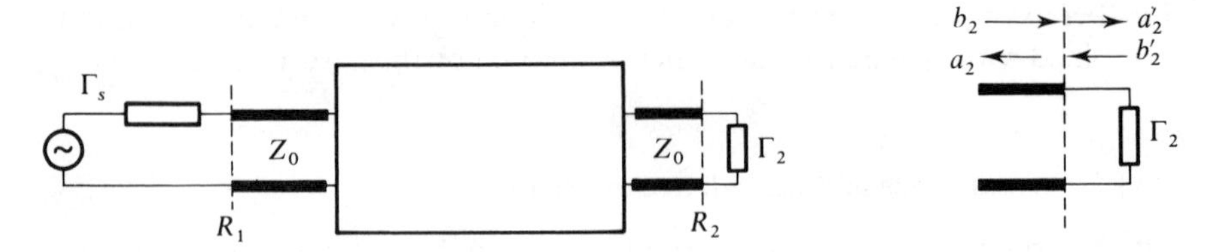

Figure 2.22 A 2-port terminated in a load of reflection coefficient Γ_2 and fed from a source of reflection coefficient Γ_s. The two connecting lines may be of very small length. The inset shows notation at the load.

Substitution in equations (2.36) then yields

$$b_1 = \left[S_{11} + \frac{S_{12}S_{21}}{1 - S_{22}\Gamma_2} \right] a_1 \tag{2.45}$$

$$b_2 = \frac{S_{21}}{1 - S_{22}\Gamma_2} a_1 \tag{2.46}$$

Hence the reflection coefficient of the loaded 2-port at the source output can be determined as $\Gamma_1 = (b_1/a_1)$.

The source has to be characterized by two quantities. Its internal impedance defines a reflection coefficient, Γ_s. The second parameter may be taken as the complex amplitude b_s, which occurs when the source feeds a matched line, as indicated in Figure 2.23(a). (Note that this does not correspond to maximum power output, which will occur with conjugate matching.) The situation when the source is connected to a load of reflection coefficient Γ_1 is shown in Figure 2.23(b): successive reflections occur at both load and source. The resultant may be obtained by summation. For the forward waves in the line

$$\begin{aligned} b &= b_s(1 + \Gamma_s\Gamma_1 + (\Gamma_s\Gamma_1)^2 + \ldots) \\ &= b_s(1 - \Gamma_s\Gamma_1)^{-1} \\ a &= \Gamma_1 b = \Gamma_1 b_s(1 - \Gamma_s\Gamma_1)^{-1} \end{aligned} \tag{2.47}$$

The power delivered by the source is given by

$$\begin{aligned} P &= \tfrac{1}{2}(|b|^2 - |a|^2) \\ &= \tfrac{1}{2}|b_s|^2 \frac{1 - |\Gamma_1|^2}{|1 - \Gamma_s\Gamma_1|^2} \end{aligned} \tag{2.48}$$

It will be found that for constant b_s this expression has a maximum value

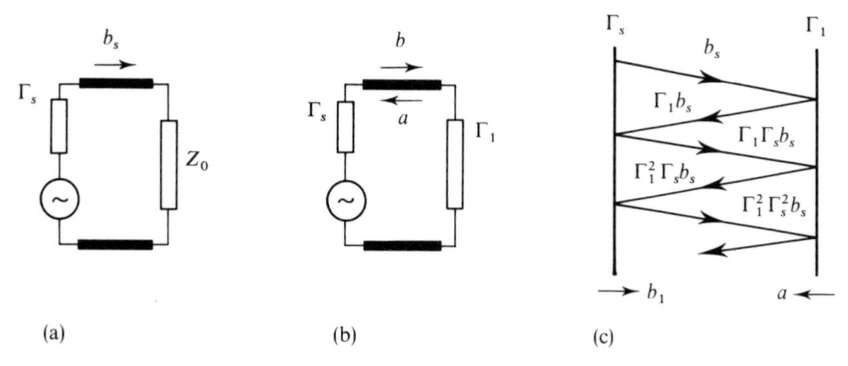

Figure 2.23 Waves produced by the source: (a) definition of source strength by means of the Z_0-available power, when the source is connected to the matched line, not to its conjugate complex impedance; (b) connection of the source to an arbitrary load through a very short length of Z_0 line; (c) successive reflections.

when $\Gamma_1 = \Gamma_s^*$, defining available power by

$$P_a = \frac{1}{2} \frac{|b_s|^2}{1 - |\Gamma_s|^2} \tag{2.49}$$

The quantity

$$P_s = \tfrac{1}{2} |b_s|^2 \tag{2.50}$$

may be termed **Z_o-available power**. It is this quantity that is usually measured.

From the relationships derived in this chapter, the power delivered to the load Γ_2 can be calculated. If the 2-port is lossless it is of course the same as that calculated for Γ_1.

2.12 Summary

This chapter has been concerned with transmission in lines and waveguides. Formulae describing field patterns and losses have been given for the specific cases of coaxial transmission line and waveguides of rectangular and circular cross-section. Transmission in microstrip has been discussed, with references for detailed design data.

Transmission line problems have been formulated in terms of complex wave amplitudes, and the use of the Smith Chart illustrated. Finally, scattering parameters have been introduced both in 2-port and multi-port situations.

2.13 Summary of formulae

(See also Tables 2.1, 2.4 and 2.7.)

Wave propagation on lossless transmission lines and waveguides:

transverse electric field $\boldsymbol{E}_t = \boldsymbol{e}(x, y)[A\exp(-\mathrm{j}\beta z) + B\exp(\mathrm{j}\beta z)]$

transverse magnetic field $\boldsymbol{H}_t = Z_g^{-1}\mathbf{a}_z \times \boldsymbol{e}[A\exp(-\mathrm{j}\beta z) - B\exp(\mathrm{j}\beta z)$

where $\boldsymbol{e}(x, y)$ is the modal pattern vector

Z_g is the wave impedance

phase velocity $c_p = \omega/\beta$

group velocity $c_g = \mathrm{d}\omega/\mathrm{d}\beta$

guide wavelength $\lambda_g = 2\pi/\beta$

In the presence of small losses the formulae are to the first order unchanged except that $\mathrm{j}\beta$ is replaced by the propagation constant

$$\gamma = \alpha + \mathrm{j}\beta$$

in which α is the attenuation constant

Power flow $P = \frac{1}{2}Z_g^{-1}\langle e^2\rangle(|A|^2 - |B|^2)$

where $\langle e^2\rangle = \iint e^2(x, y)\,\mathrm{d}x\,\mathrm{d}y$

Attenuation constant $\mathrm{d}P/\mathrm{d}z = -2\alpha P$

$$P = P_0\exp(-2\alpha z)$$

Complex wave amplitudes:

$$a = (\langle e^2\rangle/Z_g)^{\frac{1}{2}}A\exp(-\mathrm{j}\beta z)$$

$$b = (\langle e^2\rangle/Z_g)^{\frac{1}{2}}B\exp(\mathrm{j}\beta z)$$

$$P = \tfrac{1}{2}(|a|^2 - |b|^2)$$

normalized voltage $v = a + b$

normalized current $i = a - b$

reflection coefficient $\Gamma = b/a$

voltage standing wave ratio $S = \dfrac{1 + |\Gamma|}{1 - |\Gamma|}$

Transmission line:

characteristic impedance $Z_0 = (\eta/\langle e^2\rangle)\left(\int_1^2 \boldsymbol{e}\cdot\mathrm{d}\boldsymbol{l}\right)^2$

impedance transformation $Z_1 = Z_0\dfrac{Z_2 + \mathrm{j}Z_0\tan\beta l}{Z_0 + \mathrm{j}Z_2\tan\beta l}$

Scattering parameters: for an n-port network

$$b_i = \sum_{j=1}^{n} S_{ij} a_j \quad i = 1, 2, \ldots, n$$

for a lossless network

$$S^{-1} = \tilde{S}^*$$

A source is characterized by

reflection coefficient Γ_s

complex wave amplitude into Z_0 b_s

Z_0-available power $\frac{1}{2}|b_s|^2$

available power $\frac{1}{2}|b_s|^2/(1 - |\Gamma_s|^2)$

Source feeding load of reflection coefficient Γ_1

wave incident on load $b = b_s(1 - \Gamma_s \Gamma_1)^{-1}$

power to load $P = \frac{1}{2}|b_s|^2 \dfrac{1 - |\Gamma_1|^2}{|1 - \Gamma_s \Gamma_1|^2}$

EXERCISES

2.1 An air spaced coaxial line carrying transmitter power at 2 GHz to a matched antenna is made of copper. The diameter of the inner is 4 mm, the inner diameter of the outer is 12 mm, and it is 2 m long. Determine, for a matched line

(a) the characteristic impedance of the line;
(b) the attenuation;
(c) the maximum electric field when the line carries 1 MW pulses;
(d) the power dissipated in the line if the average power is 10 kW.

What is the maximum electric field in (c) if the VSWR is 1.2? (Skin resistance for copper is 16 mΩ per square at 4 GHz.)

2.2 It is customary to operate waveguide of cut-off frequency f_0 at a nominal frequency of $1.5f_0$. Calculate the dimensions of

(a) a rectangular waveguide whose sides are in the ratio 2:1 and which operates in the dominant mode;
(b) a circular waveguide operating in the TE_{11} mode;

both suitable for working at 6 GHz. For each give the next four modes and their cut-off frequencies.

Suggest a design for a transition from rectangular to circular waveguide.

2.3 Assuming that the power which can be transmitted is limited by the breakdown of air ($3 \times 10^6\,\mathrm{V\,m^{-1}}$), determine the maximum power which can be transmitted by the rectangular waveguide of question 2.2, assuming operation at 6 GHz.

2.4 Determine the modes that can propagate at 3 GHz in a circular waveguide of diameter 8 cm. Determine the values of the components of electric field strength at the centre of the waveguide when it carries a power of 2 MW.

2.5 Discuss the causes of loss in transmission lines and waveguides, with particular reference to the way losses change as frequency is varied. The TE_{01} mode in a circular waveguide is perfectly symmetrical around the guide. The magnetic field at the walls is purely longitudinal, and, for a travelling wave in an air filled guide, its amplitude H_0 is related to the power flow P by the equation

$$P = \frac{\pi}{2}\left(\frac{\mu_0}{\varepsilon_0}\right)^{\frac{1}{2}} \frac{f(f^2 - f_0^2)^{\frac{1}{2}}}{f_0^2}(aH_0)^2$$

in which a is the guide radius in metres and $f_0(= 0.183/a\,\mathrm{GHz})$ is the cut-off frequency for the mode.

A particular guide is 6 cm in diameter. The wall material is such that at 1 GHz a magnetic field at the wall of peak value $1\,\mathrm{A\,m^{-1}}$ gives rise to a loss of $4\,\mathrm{mW\,m^{-2}}$. Estimate the attenuation coefficient, in $\mathrm{dB\,km^{-1}}$, at a frequency of 8 GHz.

2.6 A rotating joint incorporates a length of circular waveguide, and uses a symmetrical mode. Designate this mode and calculate a suitable diameter for the waveguide to operate at 2.5 GHz at 50% higher than the cut-off for the mode.

What other modes can be propagated in the waveguide?

2.7 The narrowest strip which can reliably be used as the conductor in a microstrip transmission line is 10 μm. If this is used on an alumina substrate 1 mm thick and of relative permittivity 9.8, estimate the highest characteristic impedance realizable. Estimate also the ratio of free space to strip wavelength. Use the equations of Section 2.7.1.

2.8 A matching network to be used on the board of question 2.7 at 3 GHz requires an inductance of 5 nH. Estimate the shortest length of short-circuited line that will give this inductance.

2.9 A microwave mixer operating at 3 GHz presents an impedance of 250 Ω in parallel with a 0.4 pF. It is connected to a transmission line of 50 Ω characteristic impedance. Calculate the VSWR and electrical position of the maximum or minimum nearest the mixer. Find also the electrical position where the admittance of the mixer is transformed to have a normalized real part of unity, and the reactance to be placed in parallel in order to match the mixer to the transmission line. The ratio of free space wavelength to wavelength on the line is 2.6. Work out dimensions for the matching arrangement, including an open circuit stub to give the correct reactance.

2.10 Obtain expressions for the components of the scattering matrices of the following systems in which all the transmission lines have the same characteristic impedance:

(a) two lines joined by a series impedance zZ_0;

(b) a 2-port formed by a line shunted by an admittance yY_0;

(c) an ideal transformer of ratio $n\!:\!1$.

Take reference planes at the plane of discontinuity, assuming all components are of negligibly small electrical length.

2.11 Inserting a coaxial connector into a line is equivalent to connecting a small capacitive susceptance, b, across the line. Show that the VSWR at the input when the output is matched to the line impedance is approximately equal to $1 + |b|$.

A line contains two such connectors separated by a length of cable. Show that the VSWR when the output is matched must have a value between the product and the quotient of the two VSWR values for the individual connectors.

2.12 Two lines, one of characteristic impedance Z_0 and the other n^2Z_0, are joined by a section of line of impedance nZ_0 and electrical length θ. Referring to input and output planes at the line junctions show that the scattering matrix is given by

$$((n^2+1)\cos\theta + 2jn\sin\theta)^{-1}\begin{pmatrix}(n^2-1)\cos\theta & 2n \\ 2n & -(n^2-1)\cos\theta\end{pmatrix}$$

Sketch the locus of S_{11} as the frequency varies on a Smith Chart, taking $n = 2$, $\pi/8 < \theta < 3\pi/8$.

2.13 Power is fed into a load by a transmission line. Find the level of VSWR that can be tolerated before the power into the load falls below 90% of the available power.

References

General reading

The electromagnetic theory of transmission lines and waveguides is treated in numerous standard texts, such as

Ramo, S., Whinnery, J. R. and van Duzer, T. (1984). *Fields and Waves in Communication Electronics.* New York: Wiley.

Sander, K. F. and Reed, G. A. L. (1986). *Transmission and Propagation of Electromagnetic Waves.* Cambridge: Cambridge University Press.

Text references

Edwards, T. C. (1981). *Foundations for Microstrip Circuit Design.* New York: Wiley.

Schneider, M. V. (1969). 'Microstrip Lines for Microwave Integrated Circuits'. *BSTJ*, **48**, 1421–44.

Sobol, H. (1971). 'Applications of Integrated Circuit Techniques to Microwave Frequencies'. *Proc. IEEE*, **59**, 1200–11.

Wheeler, H. A. (1977). 'Transmission Line Propagation of a Strip on a Dielectric on a Plane'. *IEEE MTT*, **25**, 631–47.

CHAPTER 3

ANTENNAS AND PROPAGATION

OBJECTIVES

This chapter considers in detail the factors involved in the transmission of microwaves, propagation between two antennas in free space, and reception. Antennas will be characterized in both transmission and reception, and a power budget equation set up. Methods of calculation for aperture antennas are introduced and a number of different reflector antennas described. Examples are given of linear arrays of radiating elements, and the concept of phased-array antennas is briefly discussed. The chapter ends by considering the measurement of antenna properties.

3.1 Introduction

That part of a microwave system which involves the launching of a beam of electromagnetic radiation by an antenna, the propagation of the beam and subsequent reception is central to the performance of the system. The special properties of the beam may in fact be the reason for operating at microwave frequencies in the first place. It is therefore necessary to consider the general properties of the radiated electromagnetic wave and the way in which these properties are influenced by the radiating antenna. In reception, the concern is the power which the antenna can absorb from the radiation falling on it.

In most systems the antennas are a long way apart, so that the receiving antenna does not interact with the transmitting antenna. In this chapter the properties of radiation and of antennas in free space will be considered. The effects of the constituents of the atmosphere will be taken into account when discussing terrestrial links in Chapter 9.

3.2 Radiation from an antenna

The general theory of electromagnetic radiation (e.g. Sander and Reed, 1986) provides the means to characterize the radiation emanating from a source confined within a limited volume, such as from an antenna. The antenna will be supplied with power through one port, which will be transmission line or waveguide. Radiation from the currents induced in the various parts of the structure forms the overall beam. The characteristics of most importance in the present application concern the **far-field**. This is usually taken as the region beyond a radius of $2D^2/\lambda$ centred on the antenna (Rudge, 1982), in which as before λ is the free space wavelength and D is a typical linear dimension of the antenna. The region inside this radius is referred to as the **near-field**. In the far-field the electric and magnetic vectors can be expressed in the following form:

$$\boldsymbol{E} = A\frac{\exp(-jkr)}{kr}\boldsymbol{e}(\theta, \phi) \tag{3.1}$$

$$\boldsymbol{H} = \eta^{-1}\boldsymbol{a}_r \times \boldsymbol{E} \tag{3.2}$$

$$\boldsymbol{a}_r \cdot \boldsymbol{e} = 0$$

in which $\eta = (\mu/\varepsilon)^{\frac{1}{2}}$, $k = 2\pi/\lambda$ and $\boldsymbol{a}_r$ is the unit vector in the radial direction, as shown in Figure 3.1. The vector pattern function $\boldsymbol{e}(\theta, \phi)$ is perpendicular to the radial vector and is a function of direction only, as indicated by the explicit dependence on the angles θ, ϕ. In equation (3.1) A is an arbitrary constant of dimensions volts per metre; the term kr is dimensionless, so that $\boldsymbol{e}$ is also dimensionless. As usual $\boldsymbol{E}, \boldsymbol{H}$ are peak phasor representations of the electric and magnetic fields. Such a wave is termed **quasi-plane**.

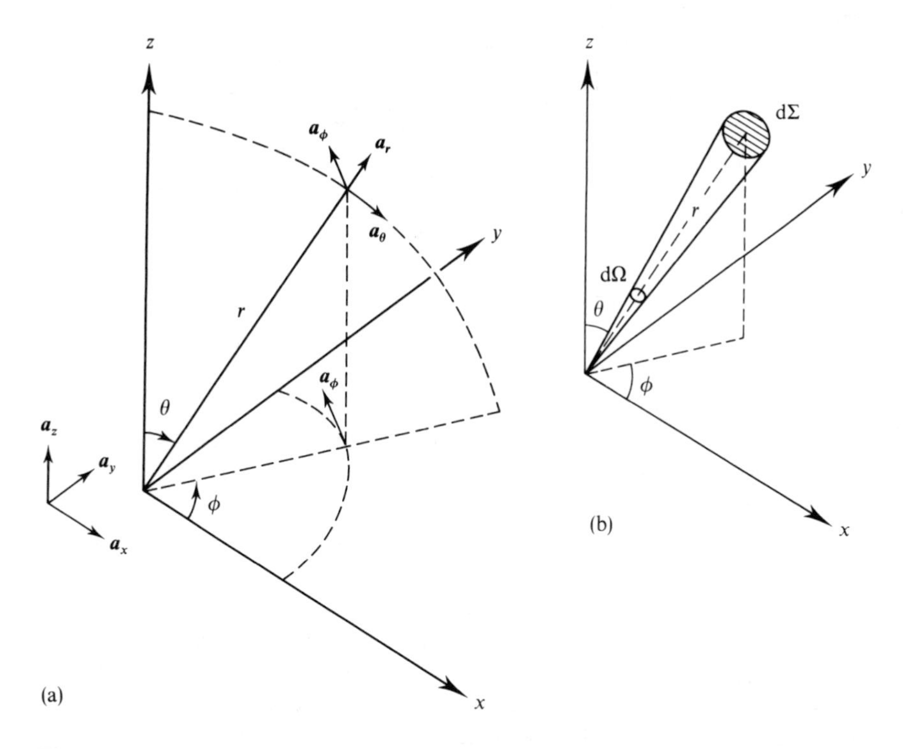

Figure 3.1 Definition of the spherical coordinate system: (a) shows the polar angles θ, ϕ and the unit vectors $\boldsymbol{a}_r$, $\boldsymbol{a}_\theta$, $\boldsymbol{a}_\phi$ of the system; (b) illustrates the definition of solid angle: $\mathrm{d}\Sigma$ is the area of the cap of the cone of height r and of solid angle $\mathrm{d}\Omega = \mathrm{d}\Sigma/r^2$.

The power flux density is given through the Poynting vector by the expression

$$\boldsymbol{p} = \tfrac{1}{2}Re(\boldsymbol{E} \times \boldsymbol{H}^*)$$

The radial power flux density is therefore given by

$$p = \tfrac{1}{2}\eta^{-1}\,|A/kr|^2 \boldsymbol{e}\cdot\boldsymbol{e}^* \tag{3.3}$$

For the total radiated power we have the expression

$$P_s = \tfrac{1}{2}\eta^{-1}\,|A/k|^2 \int_0^{2\pi} \mathrm{d}\phi \int_0^{\pi} \boldsymbol{e}\cdot\boldsymbol{e}^* \sin\theta \,\mathrm{d}\theta \tag{3.4}$$

This, correctly, is not dependent on the radial distance. The energy density in the far-field corresponds solely to the energy flow required to give the power flux. In the near-field the structure of the electromagnetic field is more complex and additional energy is stored.

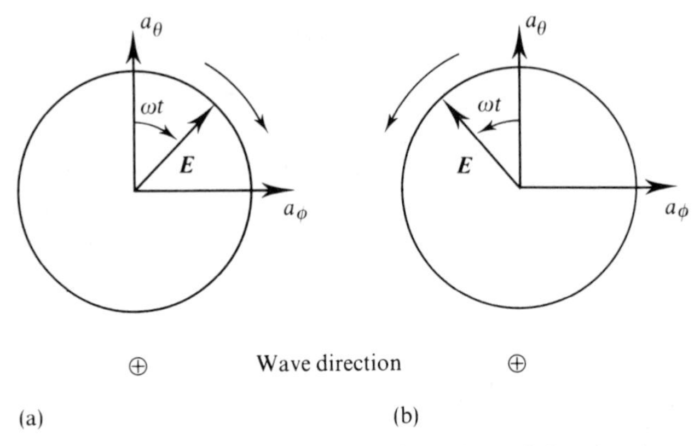

Figure 3.2 The diagrams show the direction of rotation of the electric vector of a circularly polarized wave propagating into the paper: (a) right-handed or clockwise, (b) left-handed or anticlockwise.

3.2.1 Polarization

The directional characteristics of the electric field are determined by the vector pattern function $\boldsymbol{e}$, which may be complex. If $\boldsymbol{e}$ is real, then the electric field has a direction fixed in space along any given radial line. Such a wave is said to be **plane polarized**. The direction of polarization may be different along different radial directions, and must be specified with the aid of a coordinate system. (In the case of earth based antennas the polarization vector might be referred to as vertical or horizontal.) If $\boldsymbol{e}$ is complex a different situation may arise: for a given direction θ, ϕ we may write

$$\boldsymbol{e} = e_1 \boldsymbol{a}_\theta + e_2 \boldsymbol{a}_\phi$$

Consider the case when e_1 is real and $e_2 = -je_1$. The instantaneous electric field is then given by

$$\begin{aligned} \boldsymbol{E} &= (Ae_1/kr)Re[(\boldsymbol{a}_\theta - \mathrm{j}\boldsymbol{a}_\phi)\exp(\omega t - kr)] \\ &= (Ae_1/kr)[\boldsymbol{a}_\theta \cos(\omega t - kr) + \boldsymbol{a}_\phi \sin(\omega t - kr)] \end{aligned}$$

This equation is illustrated in Figure 3.2(a), which shows that the locus of the tip of the E–vector describes a circle in a clockwise direction when looking along the direction of propagation of the wave. Such a wave is said to be **circularly polarized** in a **right-handed** or **clockwise** sense. If $e_1 = +je_2$, then as Figure 3.2(b) shows **left-handed** or **anticlockwise** polarization results. In general, the state of polarization of a wave will be elliptical in right- or left-handed sense. It will be noticed from the above derivation that a circularly polarized wave can be regarded as the sum of two orthogonal plane waves in phase quadrature. Alternatively, a plane wave can be regarded as the sum of right-handed and left-handed circularly polarized waves. The state of polarization will in general be different along different radial directions.

3.2.2 Antenna phase centre

Since the radial distance r enters into the expression of equation (3.1) a specific origin is presupposed that will lie somewhere within the antenna structure. A change of origin will alter the value of r to a new value, differing by an amount depending on direction, which in turn will modify the argument (phase) of the complex vector $\boldsymbol{e}$. If the origin can be chosen so that the argument of $\boldsymbol{e}$ does not vary with direction, or at least over a limited range of directions, that origin is said to be at the phase centre of the antenna. An alternative definition is to say that with the origin at the phase centre, the distant surfaces of constant phase are spherical. An antenna may not have a true phase centre, but frequently a point can be located which serves as an apparent phase centre for the range of directions of interest. It may on occasion be necessary to determine this position to within 0.1λ or less.

3.3 Antenna gain

The power fed into an antenna is distributed with an angular pattern characteristic of the antenna, as shown by equation (3.3). It is convenient to relate the actual power density to the value it would have with a standard antenna, usually taken as **an isotropic radiator**. If a source delivers power P_s to an isotropic radiator, the power flux density at a point distance r is given by

$$p_s = P_s/(4\pi r^2)$$

The **gain** of an antenna is defined to be the ratio of actual power density at a point to that which would be given by an isotropic radiator radiating the same total power. Thus

$$p = p_s G(\theta, \phi) = \frac{P_s}{4\pi r^2} G(\theta, \phi) \tag{3.5}$$

The total power flowing across a distant sphere must be equal to the power P_s, placing a condition on the function $G(\theta, \phi)$. Following Figure 3.1 we may write this condition in the form

$$\oint \frac{1}{4\pi r^2} G(\theta, \phi)\, \mathrm{d}\Sigma = 1$$

or

$$\oint G(\theta, \phi)\, \mathrm{d}\Omega = 4\pi \tag{3.6}$$

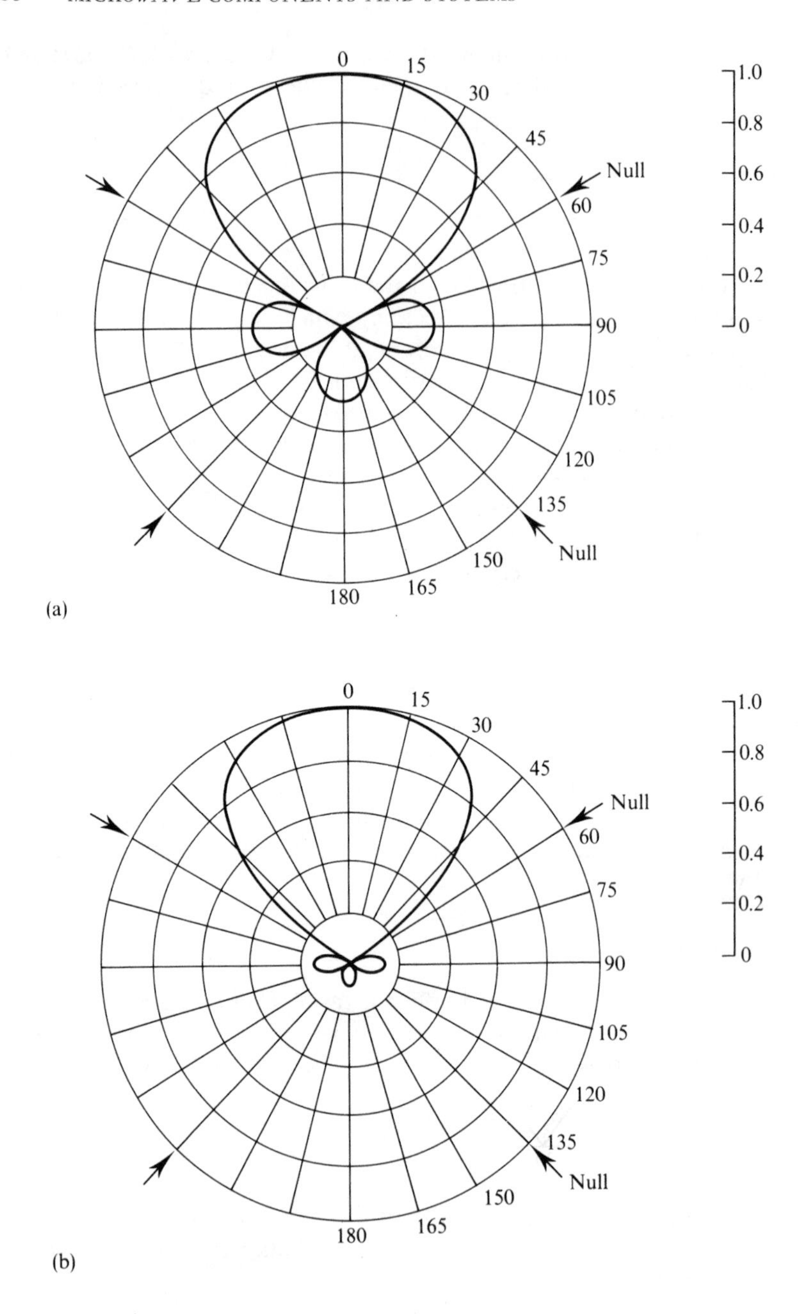

Figure 3.3 Radiation patterns in different formats: (a), (b) and (c) represent the same pattern on a polar plot but with radius proportional to (a) magnitude of electric field, (b) power flux, (c) decibels below maximum. (d) and (e) are Cartesian plots suitable for highly directional patterns. The horizontal scale is in degrees of azimuth, vertical scale: (d) magnitude of electric vector, (e) decibels below maximum.

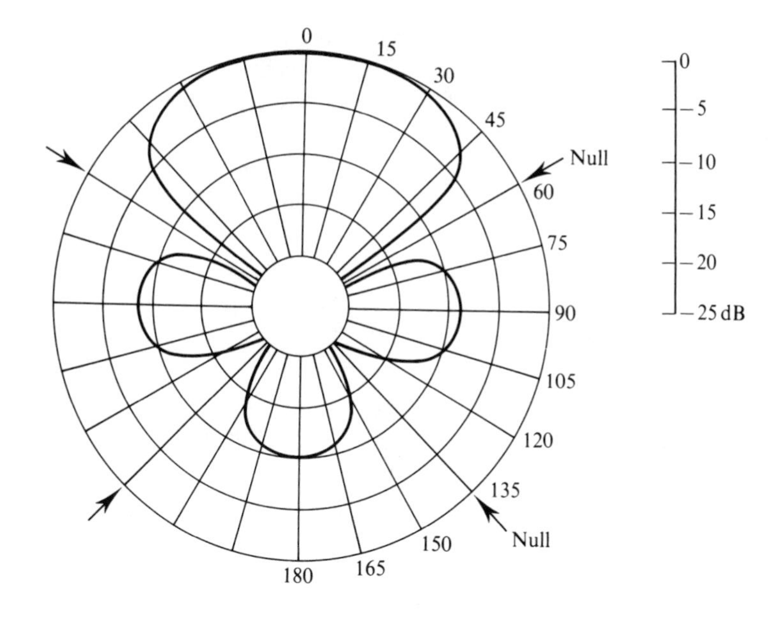

(c)

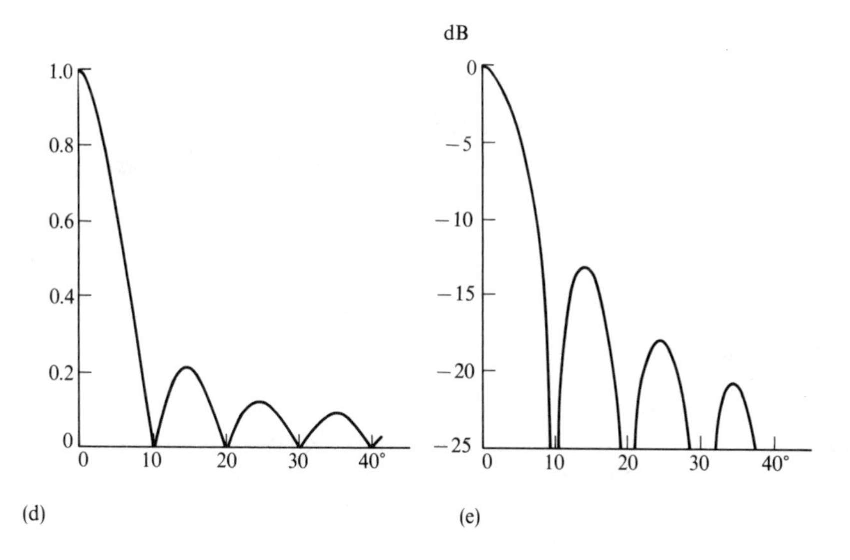

(d)

(e)

in which $d\Omega$ is the element of solid angle subtended by $d\Sigma$ at the antenna. Microwave antennas are often highly directional and therefore the solid angle subtended will be small. Equation (3.6) then shows that the gain will be high. In this instance the gain will have a well defined maximum value which is often quoted as the gain of the antenna.

Another way of looking at equation (3.5) is that the actual power density in direction θ, ϕ is that which would be obtained with an isotropic source of strength $P_s G(\theta, \phi)$. In the case above when G has a maximum G_m, the product $P_s G_m$ is referred to as the **effective isotropic radiated power** (EIRP).

Equation (3.5) can be used in conjunction with equations (3.3) and (3.4) to obtain an expression for the gain in terms of the vector pattern function $\boldsymbol{e}(\theta, \phi)$. We find

$$G(\theta, \phi) = 4\pi \boldsymbol{e} \cdot \boldsymbol{e}^* / \langle \boldsymbol{e} \cdot \boldsymbol{e}^* \rangle \tag{3.7}$$

where

$$\langle \boldsymbol{e} \cdot \boldsymbol{e}^* \rangle = \oint \boldsymbol{e} \cdot \boldsymbol{e}^* \, \mathrm{d}\Omega = \int_0^{2\pi} \int_0^{\pi} \boldsymbol{e} \cdot \boldsymbol{e}^* \sin\theta \, \mathrm{d}\theta \, \mathrm{d}\phi$$

It is useful to note that the magnitude of the electric field can be expressed in the form

$$|\boldsymbol{E}| = (\boldsymbol{E} \cdot \boldsymbol{E}^*)^{\frac{1}{2}} = (\eta P_s G / 2\pi)^{\frac{1}{2}} \frac{1}{r} \ \mathrm{V\,m^{-1}}$$

The quantity P_s in equation (3.5) is the radiated power, equal to the input power less any power absorbed by the imperfectly conducting antenna structure. Usually this lost power is small, but if a distinction has to be made, equation (3.5) technically defines directivity: gain would be defined using the same expression but with the actual input power in place of P_s.

3.4 Radiation pattern

When the antenna is designed to have a maximum gain in one direction it becomes convenient to talk about the gain normalized to this maximum. The resultant function of direction is usually presented graphically as the **radiation pattern**. A complete presentation would be three dimensional, with the radius vector in a given direction being equal to the normalized gain in that direction. For most purposes, one or more cross-sections of a projection will be used to display the radiation pattern, the format being determined by the type of pattern in question. Further variants are obtained by using relative field strength rather than relative power gain as the radius vector or, alternatively, a decibel scale. Figure 3.3 shows the form of some of these alternatives. A more sophisticated approach is to use computer graphics to represent the complete pattern as a projection. The majority of microwave antennas have a directional pattern of the form shown in Figure 3.3(d) and (e). One characteristic of such a pattern may be taken as the beam width

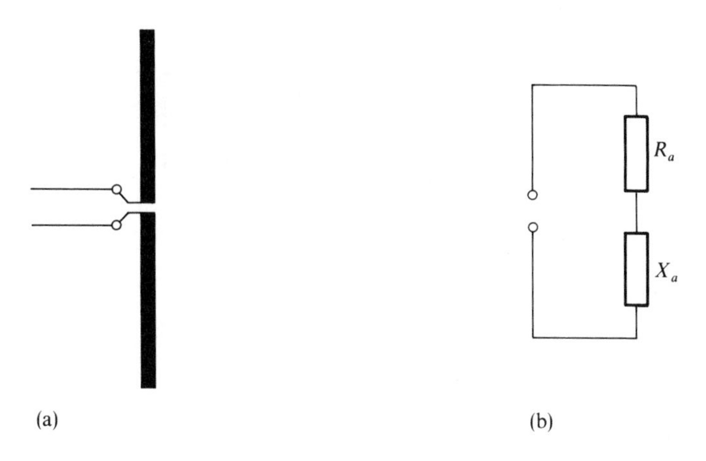

Figure 3.4 Radiation impedance: (a) shows a representation of an electric dipole fed by a transmission line; (b) shows the electrical representation of the load presented to the line by the dipole.

to half power, which in the example illustrated would be about 9°. Also of importance is the magnitude of subsidiary maxima, the **side lobes**, which can cause interference from strong signals on an 'off axis' direction. The magnitude is quoted relative to the main peak, and in the example the first side lobe is 13 dB down.

3.5 Radiation impedance

As was mentioned earlier, power is supplied to an antenna through an input port by either transmission line or waveguide. Once a reference plane is defined, this port can be characterized by a reflection coefficient or by an impedance, as discussed in Section 2.4. An antenna such as the dipole illustrated in Figure 3.4(a) which is fed by a transmission line may be said to have an impedance $R_a + jX_a$, as indicated in Figure 3.4(b). An antenna such as a waveguide horn, shown in Figure 3.5, is better characterized by a reflection coefficient, although it can also be given an equivalent circuit of the type shown in Figure 3.4(b). It would be matched to the waveguide by suitable components, as will be discussed in Section 5.3.

The resistive component R_a is a representation of the power lost to the source by radiation. To represent losses in the antenna structure, an additional resistive component would be added. The reactance X_a represents energy stored in the near-field of the antenna. In principle, the effect of X_a can be compensated by another reactive element, but such compensation is limited in frequency range and affects the bandwidth of the antenna. Both R_a and X_a will be functions of frequency.

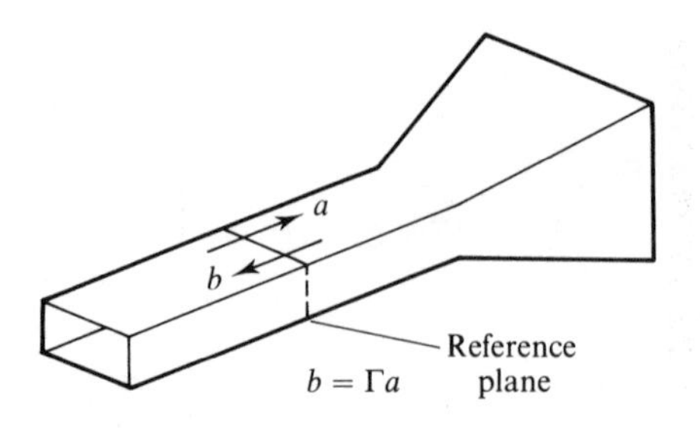

Figure 3.5 A small radiating horn with waveguide feed, best characterized by a reflection coefficient at a reference plane.

3.6 The antenna in reception

The previous sections have considered an antenna radiating power supplied to its input port. In reception the 'input' port is connected to a load and we are concerned with the power delivered to that load when a wave falls on the antenna. This situation is shown schematically in Figure 3.6(a). The performance of the antenna is most conveniently described by an effective receiving area, which is defined as the ratio of the power delivered to the load to the power flux density of the incoming wave. This effective area obviously depends on the degree of mismatch between the load and the antenna; in what follows it will be supposed that matching has been effected. A reciprocity theorem exists (Brown and Clarke, 1980) relating the antenna gain in a specific direction to the effective receiving area to a wave coming from that same direction. This theorem can be stated in two parts:

(1) The received signal amplitude in the (matched) output line arising from a plane wave with electric vector $\boldsymbol{E}_i$ approaching from the direction θ, ϕ (Figure 3.6) is proportional to

$$\boldsymbol{E}_i \cdot \boldsymbol{e}(\theta, \phi)$$

The signal is therefore maximized if

$$\boldsymbol{E}_i \propto \boldsymbol{e}^*(\theta, \phi)$$

This implies that the state of polarization of the incoming wave should be the same as that of the wave which would be transmitted in the same direction.

(2) If polarization matching is assumed, the effective receiving area is given by

$$A(\theta, \phi) = \frac{\lambda^2}{4\pi} G(\theta, \phi) \tag{3.8}$$

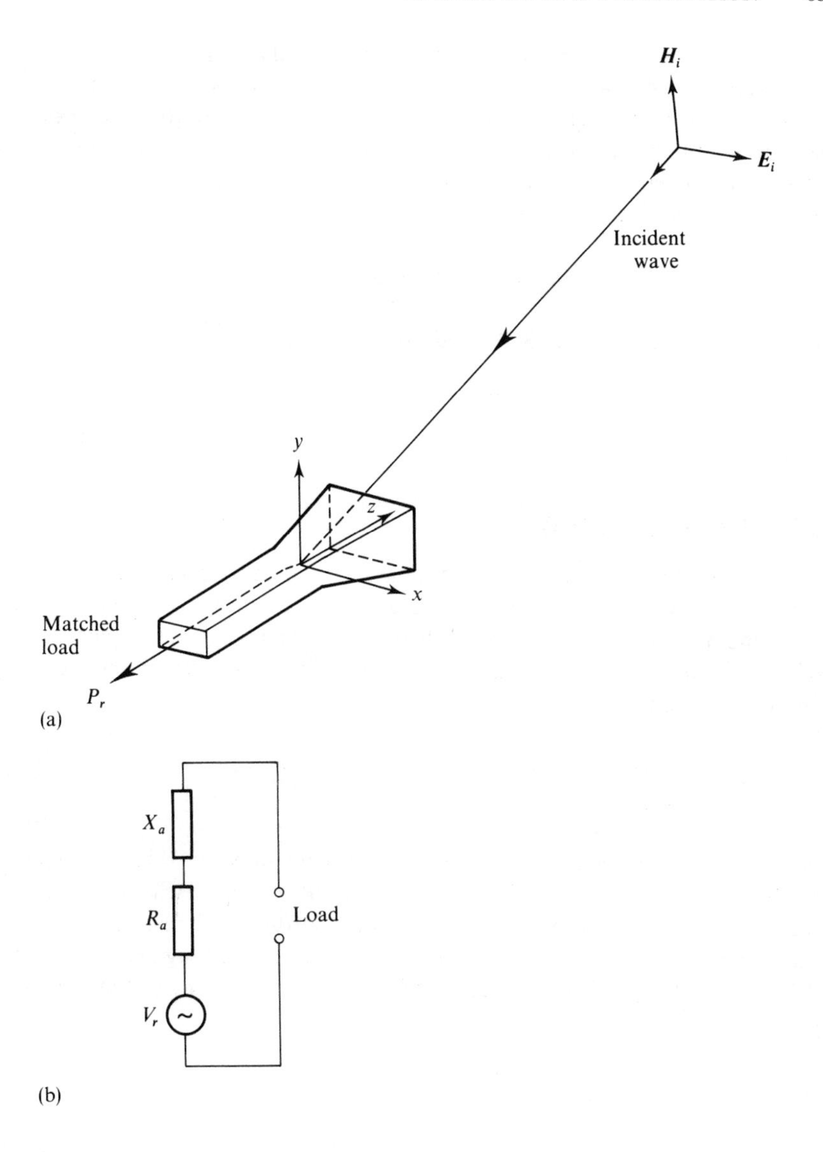

Figure 3.6 An antenna in reception: (a) the power delivered to a load by a wave coming from direction θ, ϕ is related to gain in that direction when the antenna is transmitting; (b) equivalent circuit.

Thus, with the notation of Figure 3.6, the power delivered to a matched load is given by the expression

$$P_r = \frac{1}{2\eta} \boldsymbol{E}_i \cdot \boldsymbol{E}_i^* A(\theta, \phi) \tag{3.9}$$

The power delivered by an antenna to a load appears, as far as the load is concerned, to come from a source of internal impedance equal to the radiation impedance, $R_a + jX_a$, as shown in Figure 3.6(b). In wave terms the source would have the same reflection coefficient as the antenna had in transmission. An expression for the source emf in the circuit of Figure 3.6(b) which agrees with the results given in (1) and (2) above is

$$V_r = (4R_a A(\theta, \phi)/(\eta \boldsymbol{e} \cdot \boldsymbol{e}^*))^{\frac{1}{2}} \boldsymbol{E}_i \cdot \boldsymbol{e}(\theta, \phi)$$

Using equations (3.7) and (3.8), an alternative form is

$$V_r = (4R_a \lambda^2/\eta \langle \boldsymbol{e} \cdot \boldsymbol{e}^* \rangle)^{\frac{1}{2}} \boldsymbol{E}_i \cdot \boldsymbol{e}(\theta, \phi)$$

3.7 A free space link

As was stated in Chapter 1, the propagation segment of a microwave system frequently consists of a transmitting antenna radiating towards a distant receiving antenna. This is shown schematically in Figure 3.7. In accordance with the definition of antenna gain, the power flux density at the receiving antenna is given by

$$p = P_t \frac{G_{12}}{4\pi R^2}$$

in which G_{12} denotes the gain of the transmitting antenna (1) in the direction of the receiving antenna (2). The distance R is supposed so large that radiation scattered from the receiving antenna does not react back on the transmitting antenna. By the definition of effective receiving area, the power delivered to the load, presumed matched, on the receiving antenna will be given by

$$\begin{aligned} P_r &= pA_{21} \\ &= P_t \frac{A_{21} G_{12}}{4\pi R^2} \end{aligned} \tag{3.10}$$

in which A_{21} is the receiving area of the receiving antenna from the direction of the transmitting antenna. If, as is usual, the antennas are designed and oriented so that a polarization match exists between the receiving antenna and the oncoming radiation, the effective receiving area is expressed in terms of antenna gain by equation (3.8). Equation (3.10) may then be written

$$P_r = P_t G_{12} G_{21} (\lambda/4\pi R)^2 \tag{3.11}$$

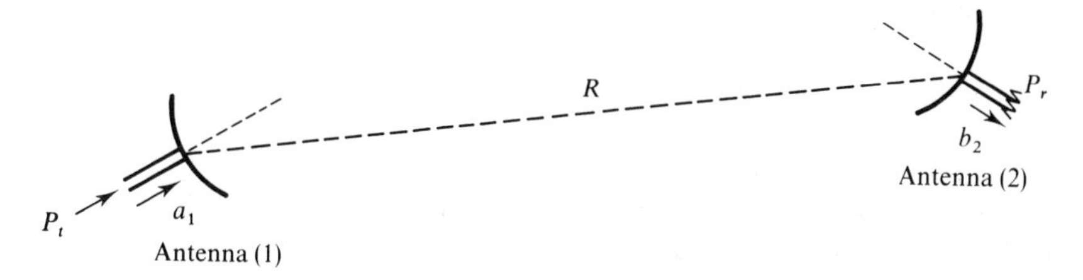

Figure 3.7 A link formed by two antennas. It is convenient to assume that both antennas together with load and source are matched to the feeders.

The equation can equally well be expressed in the form

$$P_r = P_t A_{12} A_{21}/(\lambda R)^2$$

3.7.1 The link as a 2-port

The arrangement of Figure 3.7 can be regarded as a 2-port microwave network for which scattering parameters can be defined. With a load matched to the waveguide we can then write

$$b_2 = S_{21} a_1$$

Hence

$$P_r = |S_{21}|^2 P_t$$

By comparison with equation (3.10) we have

$$|S_{21}|^2 = G_{12} A_{21}/(4\pi R^2) \tag{3.12}$$

If the roles of transmission and reception are interchanged, a matched load on the former transmitting antenna would receive power given by

$$P'_r = |S_{12}|^2 P'_t$$

in which P'_t is the power then supplied to antenna (2). We can also write

$$|S_{12}|^2 = G_{21} A_{12}/(4\pi R^2) \tag{3.13}$$

In the absence of any effects arising from static magnetic fields the 2-port will be reciprocal, so that $S_{12} = S_{21}$. Comparison of equations (3.12) and (3.13) then implies that for two arbitrary antennas

$$G_{12}/A_{12} = G_{21}/A_{21}$$

This in turn implies that a relation of the form of equation (3.8) must exist, but does not give any value for the multiplier. This value must be derived from electromagnetic theory, as stated in Section 3.6.

3.7.2 The power budget

Equation (3.11) is fundamental to the consideration of power flow through the system. In decibel units it may be written in the form

$$(P_r)_{\text{dBW}} = (P_t)_{\text{dBW}} + (G_{12})_{\text{dB}} + (G_{21})_{\text{dB}} - L_s \tag{3.14}$$

in which dBW signifies decibels above 1 Watt. The term L_s is given by

$$L_s = 20 \log (4\pi R/\lambda) \tag{3.15}$$

and is referred to as the **free space propagation loss.** This is a misnomer, but is a way of expressing the fact that much of the power radiated is not delivered to the load on the receiving antenna. Equation (3.14) in some form plays a central role in the design of most microwave systems. There are practical limits on all the terms involved: the received power has to be above a certain minimum determined by the noise level and the permissible error rate; the transmitter power is limited by the available sources and by convenience; the antenna gain will be limited by mechanical dimensions; the distance is then frequently chosen to fit, although it may be that the distance is given as in the case of a link using a satellite in geostationary orbit. It is the trade-off between these features that dominate the design process. As it stands, equation (3.14) applies only to antennas in free space: other terms have to be added when, for example, the propagation path lies in the earth's atmosphere (see Chapter 9).

3.8 Calculation of antenna gain

At lower frequencies, antennas usually consist of arrays of dipoles, each often one-half wavelength long. Since the field radiated by a single dipole is known, the resultant for several dipoles can be found by summation (e.g. Sander and Reed, 1986), allowing for the differences in path length between each dipole and the distant observation point. Such arrays permit the design of directional patterns. It is necessary to set up the design current in each dipole, allowing for mutual coupling effects. At microwave frequencies, the length of the dipole

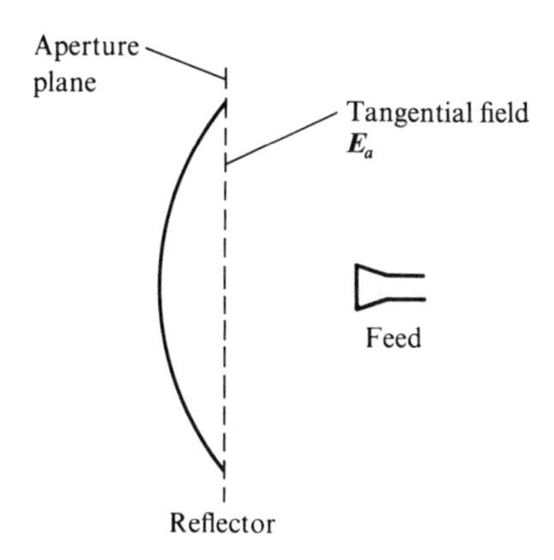

Figure 3.8 A reflector-type antenna. The reflector is a paraboloid of revolution, fed by a small waveguide horn placed at its focus. It is useful to characterize the radiation by the electric field in an aperture plane.

becomes too short for convenience in construction, and antennas take a different form. The required directional properties can still be obtained by arrays of radiating elements, as will be discussed later, but are very frequently obtained by using reflecting surfaces. An example is shown in Figure 3.8 in which the reflector is a paraboloid of revolution and is illuminated by a small waveguide horn. Although an analogy may be made with optics, it must be remembered that a large optical mirror may be 10^6 wavelengths in diameter whereas a microwave reflector is more likely to have linear dimensions of 10^2 wavelengths. The calculation of the gain and radiation pattern of such antennas is usually separated into two stages: firstly the estimation of the field pattern over the reflector surface by means of quasi-optical techniques and hence over an aperture plane, and secondly determination of the far-field associated with the distribution over the aperture plane. This two-stage process also leads to a division in the design process: the selection of a suitable aperture pattern followed by the design of an appropriate feed arrangement. The emphasis is thus shifted at microwave frequencies towards the concept of aperture distribution. Formulae to calculate the distant field from an aperture distribution will be presented in the next section. It is found to be relatively easy to estimate the main lobe and hence the beam-width, but much greater precision is needed to obtain reliable estimates of the side lobes.

3.8.1 The aperture antenna

In the simplest model for an aperture antenna it is assumed that the electric field is specified over an infinite plane: an estimation is made of the field generated by the feed over the physical aperture of the antenna; outside that aperture the field is assumed to be zero. Thus the model corresponds to the

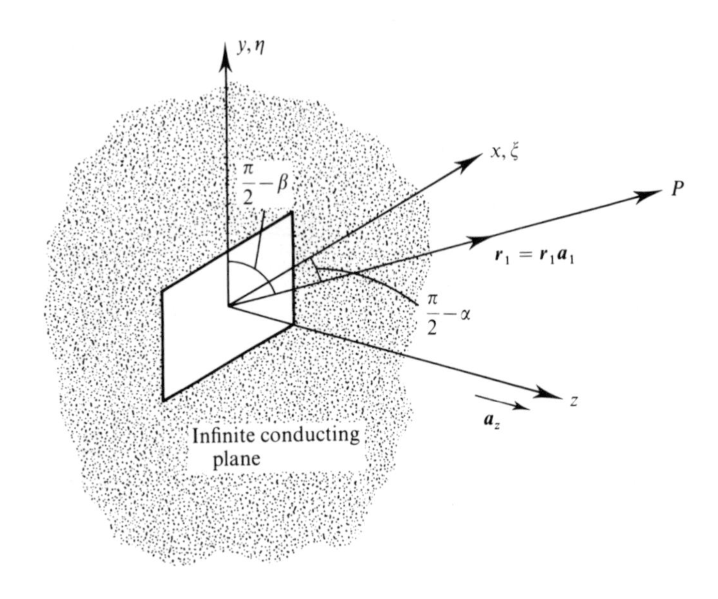

Figure 3.9 Notation for describing the radiation pattern for an aperture antenna. The angles α and β are defined by $\sin\alpha = \boldsymbol{a}_r \cdot \boldsymbol{a}_x$, $\sin\beta = \boldsymbol{a}_r \cdot \boldsymbol{a}_y$. Note that the directional changes produced by varying α and β are only orthogonal for small angles.

aperture inserted into an infinite conducting plane. It is known from electromagnetic theory that this prescription uniquely defines the radiated field elsewhere. Clearly this model can only predict forward radiation, but for highly directional antennas this is frequently sufficient. The radiated field may be calculated by means of the following formula, using the geometrical notation of Figure 3.9:

$$\boldsymbol{E}(\boldsymbol{r}_1) = -\frac{\mathrm{j}k}{2\pi r_1}\exp(-\mathrm{j}kr_1)\boldsymbol{a}_1 \times (\boldsymbol{a}_z \times \boldsymbol{F}) \tag{3.16}$$

where

$$\boldsymbol{F} = \int_{-\infty}^{\infty}\int_{-\infty}^{\infty} \boldsymbol{E}_a(\xi,\eta)\exp(\mathrm{j}k(\xi\sin\alpha + \eta\sin\beta))\,\mathrm{d}\xi\,\mathrm{d}\eta \tag{3.17}$$

and $\boldsymbol{E}_a$ is the tangential component of the electric field in the plane of the aperture. If only directions near the axis are of importance, as in the case of highly directional antennas, equation (3.16) may be simplified to the form

$$\boldsymbol{E}(\boldsymbol{r}_1) = \frac{\mathrm{j}k}{2\pi r_1}\exp(-\mathrm{j}kr_1)\boldsymbol{F} \tag{3.18}$$

It will be noticed that in this approximation the electric field is not quite perpendicular to the radius vector, which in the far-field should be the case. The derivation of these formulae will be found in Rudge (1982).

3.8.2 The uniformly illuminated aperture

Consider the case when the aperture is rectangular in shape, defined by $|x| < a$, $|y| < b$, and $\boldsymbol{E}_a = E_a \boldsymbol{a}_x$ with E_a constant. Evaluating the integral in equation (3.17) we find

$$\boldsymbol{F} = \boldsymbol{a}_x 4E_a \frac{\sin(k\, a \sin\alpha)}{k \sin\alpha} \cdot \frac{\sin(kb \sin\beta)}{k \sin\beta}$$

This function, and hence $\boldsymbol{E}(\boldsymbol{r}_1)$, is the product of two functions of the general form $\sin X/X$, from which the radiation pattern in Figure 3.3(d) was calculated. The maximum value of $|E|$ occurs on the axis, when

$$|E| = \frac{2k}{\pi r_1} E_a ab$$

Zeros occur when either $a \sin\alpha = m\lambda/2$ or $b \sin\beta = n\lambda/2$. Those zeros closest to the axis occur for $\alpha \simeq \lambda/a$, $\beta \simeq \lambda/b$. The beam spread thus depends on the size of the aperture. To evaluate gain it is necessary to know the power input. If the aperture is large compared to the wavelength it is plausible to assume that the wave in the aperture is plane, when the power crossing the aperture is equal to

$$P_s = 4ab\,|E_a|^2/2\eta$$

The actual power flux density on the axis is given by

$$\begin{aligned} p &= |E|^2/2\eta \\ &= \frac{1}{2\eta}\left(\frac{2k}{\pi r_1}\right)^2 (E_a ab)^2 \end{aligned}$$

Using these values in equation (3.5) we find

$$G_m = \frac{4\pi}{\lambda^2}(4ab)$$

This implies an effective receiving area equal to the physical aperture, which the optical analogy would lead one to expect. The realization of a feed to produce uniform illumination would be difficult and not necessarily desirable, for although maximum gain is achieved for the given aperture, the side lobes

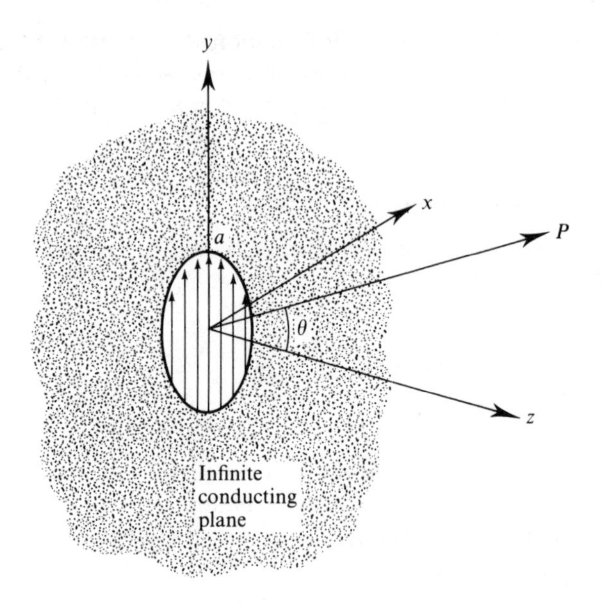

Figure 3.10 A uniformly illuminated circular aperture. Because of the axial symmetry the radiation pattern is a function only of the polar angle θ.

are too high for many applications. A distribution which tapers towards the edges produces smaller side lobes at the expense of a wider beam and lower gain. An example appears in the exercises at the end of the chapter.

The physical aperture nevertheless provides a good guide to the parameters of a microwave antenna. The effective receiving area of a well designed antenna is typically in the range 0.55 to 0.65 of the physical aperture. The beam width can be estimated from the linear dimensions of the aperture.

3.8.3 Uniformly illuminated circular aperture

It is also of interest to consider a circular aperture with uniform illumination, $\boldsymbol{E}_a = E_a \boldsymbol{a}_y$, as indicated in Figure 3.10. Evaluation (see Brown and Clarke, 1988, Section 3.2.2) of equation (3.17) yields the result

$$\boldsymbol{E} = \boldsymbol{a}_y (\mathrm{j}k^2 a^2 E_a/2) \frac{\exp(-\mathrm{j}kr_1)}{kr_1} f(\theta)$$

in which $f(\theta)$ is expressed in terms of the first order Bessel function J_1 in the form

$$f(\theta) = \frac{2J_1(ka \sin\theta)}{ka \sin\theta}$$

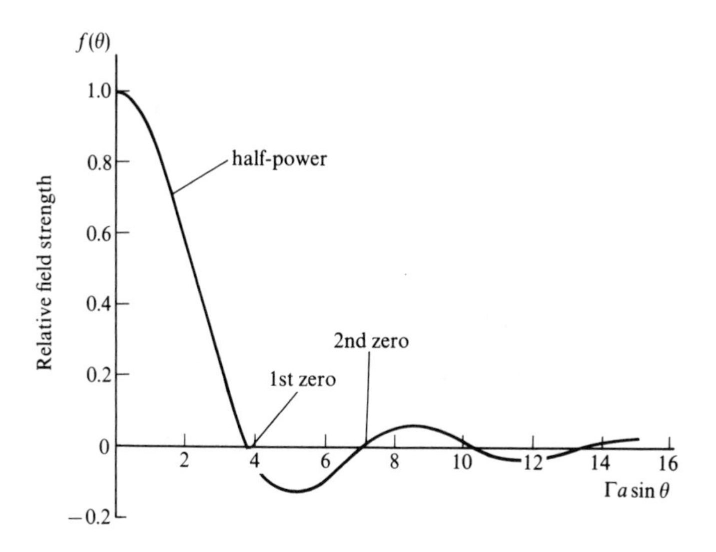

Figure 3.11 The function $f(\theta) = 2J_1(ka \sin\theta)/k \sin\theta$. The abscissa is $ka \sin\theta$ and the ordinate relative field strength in direction θ.

This function $f(\theta)$ is shown graphically in Figure 3.11, and represents a section of the axially symmetric radiation pattern. An analysis similar to that of Section 3.8.2 shows that the effective receiving area is πa^2. From Figure 3.11 the beam width to half-power may be estimated as $29\lambda/a$, and to the first zero as $70\lambda/a$, both values in degrees. The first side lobe is 17 dB down on the maximum, and the second 23 dB.

3.8.3 Illustrative calculations

A close approach to a free space link is that between an earth station and a satellite. We consider the case when the satellite is in geostationary orbit, for which $R \simeq 36\,000$ km. Plausible values for the various parameters are as follows:

Frequency of down link	4 GHz
Power delivered to antenna	5 W, or 7 dBW
Diameter of satellite antenna	1.5 m
Diameter of earth station antenna	10 m

Allowing a factor of 0.6 relating effective receiving area to physical area, we find

$$A_s = 1.06\,\text{m}^2 \qquad A_e = 47\,\text{m}^2$$

in which the suffixes s and e denote satellite and earth respectively. Given the frequency we may then estimate the gains of the two antennas. We find

$$G_s = 33.7\,\text{dB} \qquad G_e = 50.2\,\text{dB}$$

For the free space propagation loss

$$20 \log(4\pi R/\lambda) = 195.6\,\text{dB}$$

Inserting these figures into the power budget equation, (3.14), we have

$$\begin{aligned} P_r &= 7 + 33.7 + 50.2 - 195.6 \\ &= -104.7\,\text{dBW, or } 34\,\text{pW} \end{aligned}$$

It will be seen later that the noise signal at the earth station receiver would be about 0.3 pW, resulting in a signal-to-noise ratio of some 20 dB. The EIRP (Section 3.3) for the satellite transmitter is $7 + 33.7 = 41.3$ dBW, corresponding to an isotropic source of about 13.5 kW.

3.9 Frequency re-use: copolar and cross-polar fields

It has been mentioned earlier, in Section 3.6, that the signal delivered to the load on the receiving antenna depends on the polarization match. Antennas can be designed with two independent feeds which radiate with mutually orthogonal polarizations. Two such antennas form a link which can provide two completely independent channels on the same frequency. Hence the term **frequency re-use**. The aperture antennas considered in Sections 3.8.2 and 3.8.3 gave a distant electric field plane polarized in the sense of the aperture polarization, which in turn is determined by the feed arrangements. An antenna will not, in practice, have this simple property: if the aperture has electric field lines of the form shown in Figure 3.12, for example, the distant field will be polarized in the x direction on axis but, off axis, will contain a component polarized in the y direction. The use of two feeds giving orthogonal polarizations would provide independent channels when the antennas were aligned precisely on axis, but cross-talk would occur if they were misaligned. In this example the field with electric component in the x direction would be termed the **copolar field**, and that in the y direction the **cross-polar field**. This illustration is satisfactory for the case of highly directional antennas but more elaborate definition is needed for the general case (as in Ludwig, 1973).

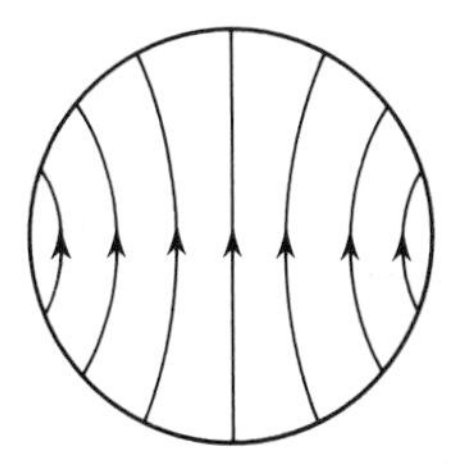

Figure 3.12 A form of non-uniform illumination over a circular aperture which will give rise to cross-polar fields off axis.

3.10 Some types of antenna

The following sections provide a brief resumé of some of the different types of antenna which are used in the microwave region of the spectrum. The features of interest to antenna designers obviously include gain and beam width, but there may also be requirements on the magnitude of side lobes on the radiation pattern and on the cross-polar fields as well as on bandwidth of operation. The precise requirements vary with application, and new applications may bring new problems. There is therefore continual development, and since any collection of conductors forms an antenna, new types of antenna appear. The following sections can only present in broad outline some of the antennas in regular use, and reference must be made to other works, especially Brown and Clarke (1980), Jasik and Johnson (1984), Rudge (1982), and Silver (1949), for detailed presentation.

3.10.1 Reflector antennas

A widely used reflector antenna has been illustrated in Figure 3.8. Two components may be distinguished, the reflector and the primary feed that controls the illumination of the aperture, and which will have a relatively broad radiation pattern.

PRIMARY FEED ANTENNAS

The design of the primary feed antenna must take into account the way in which the incident field strength varies over the reflector and the polarization pattern required. The simplest feed is the half-wave dipole of which a considerable number of variants have been produced (some are illustrated in Figure 3.13). All are designed to feed through the reflector. These are useful when only moderate performance is required. The small horn fed from rectangular waveguide in the TE_{01} dominant mode provides a useful feed when the mouth dimensions are correctly chosen to give a symmetrical

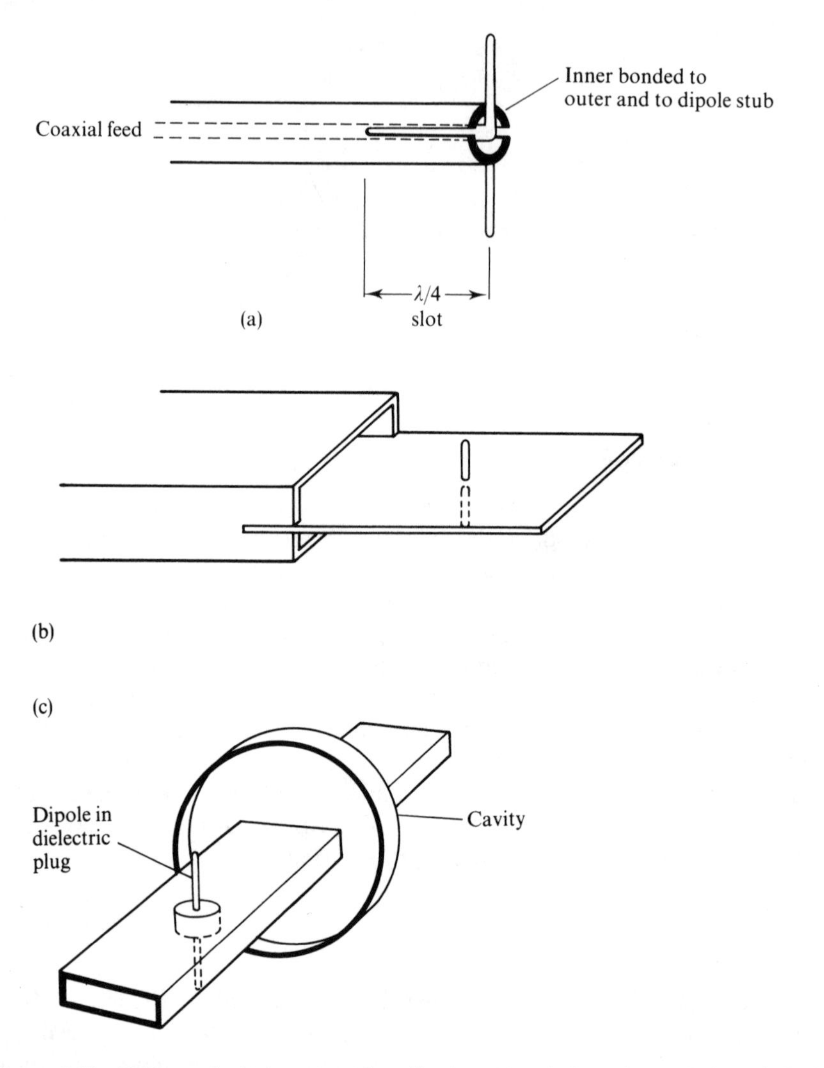

Figure 3.13 Primary feeds incorporating dipoles: (a) a balanced-to-unbalanced feed (*balun*) which avoids unsymmetrical fields between the dipole and outer edge of the coaxial line; (b) a dipole excited from a waveguide; (c) dipole backed by a cavity reflector.

pattern, but it is suitable only for linear polarization. A conical horn from circular waveguide using the TE_{11} mode is suitable for dual or circular polarization. In order to obtain a polarization pattern in circular guide similar to that in the small rectangular horn, a properly phased mixture of TE_{11} and TM_{11} can be used: the mode patterns are those shown in Figure 2.15 where it will be seen that the electric field lines curve in opposite senses. The effect may be obtained either by exciting the modes separately by means of discontinuities in the guide, or by use of structures like the corrugated horn

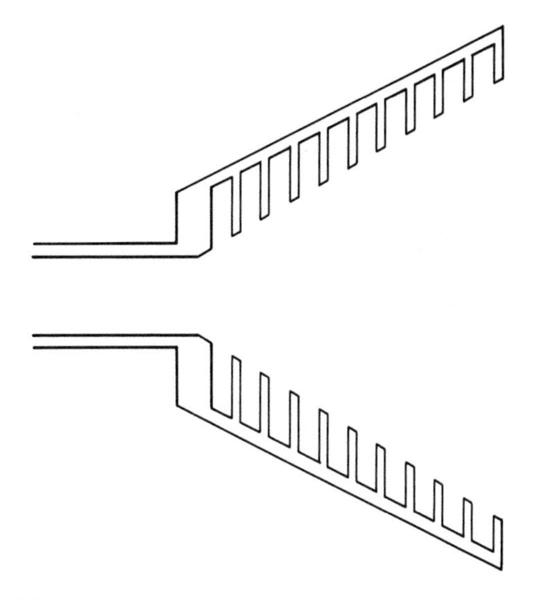

Figure 3.14 Axial section of a corrugated horn designed to propagate a hybrid mode reducing cross-polar fields.

(Figure 3.14), which ensure propagation of both modes in a hybrid combination. For further details, consult Rudge (1982).

REFLECTORS

The paraboloid of revolution illustrated in Figure 3.8 formed the earliest and still very widely used reflecting surface. The optical ray properties of such a reflector are shown in Figure 3.15. Diameter, focal length and primary feed are at the disposal of the designer. Design can be based on a circular aperture antenna, which for the case of uniform illumination was considered in Section 3.8.3. If the illumination is made to taper off towards the edges of the aperture, the side lobes are reduced relative to the main beam, at the expense of increased beam width. The tailoring of the aperture illumination by choice of primary feed and by choice of dimensions and focal length to give required performance is a major part of the design of such antennas. It is found that far-out side lobes arise from diffraction at the edge of the reflector: these are often modified by the use of absorbent material or other edging forms. A further point to be considered is polarization: in the case of the uniformly illuminated aperture the aperture field was polarized uniformly in the x direction, and the resulting far-field pure copolar. A common feed is a small rectangular horn which is also polarized uniformly in one direction, but it is found that in the process of reflection from the curved reflector the aperture field is no longer uniformly polarized, resulting in the far-field containing a

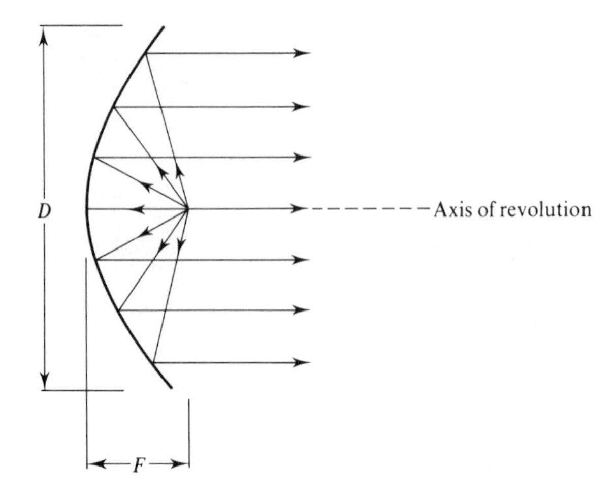

Figure 3.15 The ray properties of a paraboloid reflecting surface. The parameters are the diameter D and the focal length F.

cross-polarized component. If it is wished to use the antenna in two polarization modes, the polarization pattern of the feed must be chosen so as to reduce cross-polarization to an acceptable level. Figure 3.8 shows the feed at the focal point. This is sometimes undesirable in that it blocks part of the aperture and further may involve a long length of transmission line or waveguide to connect with the receiver. Two other variants are shown in Figure 3.16(a) and (b): the **offset reflector** antenna and the **horn reflector** antenna respectively. The aperture field for the first can be evaluated in exactly the same way as for the full paraboloid; the beam is not axi-symmetrical in this case, and it can be designed to give low cross-polar fields. In the horn-fed reflector, the horn is fed with a dominant mode waveguide, with the phase centre of the horn at the focal point of the reflector. The field in the horn is of closely dominant mode form but diverging up the horn. It is finally reflected and illuminates the aperture. This antenna is noteworthy in one respect: no radiation can take place except through the aperture. As a result, it is useful when low antenna noise must be achieved (Crawford *et al.*, 1961).

In these, as in all other antennas, it should be noted that the mechanical construction and weight must be considered in making a choice.

Another configuration which has been found to be useful takes its name from its optical counterpart, the Cassegrain telescope. This can be of symmetrical or offset form, as shown in Figure 3.17(a) and (b). All these configurations avoid the need for long transmission lines between antenna and receiver: the receiver preamplifier may be mounted immediately adjacent to the primary feed. The symmetrical version is mechanically easier to construct than the offset but suffers aperture blocking by the subreflector and

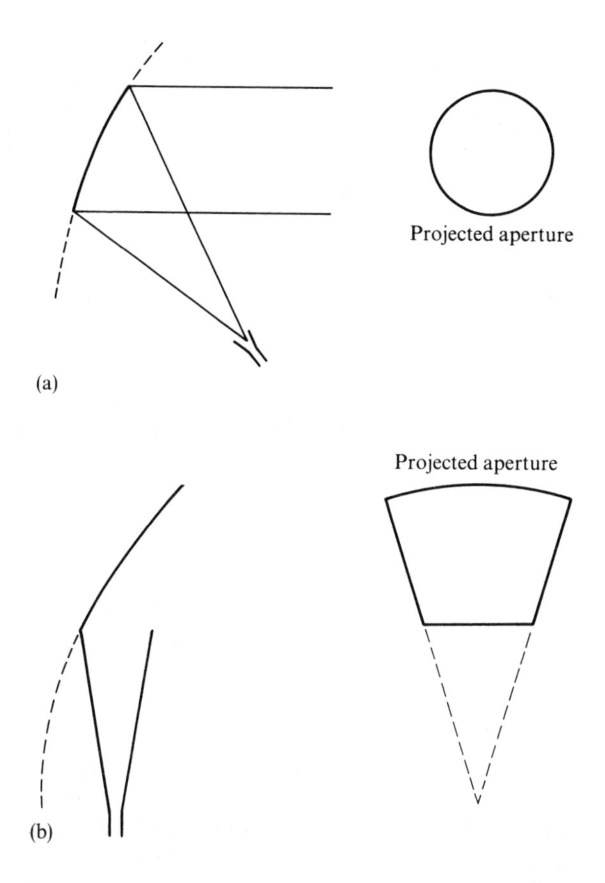

Figure 3.16 Two antennas using offset paraboloid reflectors: (a) a primary-fed reflector; (b) the horn-paraboloid antenna.

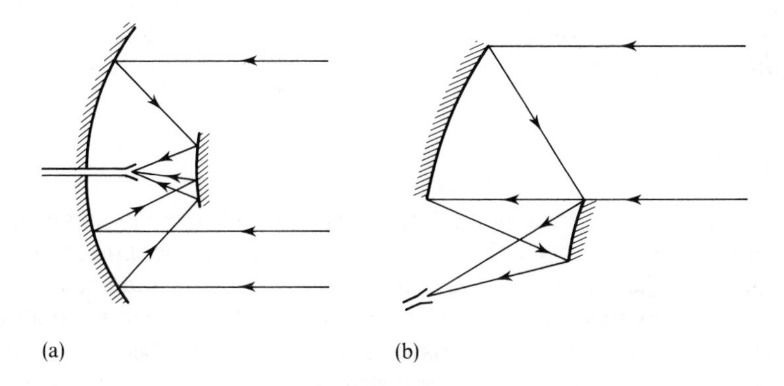

Figure 3.17 Cassegrain antennas with subreflectors: (a) symmetrical, (b) offset.

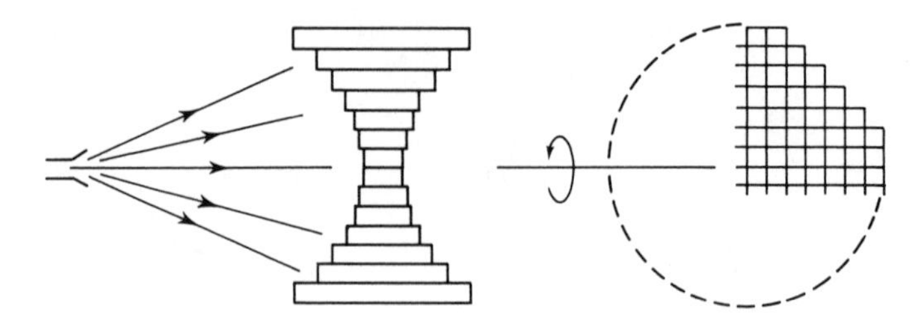

Figure 3.18 A waveguide lens: (a) section in a plane through the axis, (b) end view.

by the struts needed to support it. The offset version has been widely used in communications satellites, where a multiplicity of feeds are used to provide multiple beams or shaped beam coverage.

3.10.2 Lens antennas

In optics, collimation of light is frequently achieved by lenses. Lenses can also be made for microwaves (see Silver, 1949 and Rudge, 1982). The obvious choice is to use dielectrics, but alternatives are possible. The useful property of lens collimation is that lateral movement of a source in the focal plane produces beam swinging with less aberration than with reflectors. For this application lightweight waveguide lenses have been designed. The basic structure is indicated in Figure 3.18. The phase velocity in a waveguide is greater than in free space, so that in order to make the delay along all parts equal, the lens must take a concave form as shown. Calculation of the properties of such a system takes into account the power incident on each waveguide, the proportion transmitted and the way it is reradiated at the far end (Rudge, 1982). When many elements are used, this is clearly a substantial computational task. The lens size may be reduced by shortening waveguide elements where possible by lengths equal to integer multiples of guide wavelength. Such a lens is called a **zoned** lens. Clearly, a lens using waveguide elements is frequency sensitive and has limited bandwidth. Zoned lenses are slightly better in this respect.

3.10.3 'Flat' antennas

For certain purposes, such as in aircraft, it is desirable to use as an antenna a structure which conforms to the shape of the vehicle and does not give surface irregularities. The class can be illustrated by the microstrip antenna indicated in Figure 3.19(a): this shows a rectangular 'pad' fed by a line at the midpoint of one side. The general characteristics of microstrip have been discussed in Section 2.7: the electric fields in the structure occur only at the edges of

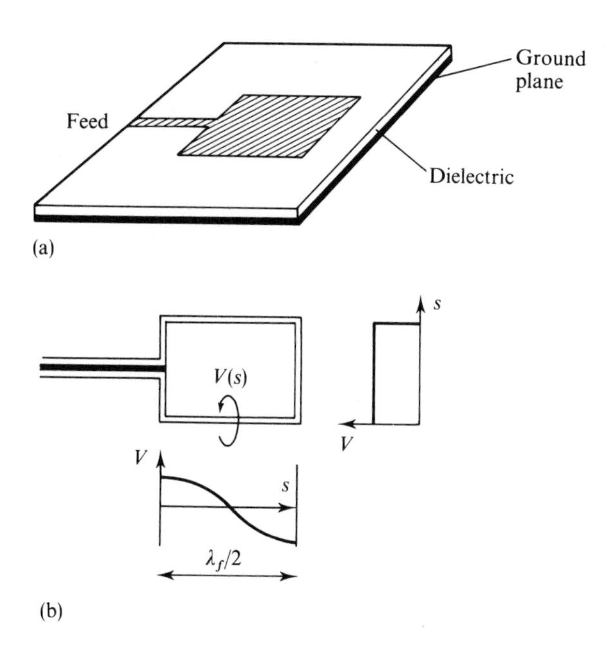

Figure 3.19 A microstrip antenna: (a) physical form, (b) slot equivalent, showing variation of voltage round the slot.

the pad, and are thus simulated closely by the planar arrangement of Figure 3.19(b). The far-field radiation of this structure can be found by use of equations (3.16) and (3.17). Since the slot width is much less than the wavelength equation (3.17) can be written in the form

$$\boldsymbol{F} = \oint V(s) \exp(\mathrm{j}k\boldsymbol{r}_2(s) \cdot \boldsymbol{a}_1)\, \mathrm{d}s$$

where s is the position along the slot and $V(s)$ is the voltage across the slot. This is equivalent to saying that the element ds of the slot is behaving like a short magnetic dipole of strength $2V(s)\mathrm{d}s$. The voltage distribution around the edge will depend on frequency and the dimensions of the patch: the situation at resonance is indicated in the diagram. Such a pad has maximum radiation normal to its plane, and has a rather broad radiation pattern.

Complete arrays of radiating elements can be made and used to give greater directivity as described by James *et al.* (1981) and Rudge (1982). In general, microstrip tends to be lossy, and it is necessary to distinguish between directivity and gain. Microstrip antennas are simple to construct and can be designed to fit curved surfaces. Flat antennas can also be fabricated out of separate components.

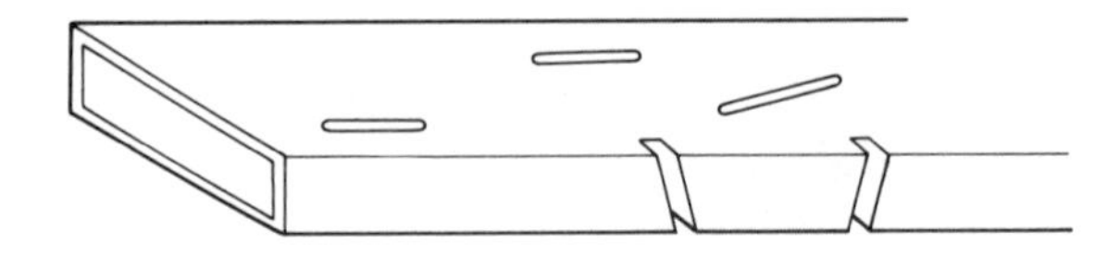

Figure 3.20 Various radiating slots in waveguide walls: broad face and edge slots are shown.

3.10.4 Antenna arrays

As discussed in Section 3.8, the field of any antenna is found by summation of the fields from the various sources comprising the antenna. In Section 3.8.1 this summation process was presented for a continuum of equivalent sources. A similar, simpler, process applies when the antenna consists of a number of discrete sources. The theory of such arrays is presented in standard texts such as Sander and Reed (1986), usually applied to somewhat lower frequencies, and where the individual source is a half-wave dipole. Such arrays are used in communications systems at the lower microwave frequencies. An example of a more specifically microwave array is the slotted waveguide: slots cut in the wall of a rectangular waveguide will act as radiators if suitably disposed. Three configurations are shown in Figure 3.20. In order to radiate, a broad wall slot must either be turned at an angle on the centre line or moved to one side; an edge slot must be inclined at an angle to the edge. The phase of the signal coupled out of the slot is 0° or 180°, depending on the sense of the inclination in the case of an edge slot, or on the direction of displacement in the case of a broad face slot. The strength of the signal depends on the angle of inclination or on the displacement. Only a small fraction of the power flowing in the waveguide is coupled out of one slot. Thus an array of sources can be fabricated as indicated in Figure 3.21. The slots are alternated in direction of inclination at intervals of about one-half wavelength, producing an array of in-phase radiators at half-wavelength intervals. This technique is highly suitable for precision automatic manufacture. Such an array can be used with a horn flare as indicated in Figure 3.22(a), or with a reflector as in Figure 3.22(b). Precision design for such an array must allow for mutual coupling between adjacent slots.

The concept of a linear array can be extended to a two-dimensional array, either planar or curved. The elements in such an array could be dipoles, or low gain devices such as horns, or even open ended waveguides. The attractiveness of such an array lies in the fact that the required source distribution can in principle be produced directly, rather than through an intermediate reflector. It is necessary to define the amplitude and phase of each source, which can be done in various ways. For example a two-dimensional array can be made of an assembly of linear arrays such as slotted waveguides: this gives a measure of independence between the design of the linear array and the power splitting between the arrays. The use of a

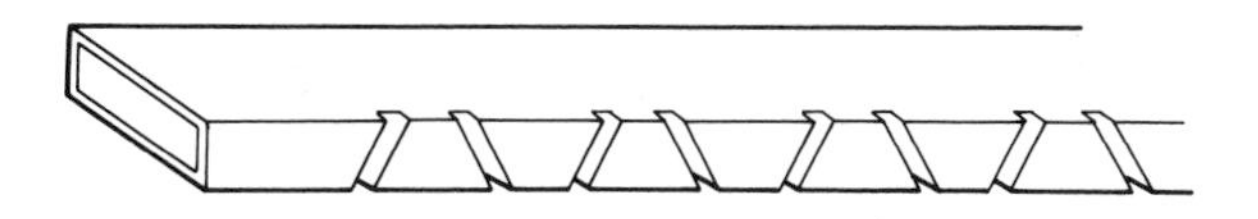

Figure 3.21 An edge slot linear array.

waveguide allows phase adjustment by frequency change, giving electronic beam swinging. The choice of the type of radiating element involves the operating frequency and the bandwidth.

The logical end point of the planar array concept is the **phased array**, consisting of a very large number of independently controlled radiators, perhaps over 10^3 in number. Such a concept has been made practically possible only by the development of small microwave components and phase shifters with low power consumption. The design of planar arrays must take into account the mutual coupling between radiating elements. A number of methods, some experimental and some involving mathematical analysis, are now available (Rudge, 1982).

It is sometimes desirable to have a non-planar phased array, perhaps to fit a curved surface or to give a wider angle of view. For the latter purpose three planar arrays might suffice; for the former the curved surface is essential. Such arrays are made, but are much less easy to design than planar ones because of the difficulty of calculating mutual effects.

Although design methods have not been discussed in detail in the above sections, it is desirable to emphasize that in practice, design is heavily computational, both in the application of electromagnetic theory and in the computer aided design of mechanical and electrical hardware.

3.11 Antenna measurements

This section considers the measurement of antenna gain and radiation patterns. Radiation impedance is measured in the same way as for any other microwave load and this is dealt with in Chapter 11.

3.11.1 Direct gain measurement

The gain of an antenna was defined in the context of the link illustrated in Figure 3.7, between two antennas of fixed orientation. The equation relating to this link was equation (3.11), repeated here for convenience:

$$P_r = P_s G_{12} G_{21} \left(\frac{\lambda}{4\pi R}\right)^2$$

It is assumed that the two antennas are polarization matched.

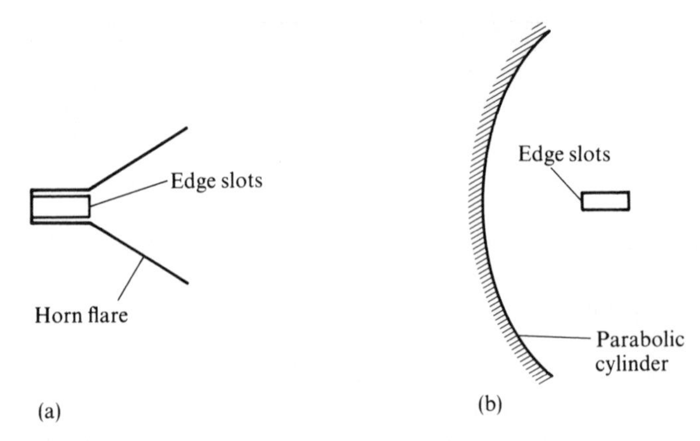

Figure 3.22 An edge slot array: (a) with a horn flare, (b) with parabolic cylinder reflector. Both configurations reduce vertical spread of radiation.

If G_{12} is the required gain, it is necessary to know (P_r/P_s), R and G_{21}. There are three ways of dealing with the last matter:

(1) use a standard horn antenna,

(2) use two identical antennas so that $G_{21} = G_{12}$,

(3) use three antennas, obtaining numerical values for the three products $G_A G_B$, $G_B G_C$, $G_C G_A$.

From these each of G_A, G_B and G_C can be determined. Since the system is reciprocal, the antenna under test can be used either in transmit or receive mode, whichever is the more convenient. Considerable care is needed to obtain an accurate measurement of the ratio P_r/P_s. By definition, P_r is the power delivered to a matched load, so that even if P_r is measured directly, matching must be checked and, if necessary, correction made. Similarly, P_s is the power accepted by the transmitting antenna, and P_s is the same as the available power from the source only if it is matched to the source. Direct power measurement may not be possible, and a more sensitive device or indeed a microwave receiver may need calibrating. Separate determination of P_r and P_s may introduce more errors than would a direct determination of the ratio (P_r/P_s): this can be done if a screened link is provided between the two antennas and suitable equipment for comparison is available. In this way phase can also be measured. An alternative procedure is to mount a reference antenna as near as possible to the test antenna: this effectively calibrates the source antenna.

A test range to carry out the measurements takes the form illustrated in Figure 3.23. The test antenna is used in the receive mode and is illuminated by a nominally 'plane' wave from a source antenna. A reference antenna

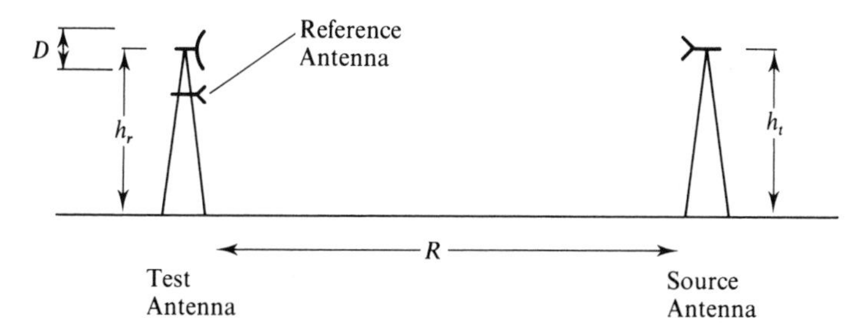

Figure 3.23 Configuration of an antenna test range, showing antenna under test, source antenna and reference antenna.

monitors the source in amplitude and phase. The principal sources of error are:

(1) multiple reflections between antennas,
(2) deviations from plane of the source wavefronts,
(3) ground (or other) reflections.

The magnitude of the multiple reflection effects depends on the directivity of both the source and test antennas and on their separation. If these parameters are chosen so as to satisfy the second criterion, errors from multiple reflections are usually negligible.

Deviations of source wavefronts occur both in phase and amplitude. Of these, phase is the more important, and non-uniformity of phase across the aperture plane has the effect of filling in the nulls of the radiation pattern. The Rayleigh criterion $R > 2D^2/\lambda$ limits phase taper to less than $\pi/8$. Amplitude deviation across the aperture plane is kept to less than 0.25 dB. Errors in gain measurement with these restrictions will be of the order of 0.2 dB. Reflections from the ground are reduced if the main lobe of the source antenna does not illuminate the ground so as to give a reflection into the test antenna. Associated with the taper of 0.25 dB, this dictates that the height of the towers must be greater than $4D$. Reflections from nearby buildings must also be considered. For example, the measurement of side lobes can be seriously affected by scattering from such obstacles into the main lobe of the antenna being measured.

For some antennas at microwave frequencies an indoor test range is feasible: unwanted reflections can be suppressed with absorbing material, enabling smaller spaces to be used. Complete anechoic rooms are in regular use.

This discussion has been primarily concerned with measurement of gain, but measurement of radiation patterns follows a similar course. It is necessary to provide suitable arrangements for obtaining variable orientation between the antennas. It may be noted that good measurement of side lobes

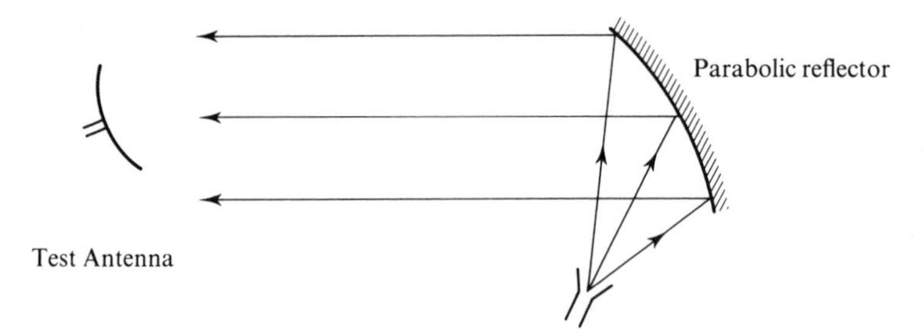

Figure 3.24 Schematic form of a 'compact' antenna range.

requires a separation of greater than twice the Rayleigh distance $2D^2/\lambda$. The measurement of radiation patterns may well be restricted to two principal planes, that containing the electric vector, or E-plane and that perpendicular to it, the H-plane.

3.11.2 Short ranges

Various artifices have been devised to shorten the length of range needed. The prime need is to illuminate the test antenna with a sufficiently uniform wave. Ray optics suggest the use of arrangements such as the compact range illustrated in Figure 3.24. Design of such a system is evidently not dissimilar to designing an antenna. The distance between test antenna and reflector is not restricted since to a first order a parallel beam is produced. The actual field present in the test position can be explored with probes, and effects on the accuracy of test measurements assessed. It is aimed to produce uniformity over an aperture substantially greater than the aperture of the test antenna.

The plane wave in the above situation can be thought of as produced by a distribution of sources over the reflector. An approximation to a plane wave can also be made from a finite number of sources. Such a source distribution could actually be made or, more practically, synthesized by summation in a computer of the responses of the test antenna as a single source is moved to the various positions in the simulating array.

3.11.3 Near-field measurements

Attention has been confined in this chapter to the far-field region of an antenna, and endeavours are usually made to operate in this region. However, as exemplified by the treatment of aperture antennas, knowledge of the near-field values can be used to calculate behaviour in the far-field. These near-field values can be measured given suitable equipment. A system to make use of such measurements is highly sophisticated: a mechanical system is needed for accurately positioning a probe antenna over the surface to be scanned, which may be planar, cylindrical or spherical, depending on the

(a)

Plate 1 Satellite links. (a) An Intelsat 'Class A' Earth Station at Goonhilly Down, Cornwall, operated by British Telecom. The 30 m diameter reflector is fitted with a subreflector and rear feed. (b) An 'earth' station on a North Sea oil rig, providing communication via satellite. The reflector is 3.7 m in diameter, and also uses subreflector and rear feed. (*Photographs courtesy of British Telecom.*)

(b)

method adopted; there is a very large amount of data to be handled and a large amount of numerical manipulation. Such specialist facilities may be the best way of dealing with certain large antennas.

3.12 Summary

Emphasis has been placed on the importance of the 'propagation' segment of a microwave link. The need to have a quantitative estimation of received power requires characterization of antennas both in transmission and reception, leading to the concepts of radiation pattern, antenna gain and effective receiving area. The correlation of the dimensions with gain and beam width have been demonstrated for aperture antennas. Various practically important antennas have been described.

3.13 Summary of formulae

Antenna far-field
$$\boldsymbol{E} = A\frac{\exp(-jkr)}{kr}\boldsymbol{e}(\theta,\phi), \quad \boldsymbol{a}_r\cdot\boldsymbol{e} = 0$$
$$\boldsymbol{H} = \eta^{-1}\boldsymbol{a}_r \times \boldsymbol{E}$$

If $\boldsymbol{e} = e_1\boldsymbol{a}_\theta + e_2\boldsymbol{a}_\phi$,

for linear polarization $e_1 = e_2$
for circular right-handed polarization $e_2 = -je_1$
for circular left-handed polarization $e_2 = je_2$

Power
$$P_s = \tfrac{1}{2}\eta^{-1}|A/k|^2\langle\boldsymbol{e}\cdot\boldsymbol{e}^*\rangle$$

where
$$\langle\boldsymbol{e}\cdot\boldsymbol{e}^*\rangle = \int_0^{2\pi}\int_0^{\pi}\boldsymbol{e}\cdot\boldsymbol{e}^*\sin\theta\,d\theta\,d\phi$$

Power flux density $p = \tfrac{1}{2}\eta^{-1}|A/kr|^2\boldsymbol{e}\cdot\boldsymbol{e}^*$
For an isotropic radiator $p = P_s/4\pi r^2$
Antenna gain $G(\theta,\phi)$ $p = (P_s/4\pi r^2)G(\theta,\phi)$

$$G(\theta,\phi) = 4\pi\boldsymbol{e}\cdot\boldsymbol{e}^*/\langle\boldsymbol{e}\cdot\boldsymbol{e}^*\rangle$$
$$|\boldsymbol{E}| = (\eta P_s G/2\pi)^{\frac{1}{2}}\frac{1}{r}$$
$$\int G\,d\Omega = 4\pi$$

Effective isotropic radiated power P_sG

Effective receiving area with polarization and impedance matching

$$A(\theta, \phi) = \frac{\lambda^2}{4\pi} G(\theta, \phi)$$

Link length R in free space, transmitter power P_t, antenna gains G_t, G_r and wavelength λ, received power P_r

$$\begin{aligned} P_r &= P_t G_t G_r (\lambda/4\pi R)^2 \\ &= P_t A_t A_r/(\lambda R)^2 \end{aligned}$$

Free space 'loss' $20 \log (4\pi R/\lambda)$ dB

Aperture antennas: far-field at $\boldsymbol{r}_1$

$$\begin{aligned} \boldsymbol{E}(\boldsymbol{r}_1) &= \frac{-\mathrm{j}k}{2\pi r_1} \exp(-\mathrm{j}kr_1)\boldsymbol{a}_1 \times (\boldsymbol{a}_z \times \boldsymbol{F}) \\ &\simeq \frac{\mathrm{j}k}{2\pi r_1} \exp(-\mathrm{j}kr_1)\boldsymbol{F} \end{aligned}$$

for highly directional antennas

$$\boldsymbol{F}(\theta, \phi) = \iint \boldsymbol{E}_a(\xi, \eta) \exp(\mathrm{j}k(\xi \sin\alpha + \eta \sin\beta))\, \mathrm{d}\xi\, \mathrm{d}\eta$$

Uniformly illuminated rectangular aperture $a \times b$

$$\boldsymbol{F} = 4E_a \boldsymbol{a}_x \frac{\sin(ka \sin\alpha)}{k \sin\alpha} \cdot \frac{\sin(kb \sin\beta)}{k \sin\beta}$$

Uniformly illuminated circular aperture, radius a

$$\boldsymbol{E} = \boldsymbol{a}_x (\mathrm{j}k^2 a^2 E_a/2) \frac{\exp(-\mathrm{j}kr_1)}{kr_1} \cdot \frac{2J_1(ka \sin\theta)}{ka \sin\theta}$$

Beam width to half-power $29\lambda/a$ degrees

to first nulls $70\lambda/a$ degrees

Antenna specification relates to:

- gain;
- receiving area;
- radiation impedance, or reflection coefficient;
- polarization;
- bandwidth of operation.

EXERCISES

3.1 A quasi-plane wave carries locally a power density of $1\,\mathrm{mW\,m^{-2}}$. Determine the peak electric field strength. What would be the power of an isotropic source that would give this power density at a distance of 5 km?

3.2 An antenna operates at a wavelength of 6 cm, and has a gain of 35 dB. Calculate its effective aperture at this wavelength.

3.3 The antenna of question 2 is fed at the design frequency with a total power of 5 W. What is the effective isotropic radiated power? The maximum permissible radiation density to which the human body should be exposed is put at $5\,\mathrm{mW\,m^{-2}}$. How closely can the antenna be approached?

3.4 The antenna of question 3 forms the transmitter of a transmit–receive link 35 km in distance. The receiving antenna has a gain of 25 dB. Assuming free space propagation, estimate the received signal power. What adjustments to the two antennas would be needed to realize this?

3.5 A communication system is designed to operate at 90 GHz. The physical aperture of the antenna is circular with diameter of 0.4 m, and it is estimated that the effective aperture is one-half of the physical area. Estimate the transmitter power required to give 1 nW received signal over a 15 km link, assuming free space propagation.

3.6 The far-field radiation pattern of an aperture is chiefly determined by the function $\boldsymbol{F}$ defined in equation (3.17). Show that if $\boldsymbol{E}_a(\xi,\eta) = f_1(\xi)f_2(\eta)\boldsymbol{a}_x$ then

$$\boldsymbol{F} = \boldsymbol{a}_x F_1(\alpha)F_2(\beta)$$

Thus investigate the far-field radiation pattern when

(1) $$E_a = E_0 \cos\frac{\pi\xi}{2a}\cos\frac{\pi\eta}{2b} \quad |\xi| < a, |\eta| < b$$

$$= 0 \qquad |\xi| > a, |\eta| > b.$$

(2) $$E_a = E_0 \exp\left(-\frac{\xi^2+\eta^2}{\omega^2}\right)$$

$$\left[\int_{-\infty}^{\infty} \exp(-ax^2 \pm jbx)\,\mathrm{d}x = (\pi/a)^{\frac{1}{2}} \exp(-b^2/4a)\right]$$

Compare and contrast the far-fields produced by these two distributions and by the uniformly illuminated aperture.

3.7 A linear phased array consists of N sources each radiating isotropically in the equatorial plane xz. They may be taken as located at the points $x = na$, $y = 0$, $z = 0$, $n = 0, 1, \ldots, N-1$. The strength of the source at $x = na$ is $I_0 \exp(jn\psi)$. Show that the radiation pattern in such a plane is determined by

$$\frac{1}{N}\frac{\sin\frac{1}{2}N(\psi + ka\sin\theta)}{\sin\frac{1}{2}(\psi + ka\sin\theta)}$$

In a particular case, $ka = \pi$, $\psi = 0$. Show that the beam is then broadside to the array, and the angular separation between the zeros adjacent to the main beam is approximately $4/N$ radians, $N \gg 1$. If ψ is non-zero, show that the beam is swung to one side, and that the separation between zeros becomes larger by a factor $(1 - \psi^2/\pi^2)^{-\frac{1}{2}}$, assuming $|\psi| < \pi$.

Comment in this context on the use of the slotted waveguide array of Figure 3.21.

3.8 Two identical horns are mounted in the orientation for maximum transmission 5 m apart and the attenuation at 10 GHz under matched conditions is found to be 33 dB. One horn is used as a source for a 1 m diameter parabolic reflector antenna at a distance of 75 m. In this situation the attenuation is found to be 46 dB. Estimate the gain of the parabolic reflector antenna.

Comment on the adequacy of the separation and suggest a suitable height above ground for the test antenna.

References

Brown, J. and Clarke, R. (1980). *Diffraction Theory and Antennas.* Chichester: Ellis Horwood, Sections 3.2.2 and 4.2.1.

Collin, R. E. (1985). *Antennas and Radiowave Propagation.* New York: McGraw-Hill.

Crawford, A. B., Hogg, D. C. and Hunt, L. E. (1961). 'Horn-reflector Antenna for Space Communication', *Bell System Tech. J.* **40**, 1095–116.

Harrington, R. F. (1961). *Time–Harmonic Electromagnetic Fields.* New York: McGraw-Hill.

James, J. R., Hall, P. S. and Wood, C. (1981). *Microstrip Antenna Theory and Design.* Stevenage: Peter Peregrinus.

Jasik, H. and Johnson, R. C. (1984). *Antenna Engineering Handbook.* New York: McGraw-Hill.

Jordan, E. C. and Balmain, K. G. (1968). *Electromagnetic Waves and Radiating Systems.* Englewood Cliffs, New Jersey: Prentice Hall.

Ludwig, A. C. (1973). 'The definition of cross-polarisation'. *IEEE Trans.* **AP21**, Jan, 116–9.

Ramo, S., Whinnery, J. R. and Van Duzer, T. (1984). *Fields and Waves in Communication Electronics.* New York: Wiley.

Rudge, A. W. Ed. (1982). *The Handbook of Antenna Design.* Vols. I and II. Stevenage: Peter Peregrinus, cf. Vol. I Section 1.4.

Sander, K. F. and Reed, G. A. L. (1986). *Transmission of Propagation of Electromagnetic Waves.* Cambridge: Cambridge University Press.

Silver, S. (1949). *Microwave Antenna Theory and Design.* New York: McGraw-Hill.

Stratton, J. A. (1941). *Electromagnetic Theory.* New York: McGraw-Hill.

CHAPTER 4

MICROWAVE POWER DEVICES

OBJECTIVES

This chapter is concerned with the microwave devices that are used specifically for the generation of microwave power (transistor circuits are not included here). The earlier sections deal with electron beam devices, which provide high power sources; later sections describe the solid-state devices. Where possible, a simple theory is given to show the principles of operation.

4.1 Introduction

At one end of any microwave system is the source of power at the frequency and level appropriate to the system. The required levels cover a very large range, from less than a watt continuous wave for local communications systems to megawatts pulse power for an early warning radar. The frequency spectrum in use extends from less than 1 GHz to 90 GHz, and is continually extending to higher frequencies. To match this wide range of requirements many devices are in use. A high power level may be met by direct generation or by amplification, depending on availability and efficiency. The highest power devices utilize electron beams *in vacuo*. Solid-state devices are continually increasing in power, and, for example, are now almost universally used as local oscillators in receivers.

The primary requirements in power generation are power level and efficiency. Secondary requirements are a number of other properties which have an important role in particular systems, such as the stability of frequency, noise content, effect of modulation (e.g. frequency modulation resulting from amplitude modulation), the effect of environmental conditions, and reliability.

Electron beam devices are considered in Sections 4.2–4.6, and solid-state devices in Sections 4.7–4.9. Transistors are considered in Chapter 6.

4.2 The klystron

Figure 4.1. shows a schematic diagram of a 2-cavity klystron. In this device an electron beam, which is formed by an electron gun, passes through two resonant cavities in succession and is finally collected on a 'collector' electrode. The electron beam is formed and propagates in a vacuum, requiring a metal and glass or ceramic vacuum-tight envelope, not shown in the diagram.

The electrons in this, as in other electron beam devices, originate from a heated cathode. Electrons are emitted by the process of thermionic emission from the surface of a suitable material. The emitter used differs with the type

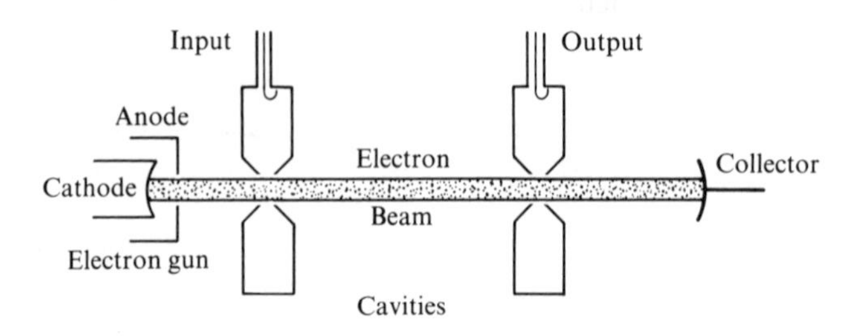

Figure 4.1 Schematic diagram of a 2-cavity klystron.

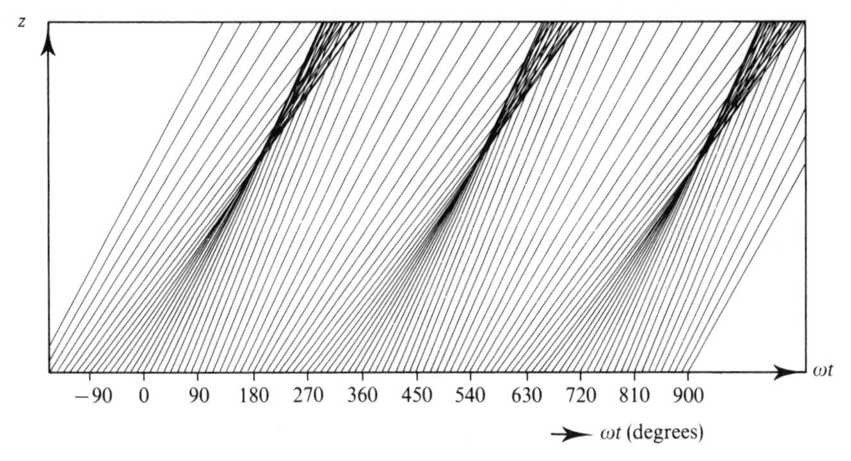

Figure 4.2 Applegate diagram. The ordinate represents distance in the drift space after velocity modulation of the beam. The formation of bunches can be clearly seen. The diagram has been calculated for $V_1/2V_0 = 0.2$.

of cathode, but usually involves barium, calcium and strontium oxides. For low current densities ($\simeq 1\,\mathrm{A\,cm^{-2}}$) the cathode layer is formed by heat treatment of a layer of the carbonates of the above elements adhering to a nickel substrate. In operation the cathode is heated to about 1100 K. For high current beams, dispenser cathodes are used: these are made with a porous body of sintered tungsten which is impregnated from a melt of barium, calcium and aluminium oxides. Following polishing and surface treatment, such cathodes are used to give emission densities of some $16\,\mathrm{A\,cm^{-2}}$ at an operating temperature of 1350 K. The cathode will have a concave surface which with the focusing electrodes forms the electron gun.

The operation of the klystron may be briefly described as follows. Excitation of the first cavity produces an alternating electric field across the gap parallel to the beam, which alternately accelerates and decelerates the electrons crossing the gap. This process is conveniently represented on the Applegate diagram shown in Figure 4.2. From this diagram the tendency for the electrons to form bunches as they progress along the beam can be seen. This process is termed velocity modulation. The second cavity is placed at a point where the bunches have developed. These bunches induce charges on the walls of the cavity, which as they change will excite an electromagnetic field. With an RF load connected to the cavity, the electric field across the gap will act in a sense so as to slow down the electrons, so that kinetic energy on the beam is transformed into RF energy in the load. The formation of bunches can be thought of as the development of an RF component of current in the beam at the signal frequency, and the action of the second loaded cavity as a resistance in series with this current.

4.2.1 Ballistic theory

An expression for the beam current may be developed as follows. We characterize the action of the cavity gap by an accelerating voltage $V_1 \sin \omega t$, acting on electrons of energy V_0 electron volts. The exit velocity, u, of an electron entering the gap (presumed short) at time t_i is given by

$$\tfrac{1}{2}mu^2 = e(V_0 + V_1 \sin \omega t_i) \tag{4.1}$$

in which e and m are the electronic charge and mass respectively. When $V_1 \ll V_0$ we have

$$u = u_0[1 + \tfrac{1}{2}(V_1/V_0) \sin \omega t_i], \qquad u_0 = (2eV_0/m)^{\frac{1}{2}} \tag{4.2}$$

If we assume a narrow gap, the time of arrival at a distant point s, further along the beam, is given by

$$\begin{aligned} t &= t_i + s/u \\ &\simeq t_i + (s/u_0)[1 - (V_1/2V_0) \sin \omega t_i] \end{aligned} \tag{4.3}$$

The charge leaving the gap in a small interval $\mathrm{d}t_i$ arrives at s during an interval $\mathrm{d}t$, whence the beam current at s is given by

$$i = I_0(\mathrm{d}t_i/\mathrm{d}t)$$

in which I_0 is the average (DC) beam current. From these equations we find

$$i = I_0(1 - X \cos \omega t_i)^{-1} \tag{4.4}$$

where

$$X = \omega s V_1/(2u_0 V_0)$$

We can also rewrite equation (4.3) in the form

$$\omega(t - s/u_0) = \omega t_i - X \sin \omega t_i$$

Although it has been assumed that $V_1 \ll V_0$, this condition does not imply that $X \ll 1$. Some graphs of i as a function of time, derived from these equations, are shown in Figure 4.3. In deriving the expression of equation (4.4), no account has been taken of the mutual repulsion of the electrons. These space charge forces are important in controlling the formation of bunches, and, for example, would prevent the apparent infinite current density indicated by equation (4.4) when $X > 1$.

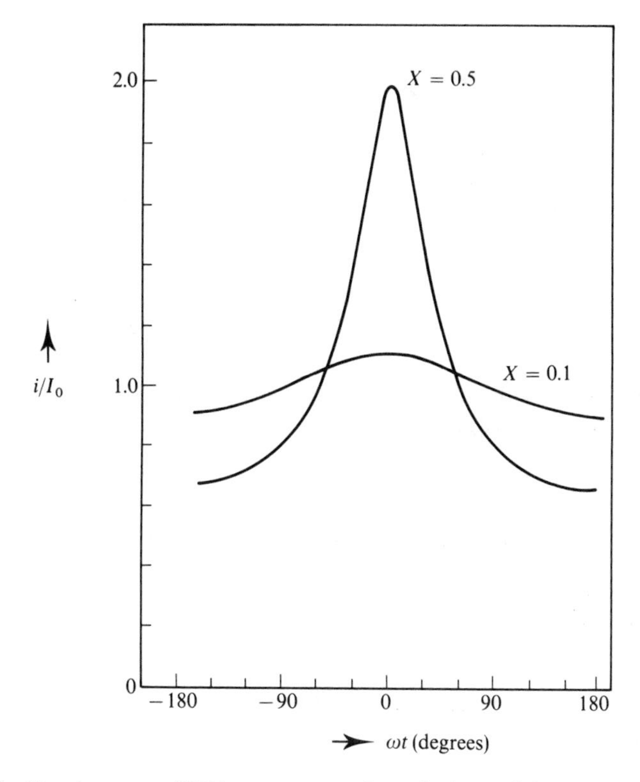

Figure 4.3 Development of RF beam current after velocity modulation. The parameter X is equal to $\omega s V_1/2u_0 V_0$, so that the two graphs can either represent current at s and $5s$, or at the same point for different depths of modulation.

4.2.2 Space charge waves

The development of RF beam current after velocity modulation can also be investigated by considering perturbations of a one-dimensional flow of electrons, as indicated in Figure 4.4. The electrons will follow straight line paths, and it may be assumed that the bulk negative charge is neutralized by a uniform, static, positive space charge. (In practice, the beam is of finite cross-section, and held by a focussing magnetic field.)

The equation of motion of an electron is given by

$$\frac{\mathrm{d}u}{\mathrm{d}t} = -\frac{e}{m}E$$

where u is the velocity of an electron, and E the electric field experienced by the electron. This is a Lagrangian description and collective action is better

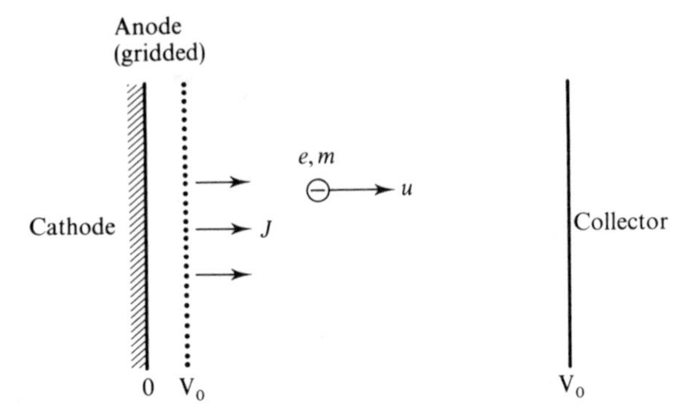

Figure 4.4 One dimensional flow of electrons. It is assumed that the paths are straight, confined by a magnetic field, and that the space between anode and collector is free of any steady electric field.

described by the velocity field $u(z, t)$, when this equation becomes

$$\frac{\partial u}{\partial t}+u\frac{\partial u}{\partial z}=\frac{e}{m}E \tag{4.5}$$

Conservation of charge requires that the current density J is related to the charge density ρ by

$$\frac{\partial J}{\partial z}+\frac{\partial \rho}{\partial t}=0 \tag{4.6}$$

We also have

$$J=\rho u \tag{4.7}$$

and

$$\varepsilon_0\frac{\partial E}{\partial z}=\rho \tag{4.8}$$

The undisturbed beam is in a region of constant potential, in which u, ρ, J, E have the values u_0, ρ_0, J_0, 0 respectively. We consider axial perturbations on the beam by writing

$$\begin{aligned} u &= u_0 + u_1 \quad u_1 \ll u_0 \\ \rho &= \rho_0 + \rho_1 \quad \rho_1 \ll \rho_0 \\ J &= J_0 + J_1 \quad J_1 \ll J_0 \end{aligned}$$

It is assumed, as mentioned above, that the bulk effect of the unperturbed electron beam is neutralized by static charges which play no other part. Thus E is in itself a perturbation due to the non-uniformity of the beam. Substituting in equations (4.5) to (4.8) and neglecting products of the perturbations, we find

$$\left.\begin{aligned}
&\frac{\partial u_1}{\partial t} + u_0 \frac{\partial u_1}{\partial z} = -\frac{e}{m} E \\
&\frac{\partial J_1}{\partial z} + \frac{\partial \rho_1}{\partial t} = 0 \\
&J_1 = u_0 \rho_1 + u_1 \rho_0; \quad J_0 = \rho_0 u_0 \\
&\varepsilon_0 \frac{\partial E}{\partial z} = \rho_1
\end{aligned}\right\} \tag{4.9}$$

These equations apply in a region where no external fields are present. In actual tubes a source field must exist somewhere. This may be very localized as in a klystron, or distributed as in a travelling wave tube (Section 4.3). Such a source field can be introduced into the first of the equations (4.9) in the form

$$\frac{\partial u_1}{\partial t} + u_0 \frac{\partial u_1}{\partial z} = -\frac{e}{m}(E + E_s) \tag{4.10}$$

Since E_s is presumed to arise from sources outside the region, the last of the equations (4.9) is unaffected. Considering a region where $E_s = 0$ and looking for a wave solution with each term proportional to $\exp \mathrm{j}(\omega t - \beta z)$, the homogeneous equations (4.9) lead to the condition

$$(\omega - u_0 \beta)^2 = \frac{1}{\varepsilon_0} \frac{e}{m} |\rho_0| = \omega_p^2$$

giving possible values of β as

$$\beta_1 = \frac{1}{u_0}(\omega + \omega_p)$$

$$\beta_2 = \frac{1}{u_0}(\omega - \omega_p)$$

The quantity ω_p is the **plasma frequency**, and is usually much smaller than the operating frequency. (In practice, because of the finite cylindrical electron beam, not all the space charge fields act in the axial direction so that the effect of these space charge fields is reduced by a numerical factor F. Then ω_p is

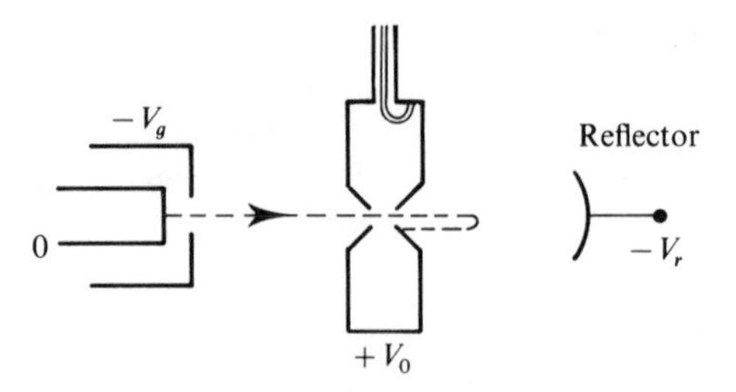

Figure 4.5 The single cavity reflex klystron oscillator. Tuning over a limited range can be obtained by variation of the reflector voltage, V_r.

replaced by $F\omega_p$. In the presence of metal walls F can be as low as 0.01.) Thus in a source free region

$$\begin{aligned} u_1 &= A\exp(-\mathrm{j}\beta_1 z) + B\exp(-\mathrm{j}\beta_2 z) \\ &= [A\exp(-\mathrm{j}\omega_p z/u_0) + B\exp(\mathrm{j}\omega_p z/u_0)]\exp(-\mathrm{j}\omega z/u_0) \end{aligned} \tag{4.11}$$

Equations (4.9) then lead to

$$J_1 = \frac{m\omega\varepsilon_0}{e}\omega_p[A\exp(-\mathrm{j}\omega_p z/u_0) - B\exp(\mathrm{j}\omega_p z/u_0)]\exp(-\mathrm{j}\omega z/u_0) \tag{4.12}$$

Thus the beam supports two forward waves, one travelling with a velocity slightly higher than that of the beam, the other slightly lower. These equations can be applied to the region of the electron beam between the two cavities: at the exit of the first gap, $z = 0$, the RF current, J_1, is zero. Thus from equation (4.12), $A = B$. Hence

$$\begin{aligned} u_1 &= 2A\cos(\omega_p z/u_0)\exp(-\mathrm{j}\omega z/u_0) \\ J_1 &= -2\mathrm{j}A\omega_p\frac{m\omega\varepsilon_0}{e}\sin(\omega_p z/u_0)\exp(-\mathrm{j}\omega z/u_0) \end{aligned} \tag{4.13}$$

At the point for which $\omega_p z/u_0 = \pi/2$, J_1 is a maximum, velocity modulation is converted to density modulation and a loaded resonant cavity at this point will extract power from the beam. Space charge wave theory concerns small perturbations of the electron beam and will not apply when the modulation is large. It can be seen from the Applegate diagram, Figure 4.2, that at large modulation the bunches tend to form closer to the gap than at weak modulation, whereas the space charge wave theory gives only one position for the maximum. This theory is useful in the design of the early stages in a klystron, but for the power output stage it is necessary to use ballistic theory corrected for the effects of space charge.

4.2.3 Multi-cavity klystrons

If the second cavity of the 2-cavity klystron is left unloaded, the induced field in the gap enhances the velocity modulation on the beam. This increases the gain to a third, output, cavity placed to extract power from the beam. The cavities in the 2-cavity klystron are resonant at the same frequency, and therefore the device is essentially narrow band. In a multiple cavity klystron, detuning the intermediate cavities alters the overall response, and valves with four or more cavities are used to produce gain over a wider bandwidth.

4.2.4 The reflex klystron

Hitherto the klystron has been treated as an amplifier, which at high power is its main use. The reflex klystron is a low power, voltage-tunable, self-oscillating tube which until the development of solid-state sources was widely used as the local oscillator in microwave superheterodyne receivers. The single cavity tube is shown schematically in Figure 4.5. The velocity modulated electron beam leaves the cavity gap and is then reversed in direction in the electric field produced by a negative 'reflector' electrode. The voltage on the reflector is adjusted so that the bunched beam reinforces the oscillations in the cavity. An X-band (10 GHz) tube might operate with a current of some 10 mA at 300 V, giving an output of the order of 50 mW.

4.2.5 Device capabilities

Klystrons are made for a very wide range of conditions. As CW power amplifiers, multi-cavity tubes are used in the UHF band as TV transmitters, at power levels of the order of 100 kW. Troposcatter systems (explained in Chapter 9) also use them as CW power amplifiers. Such devices have a signal bandwidth of the order of one or two per cent of the signal frequency, although by the use of external, tunable cavities a very wide range of operation is obtainable. Pulsed klystrons delivering megawatts of power are used in radar systems, and in the UHF band in high energy physics applications.

Physically, such klystrons are large: the tube itself may be over 2 m high, requiring a magnetic focusing system and calling for ancillary equipment appropriate to its operation at voltages as high as 100 kV.

Some details of a commercial tube are shown in Figure 4.6.

4.3 The travelling wave tube

For the klystron amplifier just discussed, the interaction of the electric field with the electron beam takes place over a distance which is short compared with a wavelength, namely the gaps in a number of re-entrant tuned cavities.

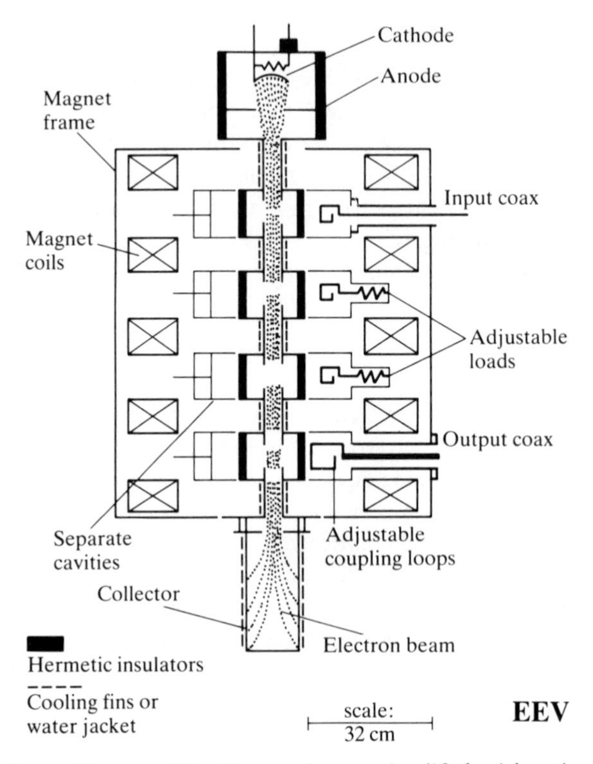

Figure 4.6 Power klystrons. The diagram shows a simplified axial section through a 4-cavity UHF klystron. Such valves are used for TV transmitters, and the following operating conditions for the English Electric K3672 indicate the capabilities of this type of valve.

Beam voltage	26.5 kV
Beam current	5.2 A
Power out (CW)	64 kW
Power input	8 W
Gain	39 dB
Efficiency	46%

Cavities can be individually tuned to give coverage 470–810 MHz, with signal bandwidth 8 MHz. Such a valve stands some 2 m high and with its magnet assembly weighs 350 kg. (*Diagram and information by courtesy of English Electric.*)

The signal bandwidth depends on the Q factor of the cavities and is generally not more than one or two per cent of the signal frequency. If the electron beam were coupled to a transmission line type of circuit so that interaction could take place on a continuous basis with the beam and line waves in approximate synchronism, a much wider bandwidth might be expected. This is the basis of the travelling wave tube, illustrated schematically in Figure 4.7. A uniform transmission line propagates electromagnetic waves at a speed much greater than the velocity of the electrons in a beam accelerated through

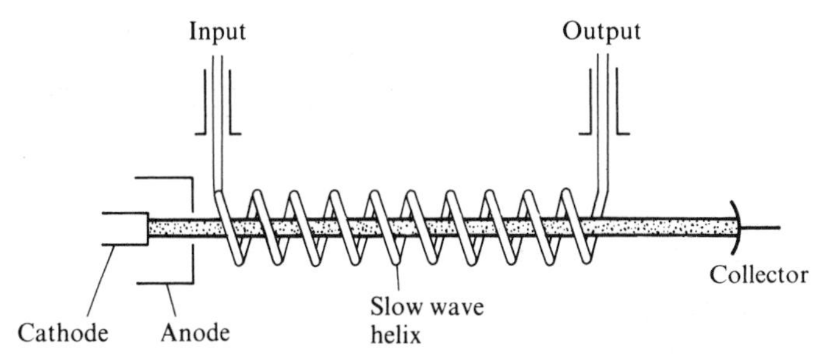

Figure 4.7 Schematic diagram of a helix-type travelling wave tube. The beam velocity and the velocity of the wave on the axis of the helix are in approximate synchronism.

'normal' voltages of some tens of kilovolts. Accordingly, to obtain synchronism, a line using a slow wave structure must be used, of which the helix shown in Figure 4.7 is a common example. It may be imagined that a wave propagates as in a uniform transmission line but bound to the wire of the helix, thus giving a much reduced speed along the axis. A simple theory of the travelling wave tube can be obtained by coupling the 'space charge wave' equations of Section 4.2.2 with suitable circuit relations. The RF current in the electron beam will excite waves in the slow wave transmission line, the electric field produced by the line will act on the electrons through the term E_s in equation (4.10). If it be assumed that each variable in equations (4.9) and (4.10) has a factor $\exp(j\omega t - \gamma z)$, a relation between J_1 and E_s can be found:

$$J_1 = \frac{J_0}{2V_0}\gamma_e[\omega_p^2/u_0^2+(\gamma_e-\gamma)^2]^{-1}E_s \tag{4.14}$$

in which $\gamma_e = j\omega/u_0$ and other symbols have their previous significance.

The mode of excitation of the slow wave structure by the beam is indicated schematically in Figure 4.8: the charge on the beam induces charge on the line, giving rise to an injected current. The current injected into a length δz of line is thus $A\delta z\partial\rho_1/\partial t$, A being the area of the beam cross-section. Using the second of equations (4.9), and denoting AJ_1 by I_1, this may be written in the form $\delta z\partial I_1/\partial z$. The line equations then take the form

$$\frac{\partial V}{\partial z} = -L\frac{\partial I}{\partial t}$$

$$\frac{\partial I}{\partial z} = -C\frac{\partial V}{\partial t} - \frac{\partial I_1}{\partial z}$$

in which L and C are respectively the inductance and capacity per unit length of the equivalent transmission line. Assumption of a factor $\exp(j\omega t - \gamma z)$ in each variable leads to the expression

$$E_s = -\frac{\partial V}{\partial z} = \frac{\gamma^2\gamma_0}{\gamma_0^2-\gamma^2}Z_cI_1 \tag{4.15}$$

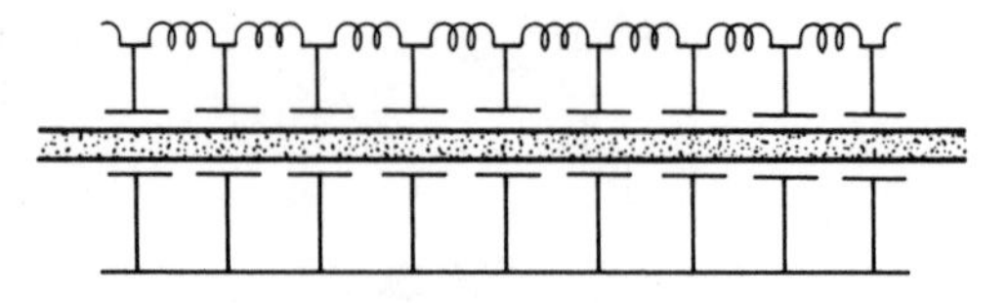

Electron beam: charge density ρ_1 C m^{-1}, current I_1 A
Line parameters: L H m^{-1}, C F m^{-1}

Figure 4.8 Showing in schematic form the mode of interaction between the electron beam and the transmission line equivalent to the helix slow-wave structure.

in which $\gamma_0 = \mathrm{j}\omega(LC)^{\frac{1}{2}}$ is the natural propagation constant of the line. $Z_c = (L/C)^{\frac{1}{2}}$ is a parameter determining the coupling between beam and line, which has dimensions of resistance. Equations (4.15) and (4.14) have to be satisfied simultaneously, fixing the value of γ in terms of the other quantities. The condition may be written in the form

$$(\gamma_0^2 - \gamma^2)(\omega_p^2/u_0^2 + (\gamma_e - \gamma)^2) = \gamma_e \gamma^2 \gamma_0 2\mathscr{C}^3 \tag{4.16}$$

where $\mathscr{C}^3 = I_0 Z_c/4V_0$ and $I_0 = AJ_0$.

Consider the case when the propagation constant of the line and of the beam are the same, $\gamma_e = \gamma_0$, and when the term $(\omega_p/u_0)^2$ can be neglected. Equation (4.16) becomes

$$(\gamma_0 - \gamma)^3(\gamma_0 + \gamma) = \gamma^2\gamma_0^2 2\mathscr{C}^3$$

For weak coupling $\mathscr{C}^3 \ll 1$, requiring that either $\gamma_0 - \gamma$ or $\gamma_0 + \gamma$ be small. If $\gamma_0 - \gamma$ is small

$$(\gamma_0 - \gamma)^3 2\gamma_0 = \gamma_0^4 2\mathscr{C}^3$$
$$\gamma = \gamma_0 - \gamma_0\mathscr{C}(+1)^{\frac{1}{3}}$$

This gives, substituting $\gamma_0 = \mathrm{j}\beta_0$

$$\gamma_1 = \mathrm{j}\beta_0(1 - \mathscr{C})$$
$$\gamma_2 = \beta_0\{\mathrm{j}(1 + \mathscr{C}/2) + 3^{\frac{1}{2}}\mathscr{C}/2\}$$
$$\gamma_3 = \beta_0\{\mathrm{j}(1 + \mathscr{C}/2) - 3^{\frac{1}{2}}\mathscr{C}/2\}$$

The fourth root, for which $\gamma_0 + \gamma$ is small, is given by

$$\gamma_4 = -\mathrm{j}\beta_0(1 - \mathscr{C}^3/4)$$

Propagation constants $\gamma_1, \gamma_2, \gamma_3$, represent waves travelling in the direction of the electron flow: γ_1 corresponds to a phase velocity slightly greater than u_0;

γ_2 and γ_3 correspond to phase velocities slightly less than u_0. The wave represented by γ_1 has constant amplitude, that represented by γ_3 increases in the direction of electron flow, whilst that represented by γ_2 is attenuated in that direction. The wave represented by γ_4 is of constant amplitude and travels against the electron flow. This simplified model implies that excitation by RF at the beginning of the helix will cause a growing wave to be propagated, allowing an amplified signal to be taken out at the far end. The interaction takes place with the slowest wave.

4.3.1 Helix tubes

The helix slow wave structure introduced in the last section has the great merit of being wide band: the phase velocity changes only slowly with frequency, and gain can be maintained over an octave change. The RF electric field within the helix is greatest near the helix wire or tape and decreases towards the centre, the more rapidly as the wavelength becomes smaller. This means that smaller diameter helices are needed as the frequency increases: at 2.5 GHz the diameter might be about 6 mm, at 9 GHz about 2 mm. Because of the decrease in area occupied by the beam, maximum power also decreases with frequency. The power is limited by current interception on the helix and by the difficulty of extracting the heat thus generated. Helix tubes are restricted to powers of the order of a few hundred watts, operating with voltages less than 10 kV. The degree of modulation on the beam is small, the ratio of RF power on the beam to the DC beam power being only of the order of 10%. This ratio is termed the electronic efficiency, and because it is small the emerging beam has only a small spread of velocities and so can be slowed down by the use of a 'depressed' collector, held at a voltage lower than that of the helix. By this means the overall efficiency can be increased to about 20%. Beam focusing in these tubes is achieved by using a periodic sequence of annular permanent magnets. Input and output connections will be by coaxial line through matching sections to the ends of the helix. The very wide bandwidth makes it difficult to achieve a good match over the entire operating range, and the consequent reflections lead to feedback and self-oscillation. This is controlled by inserting a well matched attenuating section in the helix.

Helix tubes have been used for a variety of purposes. Although now in competition with more recently developed solid-state devices, they have been widely used in microwave communications links, with tubes delivering some 10–20 W of RF power. Such tubes have been developed for flying in communications satellites, and higher power tubes giving 100–200 W output are under development for direct broadcasting from satellites. Very low power tubes provided the best noise performance before the development of microwave transistors. Broad band tubes with more than an octave bandwidth and 100 W or greater CW power are used in counter-measure systems.

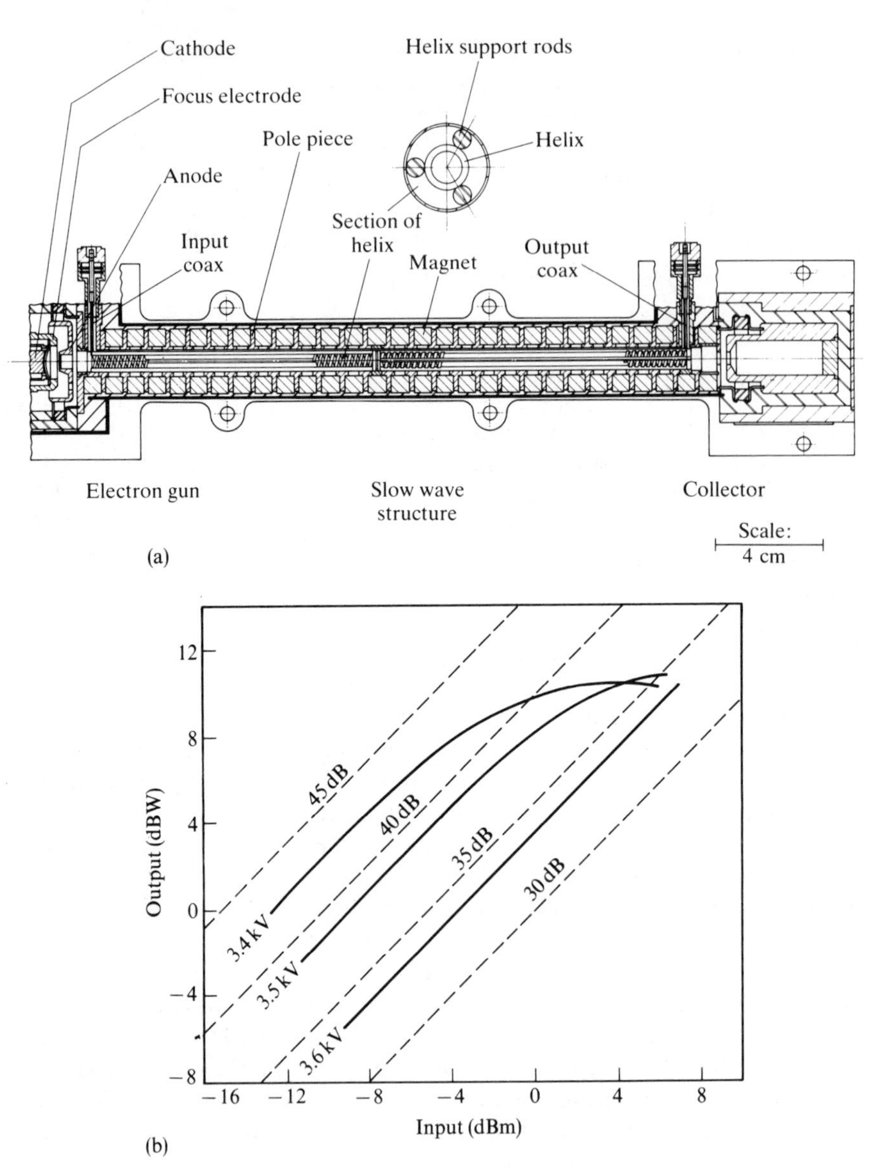

Figure 4.9 Helix-type travelling wave tubes. (a) An axial section through a tube of ceramic and metal construction, with integral focusing provided by periodic permanent magnets. To simplify presentation, full detail is shown only for the slow-wave structure. The pole pieces form part of the vacuum envelope. The section is typical of a broad-band, CW tube, operating over the range 5–10 GHz at 100 W. Approximate operating voltages: helix cathode, 6 kV; collector cathode, 3 kV. Beam current 200 mA. (b) Characteristics for a communication-type tube with a glass envelope, the English Electric N10018. This tube operates around 6.5 GHz, with a helix voltage of about 3.5 kV and beam current 30 mA. The collector is held at 1.3 kV. The gain is about 40 dB, noise figure 26 dB and power efficiency 20%. (*Diagram and information by courtesy of English Electric.*)

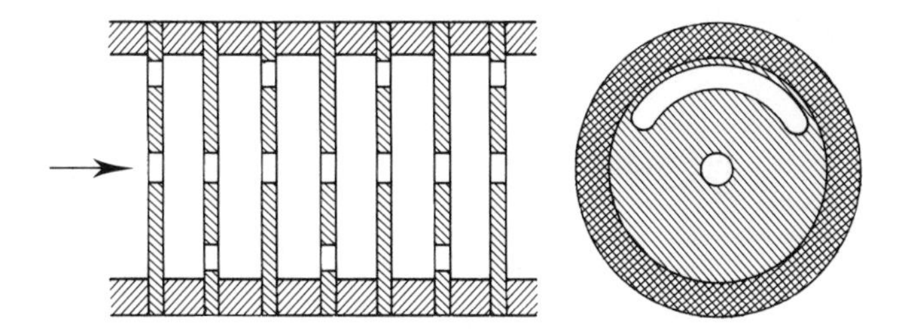

Figure 4.10 A coupled cavity slow-wave structure used in high-power travelling wave tubes.

Pulsed tubes with peak output powers of the order of 1 kW and average powers of the order of 100 W which operate at 10 kV are used in some radar systems. Some details and characteristics of one type of commercial tube are shown in Figure 4.9.

4.3.2 Coupled-cavity tubes

Alternative slow wave structures can be made using a waveguide in which some periodicity of structure has been introduced. Such structures offer the possibility of more efficient heat extraction and thus higher powers. One such coupled-cavity structure is shown in Figure 4.10. Such tubes can deliver high average powers and peak pulse powers of the order of 1 MW with a bandwidth of 10% of the mid-band frequency. A tube of this power level would operate at 100 kV with a beam current of some 30 A. Focusing can be achieved either using periodic permanent magnets or an external solenoid.

4.4 Backward wave oscillators

The backward wave oscillator (BWO) merits only a brief mention since it has been largely replaced by solid-state devices together with amplifiers. The slow wave structures used in travelling wave tubes can support 'backward waves', in which the group velocity is in the opposite direction to the phase velocity. Interaction with an electron beam leads to a wide-band voltage-tuned oscillator.

4.5 The magnetron

The magnetron is a self-oscillating device, compact and capable of producing high power. As a pulsed source it is widely used in radar systems. The cavity magnetron is cylindrical in structure, with a cross-section of the general form

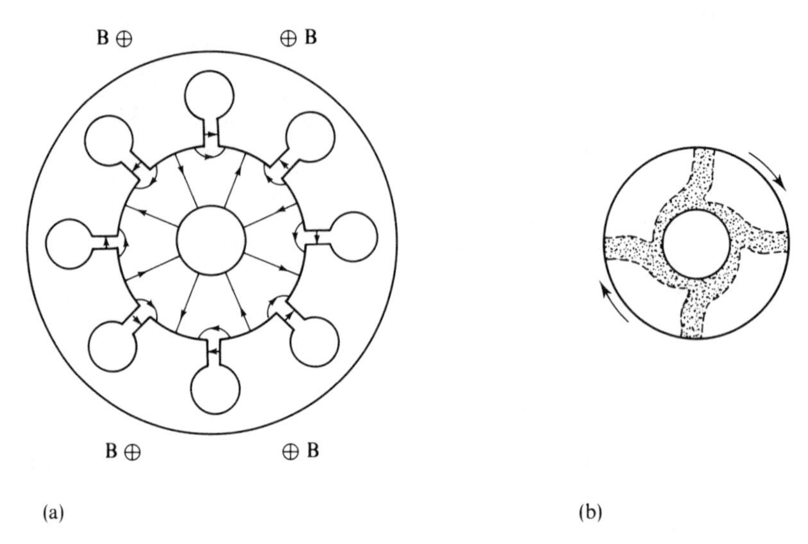

Figure 4.11 The magnetron. (a) An 8-cavity magnetron block, showing the form of the electric field lines in the π mode of oscillation. (b) Schematic representation of the electron 'spokes' formed in an oscillating magnetron.

shown in Figure 4.11(a). The surface of the centre electrode is an emitting cathode, the surrounding block forms the anode and the whole is immersed in a strong axial magnetic field. The axial length is typically of the order of one-quarter wavelength. The anode block contains a number of cavities resonant at the frequency of operation that create an electric field in the interaction space between anode and cathode. The diagram shows an 8-segment block, although the number of segments is a design parameter. The general form of the lines of electric force for the normal, π-mode of operation is indicated in the diagram. This mode can be regarded as formed by the superposition of two waves rotating in opposite directions. As will be shown below, in the absence of an RF electric field the action of the static magnetic field is to prevent electrons reaching the anode when the anode–cathode voltage is less than a certain value, the cut-off voltage. In this cut-off region the space charge is unstable to azimuthal components of the RF electric field and can be drawn into spokes as indicated in Figure 4.11(b). These spokes play a part similar to that played by bunches in the linear devices discussed earlier. Modes other than the wanted π-mode are possible, and to reduce the possibility of oscillation in these modes the plain block illustrated is modified. One modification is 'strapping', shown in Figure 4.12, in which alternate segments are connected to the same strapping ring. Another, particularly useful at high frequencies when the block becomes small, is to make alternate cavities differ in depth, forming the 'rising sun' block shown in Figure 4.13. Both these devices have the effect of increasing the frequency separation between wanted and unwanted modes.

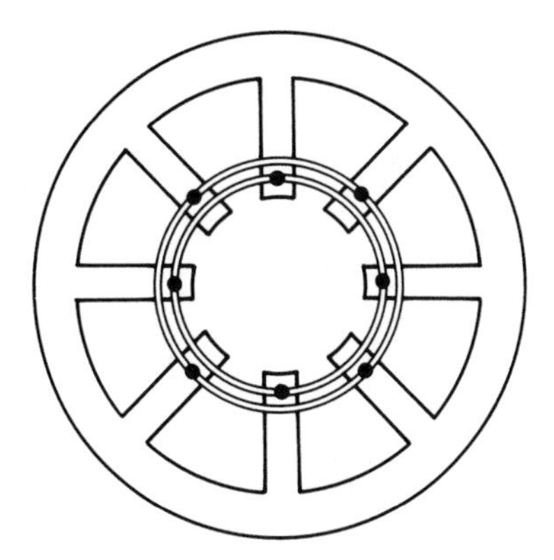

Figure 4.12 A 'strapped' magnetron block. Strapping encourages excitation in the π mode. The cavities in this block are formed by a sequence of thick vanes electrically integral with the outside wall. There may be strapping rings at both ends of the block.

4.5.1 The cut-off condition

The theory of the magnetron is complex, and no attempt will be made here to present it in detail. However, two calculations provide some indications: the 'cut-off' curve and the Hartree curve. Consider a cylindrical structure, of anode radius r_a and cathode radius r_c, in which electrons are emitted from the cathode with zero velocity. The equation of motion of an electron in the two-dimensional system has two integrals, energy and angular momentum. These take the form

$$\dot{r}^2 + r^2\dot{\theta}^2 = 2\frac{e}{m}V$$

$$r^2\dot{\theta} = \frac{1}{2}\frac{e}{m}B(r^2 - r_c^2)$$

in which the dots signify differentiation with respect to time and e and m are the charge and mass of the electron. Elimination of θ shows that the electron will only reach radius r_a if

$$V_a \geqslant \frac{1}{8}\frac{e}{m}B^2r_a^2(1 - r_c^2/r_a^2)^2$$

If the anode voltage is less than this value, no electron can arrive at the anode and no current will flow. This is the cut-off condition.

4.5.2 The threshold condition

As far as energy is concerned there is no reason why an electron should not reach the anode: in the case just presented the energy all goes into rotational

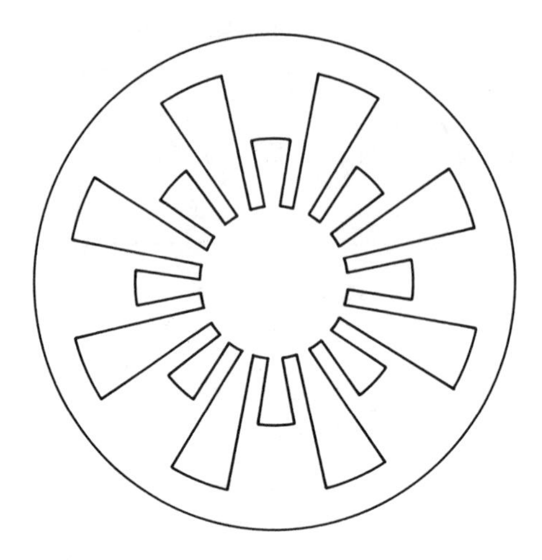

Figure 4.13 A 'rising sun' block, which also encourages operation in the π mode and is particularly useful at high frequencies.

velocity. Small perturbations such as those provided by the RF field can alter the direction of motion and allow electrons to reach the anode. Analysis based on such considerations leads to an expression for the threshold voltage at which current might just be expected to flow to the anode of an oscillating magnetron. This expression, known as the Hartree condition, takes the form

$$V_a = \tfrac{1}{2}B\Omega(r_a^2 - r_c^2) - \frac{m}{2e}\Omega^2 r_a^2$$

in which Ω is the angular speed at which both the electromagnetic wave and the electron cloud rotate. (For an N-segment block with cavities of resonant frequency ω oscillating in the π-mode, Ω is equal to $2\omega/N$.) The experimentally determined values for the threshold in practical valves are in general agreement with the values calculated from this expression. In Figure 4.14 the dependence of the cut-off voltage and the Hartree voltage on the magnetic field and other variables is shown in dimensionless form.

4.5.3 Performance

Figure 4.15(a) shows a typical high power magnetron designed for pulse working around 1.3 GHz. Characteristics are shown in Figure 4.15(b). Conduction commences around the Hartree voltage and thereafter current increases rapidly with voltage. As expected, any increase of magnetic field requires an increase of voltage for a given current, and the RF output power

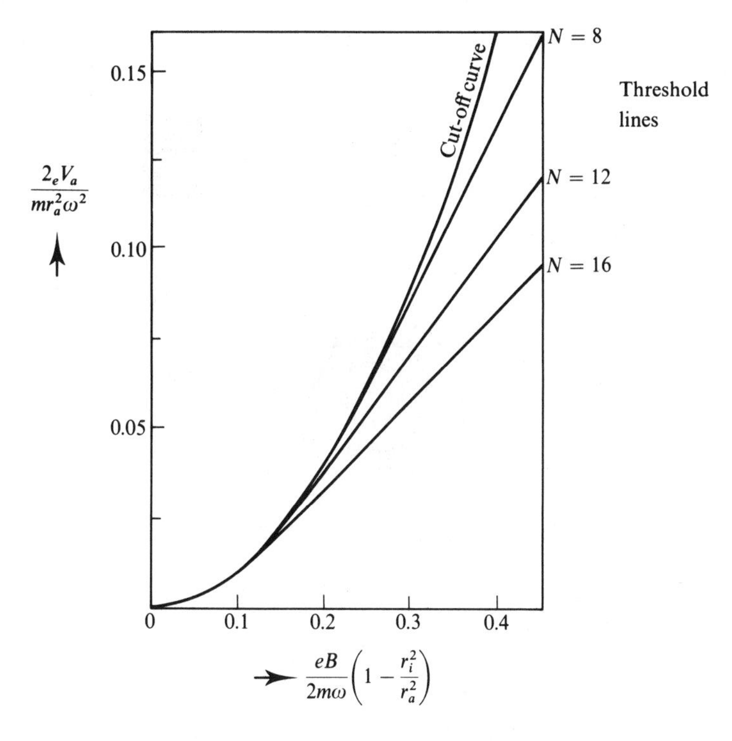

Figure 4.14 The threshold or Hartree curve. The curve relates the voltage at which current can flow to the anode, and oscillations commence, to the magnetic field and the geometrical parameters defined in the text.

increases. The maximum electronic efficiency that can be expected is given by

$$1 - m\Omega^2 r_a^2/(2eV_a)$$

Thus, high efficiency requires operation at high magnetic field. At high current the efficiency drops as electrons striking the anode do so with increasing radial velocity. Magnetrons quite commonly operate at overall efficiencies of around 50%. They are high peak power devices, and forced cooling is needed. Considerable back bombardment of the cathode by electrons takes place and, when running, external heater power may not be necessary. The average power is limited by the maximum permissible operating temperature of the cathode. Typically the ratio of peak to average power is 1000:1. The RF output is taken by coupling into one of the cavities. The connection may be through a vacuum-tight coaxial seal or through a broad band vacuum window in waveguide. The magnetic field is usually obtained with a permanent magnet. The frequency of operation is determined by the dimensions of the resonant cavities, so that the magnetron is

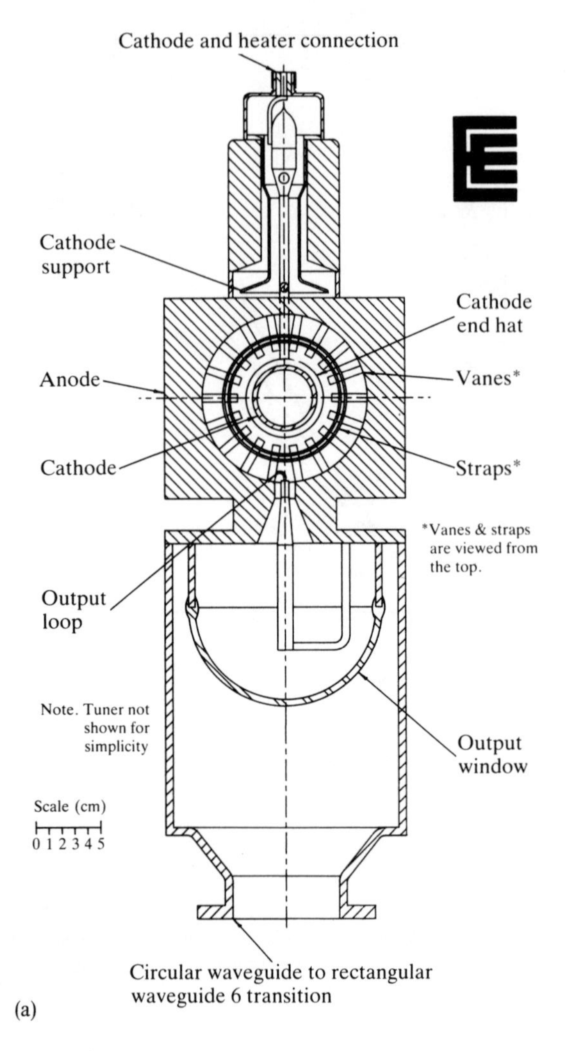

Figure 4.15 A high power D-band magnetron, English Electric M5051. (a) Simplified mid-section view. (b) Characteristics. This valve has 18 anode segments, and operates around 1.3 GHz. Mechanical tuning (not shown) is provided by the axial movement of metal slugs into the cavities, giving a tuning range of 1.25–1.31 GHz. The anode is vapour cooled. The magnetic field is provided by a permanent magnet, at about 90 mT strength. The overall weight of the valve and magnet is approximately 90 kg. (*Diagram and information by courtesy of English Electric.*)

essentially a fixed frequency device. However, tuning over a limited range may be achieved by modifying one or more of the cavities. This may be done through mechanically controlled inserts, when the variation will be continuous but at a comparatively slow rate, or by electronic means using the multipactor effect. In operation it is important to maintain a good VSWR in

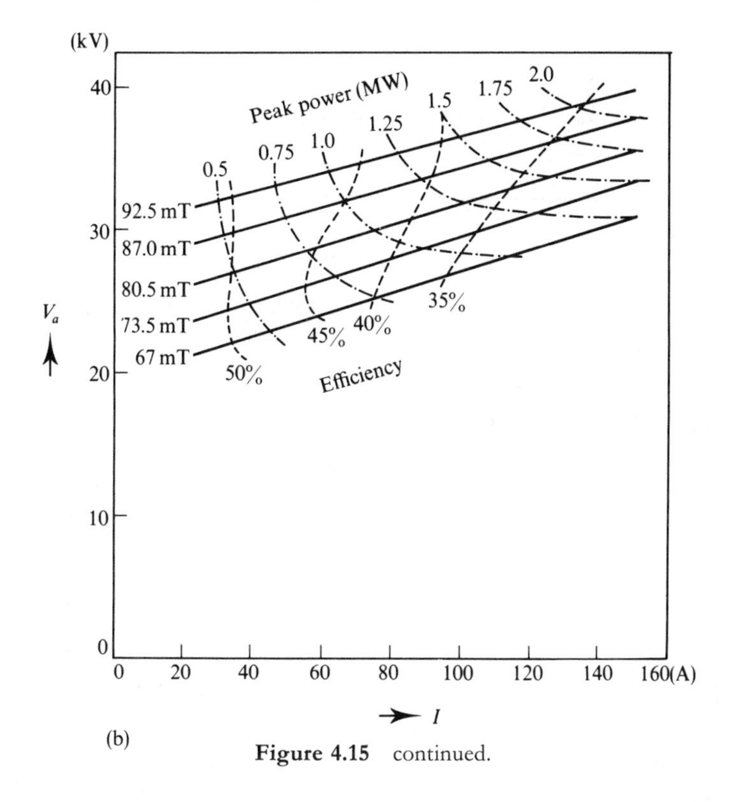

Figure 4.15 continued.

the output line over a much wider frequency band than the operating range in order to avoid oscillation in non-π modes.

A pulsed magnetron starts afresh at each pulse, and the RF pulse will be very dependent on the rate of rise of the voltage. There will generally be no pulse-to-pulse coherence, but such coherence may be obtained in suitable designs by injecting a small CW signal.

As exemplified by the characteristics shown in Figure 4.15, magnetrons can produce several megawatts of pulse power at lower microwave frequencies. As the frequency increases, dimensions become smaller, limiting the power available. However, tens of kilowatts peak power can be obtained at 40 GHz.

4.6 Crossed-field amplifiers

The high power generating properties of the magnetron can be exploited in linear form to make an amplifier, of which one type is indicated in Figure 4.16. Such devices are known as crossed-field amplifiers, and are capable of giving peak powers of the same order of magnitude as a klystron. Typically the gain is in the range 10–15 dB.

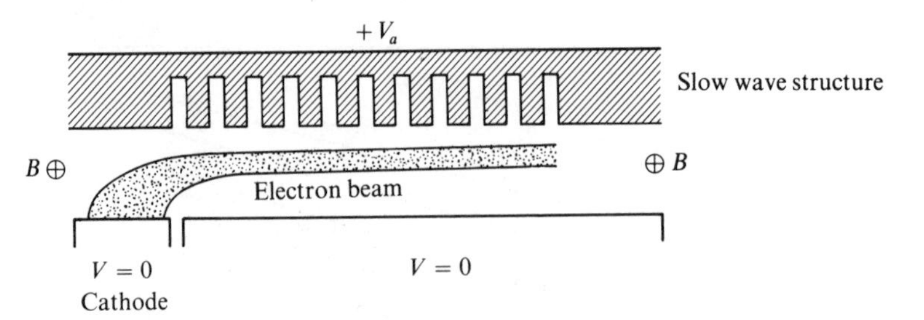

Figure 4.16 Schematic diagram of a crossed-field amplifier. Magnetron type interaction occurs in a linear structure, resulting in a medium gain amplifier with high power output.

4.7 Transferred-electron devices

The first of the solid-state devices to be considered depends on the occurrence of a bulk property, exhibited notably by gallium arsenide, which gives rise to a negative differential resistance. In this material, as the electric field is increased, the velocity of carriers at first increases but subsequently goes through a maximum to finally saturate at a lower level, as indicated in Figure 4.17. If the electric field across a sample of length L were uniform at a value E and the charge-carrier density were uniform at value n over the device area A, then the current $I = enAv(E)$ and voltage $V = EL$ would follow the same curve as shown in Figure 4.17. In the region of negative slope, the device would behave like a negative resistance which could be used to deliver power to a suitable circuit. The adjective 'transferred' when applied to an electron, comes from an awareness of the detailed structure of the conduction band of the material, whereby sufficiently energetic electrons can transfer to a state of lower mobility. The effect can be utilized in different ways depending on the interaction between device and circuit.

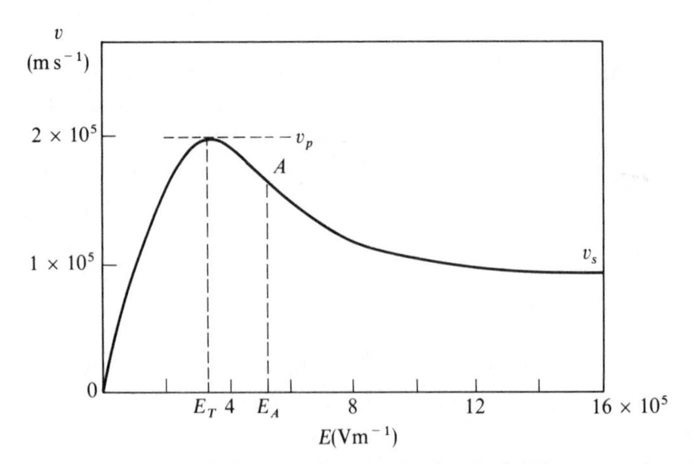

Figure 4.17 The transferred electron effect. As the electric field increases, the velocity of the electrons reaches a maximum, subsequently falling to a limiting value. The figures apply to gallium arsenide.

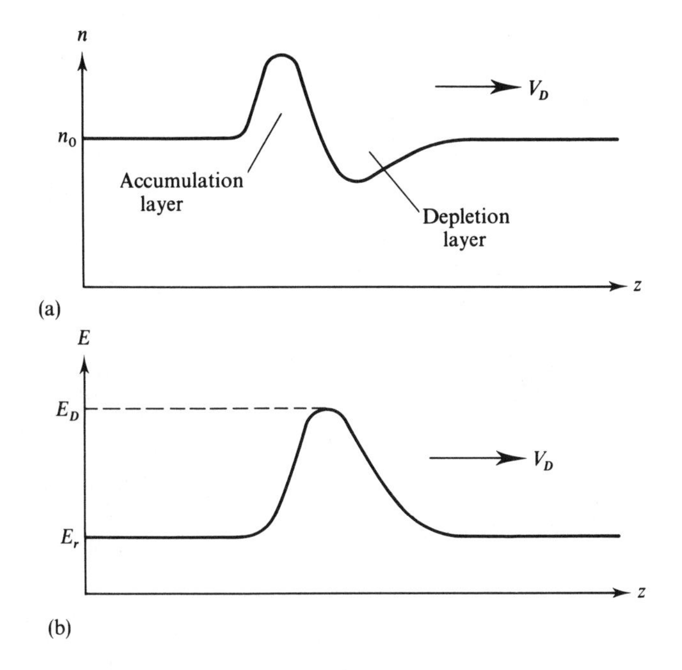

Figure 4.18 When the applied voltage would give a uniform field exceeding the threshold, non-uniformities develop: (a) electron density showing accumulation layer followed by a depletion layer, forming a domain; (b) the distribution of electric field through the specimen.

4.7.1 Domain formation

If a uniform specimen is subjected to an electric field greater than the threshold E_T (such as E_A in Figure 4.17), the state indicated by the point A is unstable. A slight perturbation or non-uniformity results in the formation of a high field domain: this can be seen by considering the result of applying a rising field. The carrier velocity steadily increases until the threshold is reached, whereupon the carriers slow down, giving rise to an accumulation of electrons near the negative electrode with a depletion further into the specimen, as indicated in Figure 4.18(a). Thus instead of a uniform field E_A across a length L, the field distribution takes the form of Figure 4.18(b). Theory shows that the domain travels towards the positive electrode with a velocity that is close to the value v_s ($\simeq \frac{1}{2}v_p$ in Figure 4.17). If the diffusion constant is independent of the electric field strength, then theory shows that the velocity of the electrons outside the domain is equal to the domain velocity. The current is then determined by the donor density and this velocity: domain formation is indicated by a drop in current when the device is operated from a constant voltage source. When one domain is extinguished at the positive electrode, another begins nucleating at the negative electrode. The domain takes time to form, of the order of 50 ps.

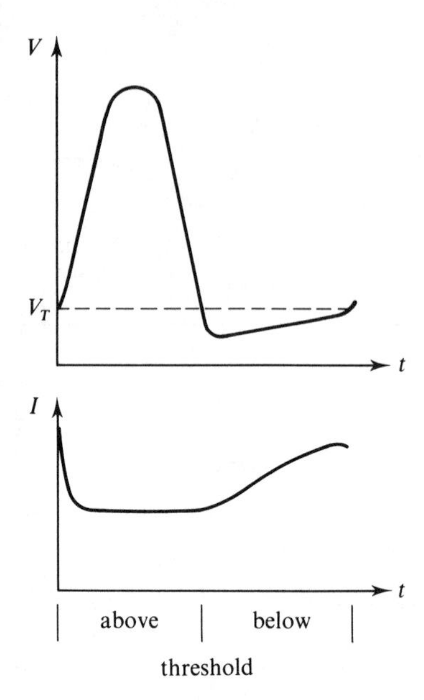

Figure 4.19 Voltage and current waveforms, when operation is in the 'Limited Spacecharge Accumulation' mode. V_T is the voltage corresponding to a uniform field at the threshold level.

4.7.2 The Gunn diode

From the above discussion it will be seen that the regular formation and extinction of domains will give rise to current pulses with a period equal to the transit time. Such oscillations are said to be in a **transit-time mode**. When the diode is incorporated in a resonant circuit, the applied voltage change may reduce the electric field below threshold before a domain has reached the anode. This is termed a **quenched mode**. The precise mode will depend on specimen doping and length and on the frequency of operation as well as on the properties of the circuit in which the device is embedded.

4.7.3 LSA modes

It is possible by using a large enough RF voltage and high frequency to decrease the time spent by the specimen in the region of negative mobility below that required for domain formation: instead, a relatively small accumulation of charge takes place, giving the acronym for 'Limited space-charge accumulation', LSA. The type of waveform in this mode is indicated in Figure 4.19. The rapid rise of voltage just above threshold can sweep the fields rapidly through the unstable region. Probably a more usual practical mode of operation is the 'hybrid' mode where the domain does not fully form but neither is the field uniform. Any small charge accumulations have to be

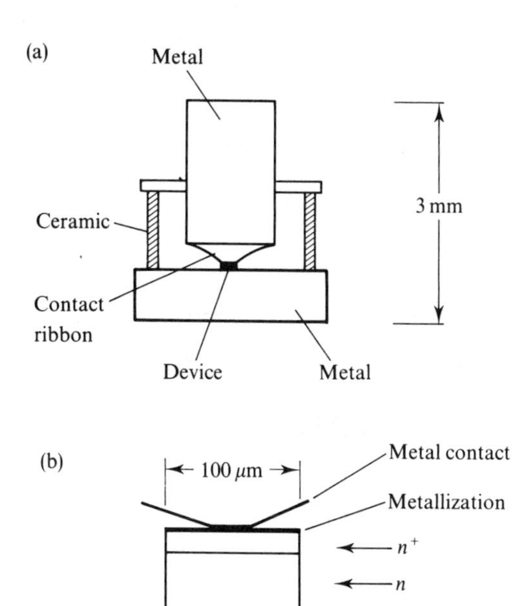

Figure 4.20 The construction of a Gunn Diode. (a) Packaging, (b) diode fabrication. Typical dimensions for the slice might be 100 μm × 100 μm with the thickness of the *n*-layer governed by the required transit time. About 10 μm at 10 GHz.

allowed to sweep through the device during each cycle in order to escape at the anode and avoid excessive build up in the system.

4.7.4 Packaging and circuit

The structural form of a Gunn diode is shown schematically in Figure 4.20. It is necessary to have good heat sinking since the power dissipation in the device can be very high, of the order 10–100 MW cm^{-3}. A diode can be incorporated into a resonant cavity structure of the form shown in Figure 4.21.

4.7.5 Device capabilities

CW Gunn oscillators are made over a wide part of the microwave frequency range between approximately 10–100 GHz, typically giving output powers of 100 mW at the lower frequencies. The power supplies are low voltage (<10 V), high current (~ 250 mA). Temperature stability is of the order of 1 MHz/°C.

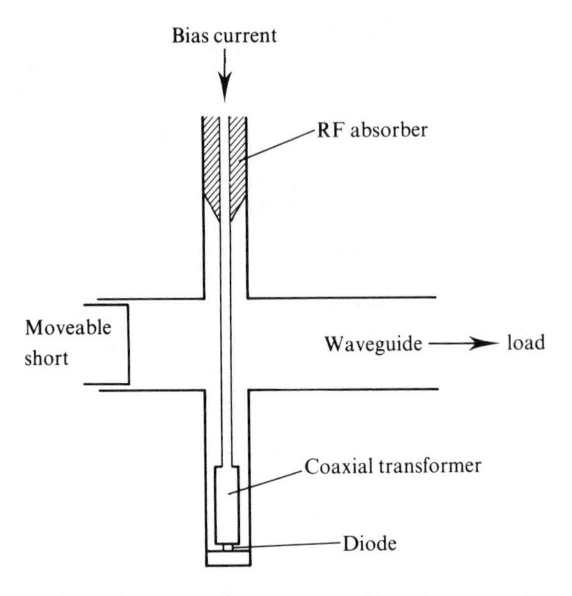

Figure 4.21 Circuit configuration for a Gunn Diode oscillator.

Pulsed devices can have peak powers of the order of tens of watts at the lower frequencies, decreasing at higher frequencies. Rise times are of the order of 25 ns. Operating voltage and current are much higher than with CW devices, typically 30 V, 10 A. Temperature rise during the pulse causes substantial frequency modulation on the output pulse.

4.8 Impatt diodes

Impatt is an acronym for 'impact avalanche and transit time' and this indicates the two aspects of the mechanism of such diodes: an avalanche zone is used to produce a suitably phased current of electrons which subsequently pass through a drift zone. The combination of injection phase and transit time act to produce a device with a negative small signal resistance. The possibility of such a source was originally postulated by W. T. Read of Bell Laboratories. The theory outlined below follows that given by him for the 'Read diode' (Read, 1958).

4.8.1 Avalanche breakdown

The passage of current through a semiconductor can bring about the ionization of neutral atoms in the lattice. For this to happen the electric field gradient must be high, of the order of $10^7\ \mathrm{V\,m^{-1}}$, and thus in a p–n junction the effect occurs in the neighbourhood of the metallurgical junction under reverse bias conditions. It is characterized by **an ionization rate** coefficient

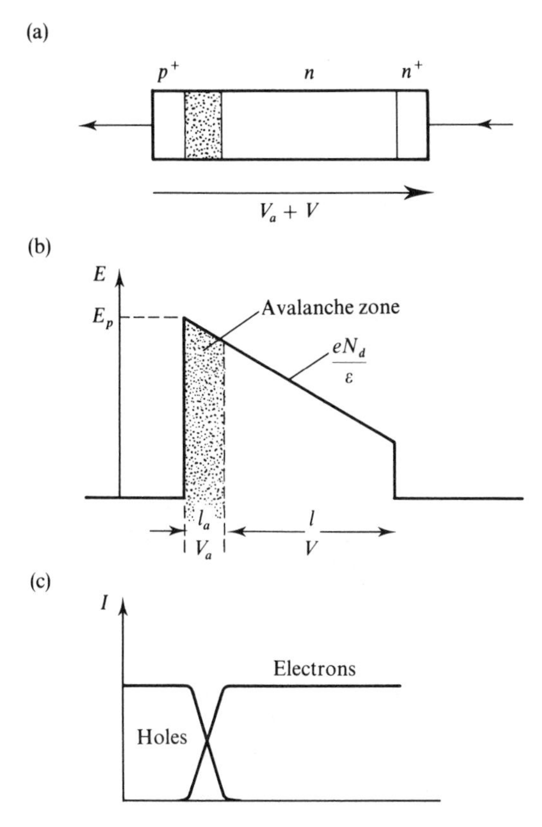

Figure 4.22 The Impatt Diode. (a) Formation of the avalanche zone in the n type material. (b) Distribution of electric field, ignoring effects of space charge. (c) Hole and electron current distribution.

defined as the probability that an electron will, in unit distance of travel, bring about one ionizing collision. The ionization coefficient depends on the electric field in the region, and increases very rapidly with increasing electric field. Every ionizing collision produces a hole as well as an electron; further, holes also can be considered to make ionizing collisions characterized by a rate coefficient.

Consider the p^+–n–n^+ structure shown in Figure 4.22(a), with a reverse bias sufficiently large to deplete the whole of the n region. With uniform doping, the form of the electric field will be as shown in Figure 4.22(b). For a sufficiently high voltage, breakdown will occur in the region of peak field, and under DC conditions the particle currents will be as shown in Figure 4.22(c), it being assumed that the total current is limited by the supply. If the current is changed, conditions in the avalanche zone readjust, but at a finite rate. It is this finite rate of readjustment which creates a phase delay between RF current and RF voltage across the avalanche zone. To simplify an analytical approach it is assumed that the electric fields are sufficiently high for both electrons and holes to travel with the same limiting velocity, v_s, and that they both have the same ionization rate coefficient, $\alpha(E)$. It can be shown under certain plausible assumptions that the density of the electron current leaving the avalanche zone satisfies the equation

$$\frac{\tau_a}{2}\frac{dJ_n}{dt} = J_n\left(\int_0^{l_a} \alpha(E)\,dz - 1\right) \tag{4.17}$$

in which $\tau_a = l_a/v_s$, the avalanche zone transit time. The form of the electric field in Figure 4.22(b) contains only one externally controlled variable, E_p: this makes it convenient to write

$$\frac{1}{l_a}\int_0^{l_a} \alpha(E)\,dz = \bar{\alpha}(E_p) \tag{4.18}$$

The externally applied voltage is equal to the area under the curve of Figure 4.22(b) and hence E_p is directly related to V. In the DC condition referred to earlier dJ_n/dt must be zero, defining a mean value E_{p0} by

$$l_a\bar{\alpha}(E_{p0}) = 1$$

If an external variation takes place so that the voltage across the avalanche region increases by V_a, E_p will increase by $E_1 = V_a/l_a$; for a small change

$$\bar{\alpha}(E_{p0} + E_1) \cong 1 + E_1\bar{\alpha}'(E_{p0}) = 1 + E_1\bar{\alpha}'_0$$

so that equation (4.18) can be written in the form

$$\frac{\tau_a}{2}\frac{dJ_{n1}}{dt} = l_a J_0\bar{\alpha}'_0 E_1 = J_0\bar{\alpha}'_0 V_a \tag{4.19}$$

in which J_0 represents the steady current and J_{n1} a small change. Considering a device of cross-sectional area A, equation (4.19) shows that the avalanching process acts like an inductance of value

$$\tau_a/(2\bar{\alpha}'_0 I_0)$$

where

$$I_0 = AJ_0$$

In addition to the particle current there is also a displacement current represented by a capacity

$$C_a = A\varepsilon/l_a$$

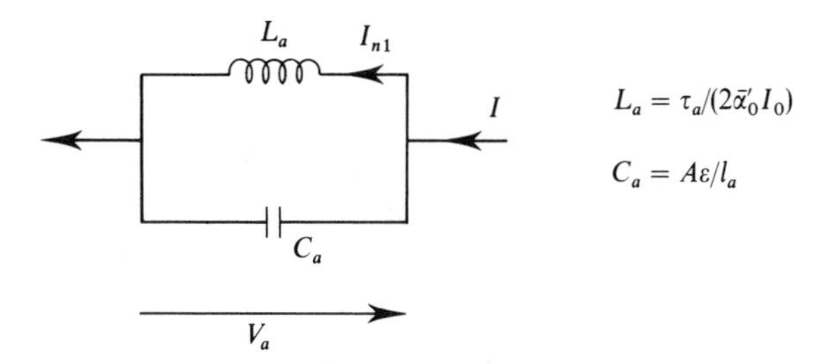

Figure 4.23 Equivalent circuit of the avalanche zone. Symbols as defined in the text.

Thus a small signal equivalent circuit for the avalanche zone is as shown in Figure 4.23.

For larger signals, remembering that $\bar{\alpha}(E_p)$ will increase very rapidly with increase of E_p, consideration of equation (4.17) shows that the emerging electron current becomes like a pulse, as indicated in Figure 4.24. The depression in mean level of E_p is necessary to maintain a steady state.

4.8.2 The drift zone

The remaining part of the n region of Figure 4.22(a) is a drift zone for the electron current injected from the avalanche zone. The current at the terminal includes a term for the rate of change of charge induced by the movement of the electrons in the drift space. Since total current is a function of time only

$$\frac{1}{A} I(t) = J(z, t) + \varepsilon \frac{\partial E}{\partial t}$$

where $J(z, t) = J_{n1}(t - z/v_s)$. Integrating over the drift space of length l

$$\frac{l}{A} I(t) = \int_0^l J_{n1}(t - z/v_s)\,\mathrm{d}z + \varepsilon \frac{\partial V}{\partial t} \tag{4.20}$$

For small sinusoidal signals represented by phasor quantities $\hat{I}$, $\hat{J}_n$, $\hat{V}$

$$\begin{aligned} \frac{l}{A} \hat{I} &= \hat{J}_n \int_0^l \exp(-\mathrm{j}\omega z/v_s)\,\mathrm{d}z + \mathrm{j}\omega\varepsilon\hat{V} \\ &= \hat{J}_n \frac{v_s}{\mathrm{j}\omega}(1 - \exp(-\mathrm{j}\omega\tau)) + \mathrm{j}\omega\varepsilon\hat{V} \end{aligned} \tag{4.21}$$

where $\tau = l/v_s$ is the transit time.

For small signals we may assume that space charge currents are small compared with the displacement currents, so that, as evident from Figure

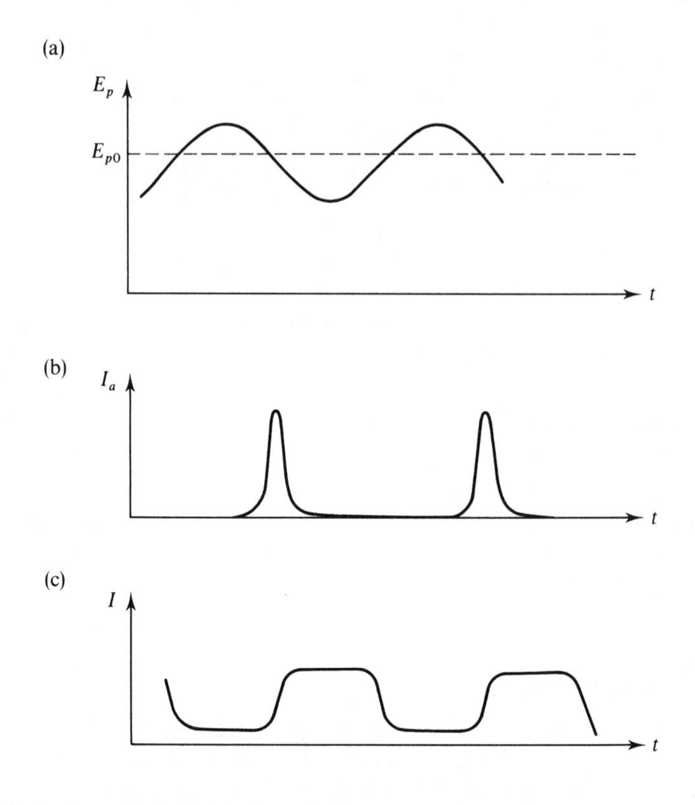

Figure 4.24 Waveforms in the Impatt diode. (a) Peak electric field strength. (b) Electron current from avalanche zone. (c) Total current. Comparison of (a) and (c) shows that the current is low when the voltage is high, so the device acts as a negative resistance.

4.22(b)

$$V_a \cong l_a V/l$$

Using this with equation (4.19) to eliminate $\hat{J}_n$ we find the admittance of the drift space is given by

$$Y = (\hat{V}/\hat{I}) = Y_e + j\omega C_d \tag{4.22}$$

where

$$Y_e = 2\bar{\alpha}_0' I_0 \frac{1 - \exp(-j\omega\tau)}{(j\omega\tau)^2} \qquad C_d = \frac{A\varepsilon}{l}$$

For large signals, the integral in equation (4.20) is almost constant over the drift space transit time, giving the electronic current shown in Figure 4.24(c).

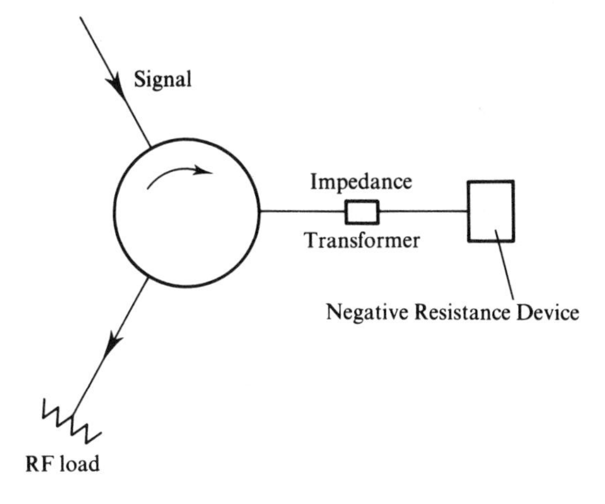

Figure 4.25 Configuration to use a negative resistance device as an amplifier.

It will be seen that the current is high when the voltage is low, and thus RF power is being delivered from the device. Equation (4.22) shows that in the small signal analysis the admittance has a negative real part over a range of transit angles, $\omega\tau$: in particular for $\omega\tau = \pi$, Y_e is real and negative.

This model ignores the effect of space charge in the transit zone. The charge carriers create their own field which can depress the overall field so that the carriers no longer travel at the scattering limited velocity. This is a saturation mechanism.

4.8.3 Packaging and circuits

Apart from the doping profile, the structural form is very similar to that of the Gunn diode of Figure 4.20. In its general form, the cavity is also similar to that of Figure 4.21. The generation of heat is even greater than that of Gunn diodes ($\sim 1\,\mathrm{GW\,cm^{-3}}$), so that heat sinking is of first importance. The junction temperature may be over 200°C.

4.8.4 Device capabilities

Impatt oscillators form at present the highest power solid-state device above 10 GHz, giving CW power of the order of hundreds of milliwatts and peak power of tens of watts. Power supplies required are typically 40 V at 100 mA for CW devices, and currents of many amps on pulse. As with a Gunn diode, frequency modulation during a pulse occurs because of temperature rise.

4.9 Negative resistance amplifiers

In the above sections both Gunn and Impatt diodes have been considered as oscillators. Negative resistance devices can also be used as amplifiers, and this is usually done by incorporating the device with a circulator, which will be considered under microwave circuits in a later section. The schematic form of such an amplifier is shown in Figure 4.25. Output powers obtainable are of the same order as those obtained as oscillators.

4.10 Transistors

Both bipolar and field-effect transistors are widely used at microwave frequencies. They are treated in Chapter 6.

EXERCISES

4.1 An electron beam formed by acceleration from rest to a potential of 1 kV carries a current of 40 mA constrained to a diameter of 2 mm. Calculate the plasma frequency.

Calculate also the wavelength of the beats between the two slow space charge waves possible on the beam (equal to u_0/f_p). Show that in general this wavelength is related to the perveance of the beam, $IV^{-\frac{3}{2}}$.

4.2 By comparison of equations (4.2) and (4.13), the constant A may be identified with $u_0V_1/4V_0$. Show that when $z = s$ is sufficiently small, the second member of Equation (4.13) agrees with the expression given in equation (4.4).

4.3 Evaluate the two smallest distances predicted by equations (4.13) at which maximum RF current will be found when the beam carries 1 A at 8 kV and is constrained to a diameter of 8 mm. (N.B. The value of ω_p calculated should really be reduced by a factor $F < 1$ to allow for finite beam cross-section.)

4.4 The simplest expression for gain of a travelling wave tube N wavelengths long is

$$(47.3N\mathscr{C} - 9.5)\,\text{dB}$$

Show that the first term corresponds to the growing forward wave. The second term is because input power is divided equally into the three possible forward waves. (N.B. This expression implies gain can be increased without limit: this is not so, since mismatch at the output and the backward wave provide feedback leading to oscillation.)

4.5 The magnetron for which the details were given in Figure 4.15 has cathode and anode diameters of 50 and 76 mm respectively. Calculate the cut-off voltage and the threshold voltage when the valve is operating at 1.3 GHz in a magnetic field of 92.5 mT. Compare with the threshold voltage indicated by the characteristics of Figure 4.15(b).

4.6 The transient behaviour of electrons flowing in a material exhibiting the transferred electron effect as illustrated in Figure 4.17 may be considered as follows: consider perturbations from a uniform state in which the electric field is E_0 and current density $J_0 = en_0u_0$. The final three equations of (4.9), with an appropriate value of ε, are valid providing that diffusion effects are small. Using these equations and expressing u_1 in the form $(\mathrm{d}u/\mathrm{d}E)_0E$, show that E satisfies the equation

$$\frac{\partial^2 E}{\partial z^2} + \frac{en_0u_0'}{u_0\varepsilon}\frac{\partial E}{\partial z} + \frac{1}{u_0}\frac{\partial^2 E}{\partial z \partial t} = 0$$

in which $u_0' = (\mathrm{d}u/\mathrm{d}E)_0$. By considering solutions of the form $\hat{E}\exp(-\gamma z + \mathrm{j}\omega t)$ show that a wave can be supported having phase velocity u_0 and attenuation coefficient $\alpha = (en_0u_0'/\varepsilon u_0)$. Discuss the significance of these results when u_0' is negative.

4.7 The data in the following question refer to Section 4.8. The ionization rate α is approximated to by the expression

$$\alpha = \kappa(E/E_0)^5\,\mathrm{m}^{-1}$$

in which

$$\kappa = 10^5\,\mathrm{m}^{-1} \quad E_0 = 2.5\times 10^7\,\mathrm{V\,m}^{-1}$$

Neglecting the change of electric field across the avalanche zone (Figure 4.22(b)) show that

$$\bar{\alpha} = \kappa(E_{p0}/E_0)^5 \quad E_{p0} = E_0(\kappa l_a)^{-\frac{1}{5}}$$

Hence show that the equivalent inductance of the avalanche zone is $(\tau_a E_{p0} l_a/10 I_0)$. Evaluate this quantity for a device operating with a current of 50 mA and with an avalanche zone 1 μm in thickness. It may be assumed that $v_s = 10^5\,\mathrm{m\,s}^{-1}$. Determine the power dissipated in the avalanche zone.

The device is of cross-section 100 μm × 100 μm. Taking the relative permittivity as 12, find the frequency at which the equivalent circuit for the avalanche zone is resonant.

References

The nature of the treatment given in the present chapter makes specific references inappropriate. Fuller treatment may be found in a number of texts, a selection of which is given below. The texts concerning electron tubes represent a period when understanding and design were well advanced, subsequent development being primarily in detailed development with new materials.

General texts on electron tubes

Dix, C. H. and Aldous, W. H. (1966). *Microwave Valves.* London: Iliffe.

Gewartoski, J. W. and Watson, H. A. (1969). *Introduction to Electron Tubes.* New York: Van Nostrand Reinhold.

Sims, G. D. and Stephenson, I. M. (1963). *Microwave Tubes and Semiconductor Devices.* London: Blackie.

Solid-state devices

Carroll, J. C. (1970). *Hot Electron Microwave Generators.* London: Arnold.

Howe, M. J. and Morgan, D. V. (1976). *Microwave Devices.* New York: Wiley.

Howe, M. J. and Morgan, D. V. (1980). *Microwave Solid State Devices and Applications.* Stevenage: Peter Peregrinus.

Milnes, A. G (1980). *Semiconductor Devices and Integrated Electronics.* New York: Van Nostrand Reinhold.

Read, W. T. (1958). 'A Proposed High-frequency Negative-resistance Diode', *BSTJ*, **37**, 401.

Watson, H. A. (1969). *Microwave Semiconductor Devices and their Applications.* New York: McGraw-Hill.

Tubes for radar

Elwell, G. W. (1981). *Radar Transmitters.* New York: McGraw-Hill.

Skolnik, M. I. (1980). *Introduction to Radar Systems.* New York: McGraw-Hill.

Travelling wave tubes

Gittins, J. F. (1965). *Power Travelling Wave Tubes.* London: E.U.P.

Hutter, R. G. E. (1960). *Beam and Wave Electronics in Microwave Tubes.* New York: Van Nostrand Reinhold.

Pierce, J. R. (1950). *Travelling Wave Tubes.* New York: Van Nostrand Reinhold.

CHAPTER 5

MICROWAVE CIRCUITS

OBJECTIVES

The fact that at microwave frequencies connections are made using transmission line or waveguides has brought about the development of many components specifically designed for these transmission media. Some have a role similar to that of lumped components at lower frequencies, others are peculiar to microwave frequencies. This chapter introduces a range of these passive components likely to be encountered in working with microwave systems.

5.1 Introduction

In all systems it is necessary to transmit and distribute signals between a variety of subsystems. Because of the high frequencies involved, the necessary microwave circuits take special forms using waveguides or transmission lines, and a range of more or less standard components has been developed to perform the various functions. The physical size of the components will vary in conformity with the appropriate size of waveguide although the mode of operation remains the same. The nearest approach to 'ordinary' circuits as used at low frequencies occurs in microstrip circuits: although these will often be designed as interconnections of transmission lines, in some cases a good approximation to 'lumped' inductors and capacitors can be realized with advantage.

5.2 Matching

Whether or not lumped realizations of a circuit design can be made, interconnections are necessarily through transmission line or waveguide. This is true for all power levels, since the distances involved are usually considerably greater than the wavelength. Care is taken to ensure that, as closely as

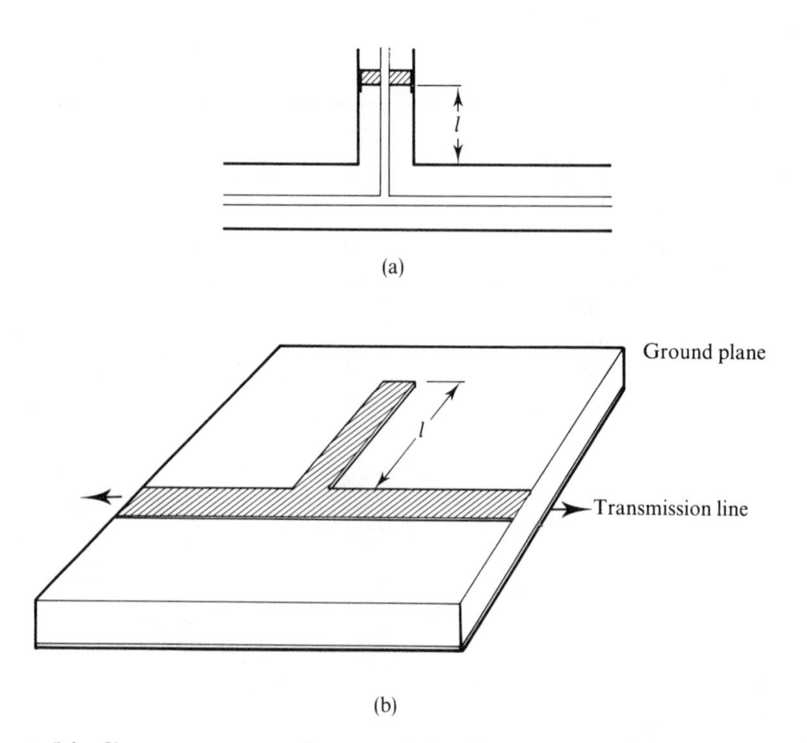

Figure 5.1 Shunt susceptances in transmission line: (a) coaxial line, normalized susceptance $b = -\cot(2\pi l/\lambda)$; (b) microstrip, $b = \tan(2\pi l/\lambda)$.

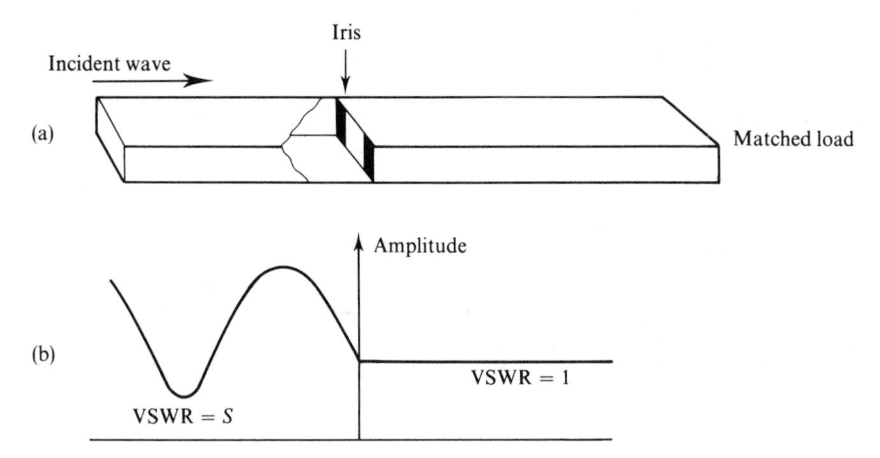

Figure 5.2 Illustrating the concept of equivalent impedance for an obstacle in a waveguide: (a) shows a diaphragm in a waveguide terminated in a matched load; (b) shows the standing wave pattern on either side of the diaphragm. By interpreting the pattern as on a transmission line terminated in a load consisting of Z_0 in parallel with a reactance, a value may be assigned to the normalized reactance of the diaphragm.

possible, each line has a reflectionless termination: in the terminology of Section 2.10.2 the standing wave ratio is kept as near unity as possible. In this condition the line or waveguide is said to be **matched**. At high powers, the breakdown condition which limits power will then be reached uniformly along the line, giving maximum power handling capacity. At low levels a matched system is relatively insensitive to the length of connections. One way in which matching can be achieved is mentioned in the discussion of Section 2.10.3, where it is shown that an appropriate inductive or capacitive susceptance across the line at the correct point will match an arbitrary load. In the case of a transmission line such admittances in parallel with line are usually realized by short-circuited stubs, of the type shown for a coaxial system in Figure 5.1(a): the length of the stub determines the value of susceptance. Such a stub can easily be made of variable length. Open circuited stubs can also be used, but the external field near the open circuit can be disadvantageous. In Figure 5.1(b) an open circuit stub in microstrip is shown. In both cases the effective length of the stub will be slightly different from the dimension l in the figure, by an amount of the order of the conductor size. Such differences can be allowed for in computer aided design methods.

5.3 Waveguide components

The previous section considered the way in which components can be placed in shunt across a transmission line. Provided the longitudinal thickness of the junction is much less than a wavelength these are accurate realizations. As the

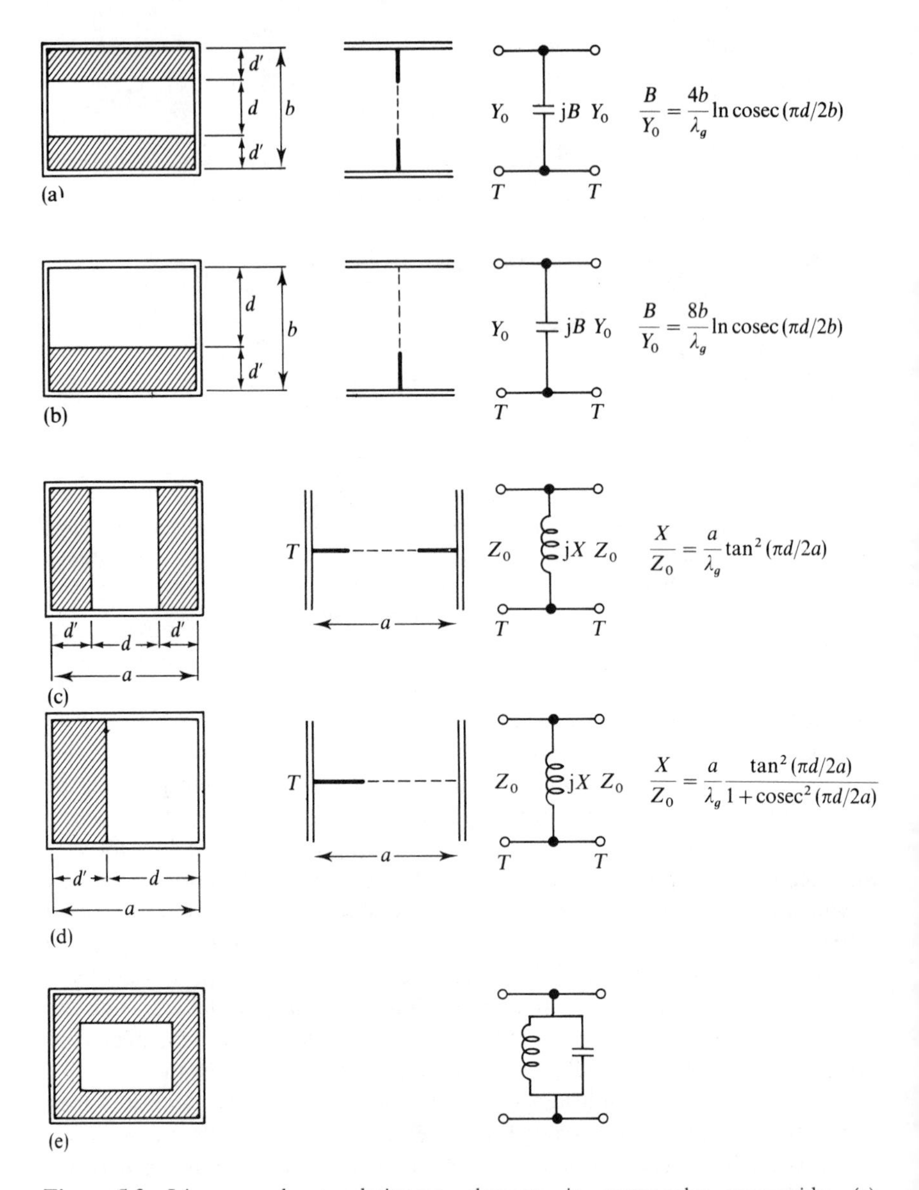

Figure 5.3 Irises as shunt admittance elements in rectangular waveguide: (a) symmetrical capacitative; (b) asymmetrical capacitative; (c) symmetrical inductive; (d) asymmetrical inductive; (e) resonant.

wavelength decreases it might become necessary to model the junction by a 2-port circuit. Similar techniques are available for use with waveguides, which for transmission are usually used in the dominant mode. Consider a thin obstacle placed in a transverse plane in a waveguide with a reflectionless termination, as indicated in Figure 5.2(a). The obstacle will bring about a

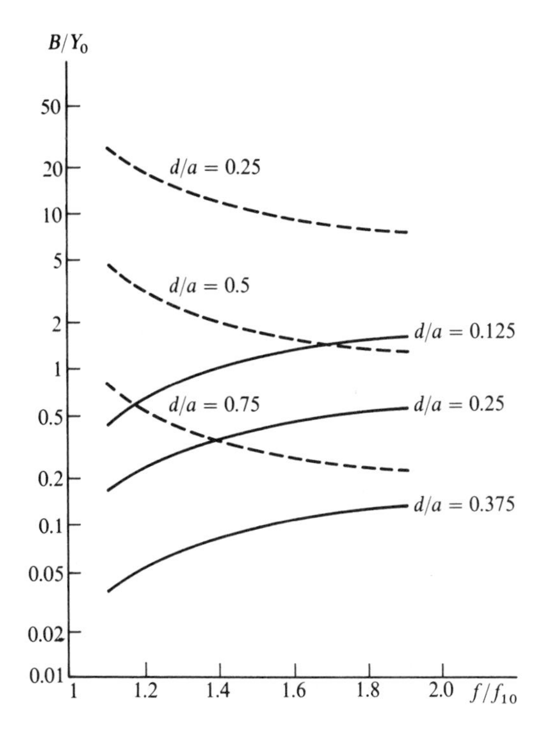

Figure 5.4 Variation with frequency of normalized susceptance for symmetrical irises. Full line denotes capacitive, Figure 5.3(a); pecked line, inductive, Figure 5.3(c).

reflection and, since the waveguide is operating in the dominant mode, the reflection will be characterized by a dominant mode field travelling in the reverse direction: all other modes needed to satisfy boundary conditions at the obstacle will be attenuated to negligible proportions within a short distance on either side of the obstacle. On the basis of the transmission line equivalent described in Section 2.4 the obstacle may be characterized by the shunt admittance which will give the same standing wave pattern. Figure 5.3 shows a number of diaphragms in rectangular waveguides, together with approximate formulae derived from field analysis. Figure 5.4 presents graphically some of these results, showing behaviour with frequency. Whether the equivalent shunt admittance is inductive or capacitive depends on the effect of the diaphragm on the electric and magnetic fields: if it causes an increase in the electric field and an increase in stored electric energy, a capacitive susceptance results; if it constricts magnetic field lines then the equivalent susceptance will be inductive. When both effects occur, a resonant diaphragm results.

In Figure 5.5 a post is shown, for which the longitudinal dimension is not negligible compared with the guide wavelength. This has to be regarded as having the equivalent circuit shown. Values of the reactances are indicated

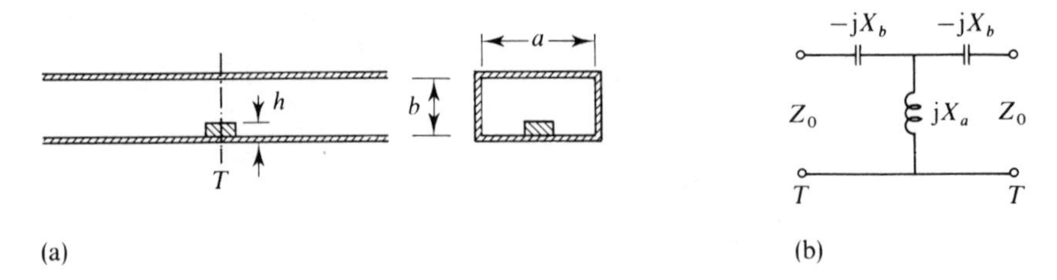

Figure 5.5 Post in waveguide. (a) Configuration. (b) Equivalent circuit.

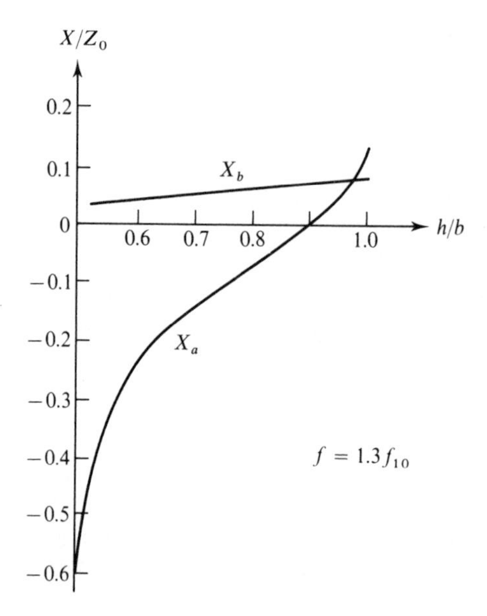

Figure 5.6 Parameters of the equivalent circuit of Figure 5.5(b). The curves show typical variation of the parameters as post height is altered at fixed frequency.

in Figure 5.6. Similar components exist in circular waveguides. All the above examples have sharp edges, giving rise to high electric fields and consequent early breakdown at high power. When high power levels are encountered a matching post of the form shown in Figure 5.7 might be used, which would have to be regarded as having the same form of equivalent circuit as the post in Figure 5.5.

Whilst a variable stub in a coaxial system is relatively simple to make, it is much more difficult to make variable components in microstrip or waveguide. In these cases a design is arrived at by calculation or experiment and components made to the resulting specifications, forming a 'preplumbed' system. In some cases, in particular for microstrip, it is possible to design for larger dimensions and then to trim for optimum results.

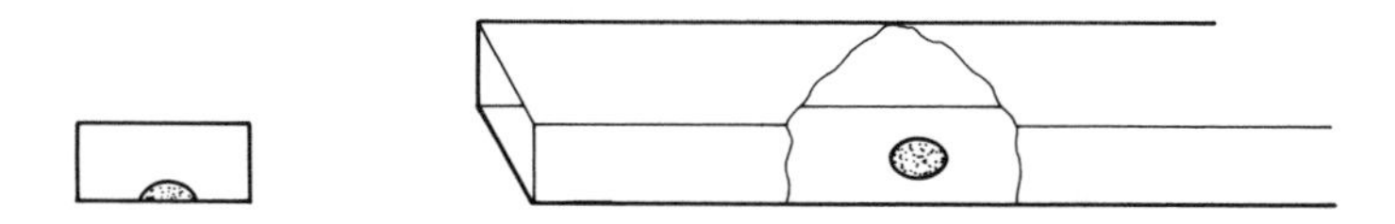

Figure 5.7 Rounded post, avoiding sharp edges and breakdown at high power.

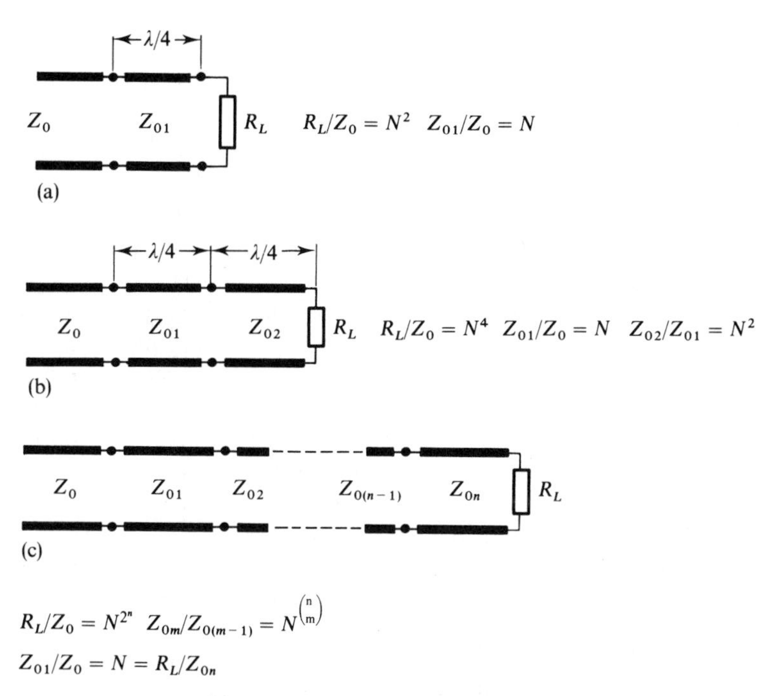

Figure 5.8 The quarter-wave transformer: (a) single stage; (b) two-stage; (c) general form for multi-stage transformer.

5.4 Transformers

The interconnections in a system will everywhere use a standard transmission line of given characteristic impedance, often 50 Ω. The devices being connected may well have a different value of conductance as well as a parallel susceptance. The susceptance can be cancelled by an equal but opposite susceptance in parallel, leaving the problem of matching a resistive load to the line impedance. A device frequently used for this purpose is the quarter-wave transformer. Application of equation (2.19) from Section 2.3.1 shows that a $\lambda/4$ length of line terminated in Z_2 presents an input impedance of Z_0^2/Z_2. Thus matching between a load R_L and a line system of characteristic impedance Z_0 can be achieved by interposing a $\lambda/4$ length of characteristic impedance $(Z_0 R_L)^{\frac{1}{2}}$. Such a transformer is correct only at the design

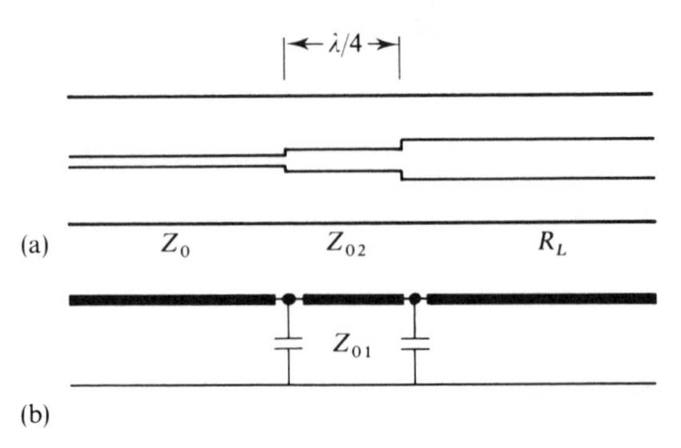

Figure 5.9 A quarter-wave transformer in coaxial line: (a) construction; (b) equivalent circuit allowing for junction effects.

frequency, but improved bandwidth may be obtained using multiple section transformers, as shown in Figure 5.8. In the general case, shown in Figure 5.8(c), the transformer is made with n sections, each of length $\lambda/4$ at the design frequency and of characteristic impedance graded according to the formula given. The increase in bandwidth is a result of the more gentle grading from one section to the next. It should be mentioned that this description of quarter-wave transformers assumes that the dimensions of the discontinuities are much less than the wavelength: at higher frequencies it may be necessary to allow for a junction effect by an additional shunt susceptance, as indicated in Figure 5.9. This will of course modify the design.

The above examples refer to transmission lines: Similar arrangements can be used in waveguides, making allowance for the more significant junction effects.

Quarter-wave transformers are also used in realizing non-contact shorts. A variable short circuit can in principle be made by a sliding plug in a waveguide. In practice, the difficulties of making consistently good contact between slider and wall are too great, and instead non-contact sliders are used. Consider the configuration of Figure 5.10(a): the plunger and waveguide may be regarded as coaxial transmission lines as shown in Figure 5.10(b). Since each section is quarter-wave

$$Z_A = Z_{01}^4/(Z_{02}^2 Z_B)$$

The characteristic impedances depend on the gaps, and $Z_{02}/Z_{01} \cong 10$ is achievable. Z_B is large, so that Z_A can be very small. A shorter version is shown in Figure 5.10(c).

A similar arrangement can be used to avoid contact problems in demountable waveguide junctions. Figure 5.11 shows a choke joint for a rectangular waveguide: the gap forms a radial waveguide terminated in a groove approximately one-quarter wave deep. Correct choice of groove radius gives a low impedance across the gap in the waveguide wall.

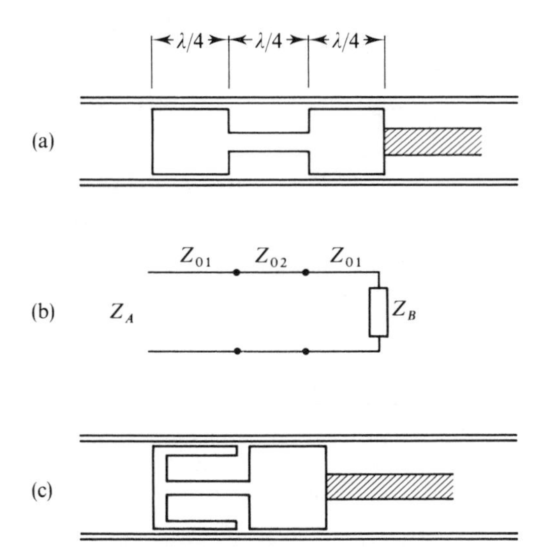

Figure 5.10 Shorts: (a) non-contact plunger; (b) equivalent circuit; (c) re-entrant form.

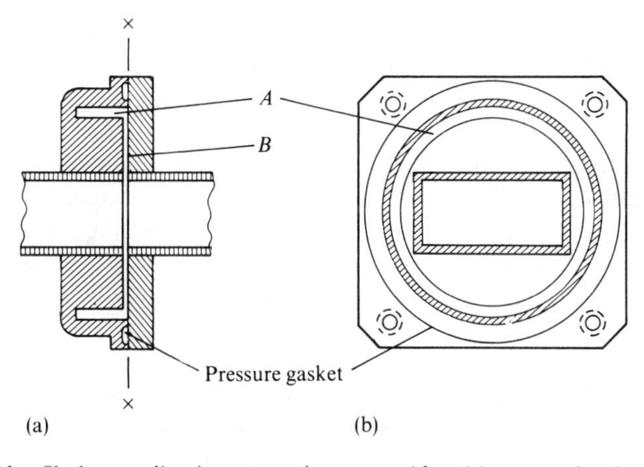

Figure 5.11 Choke coupling in rectangular waveguide. A is an annular slot $\lambda/4$ deep, B is an annular gap forming a radial transmission line $\lambda/4$ long.

5.5 Resonant cavities

Both in waveguide and transmission line systems, resonant cavities play the part of tuned circuits at low frequencies, as circuit elements and as filters. Some examples of 1-port resonators are shown in Figure 5.12. Figure 5.12(a) shows an iris coupled waveguide cavity, Figure 5.12(b) a loop coupled coaxial transmission line cavity. Both these consist of short-circuited lengths of line.

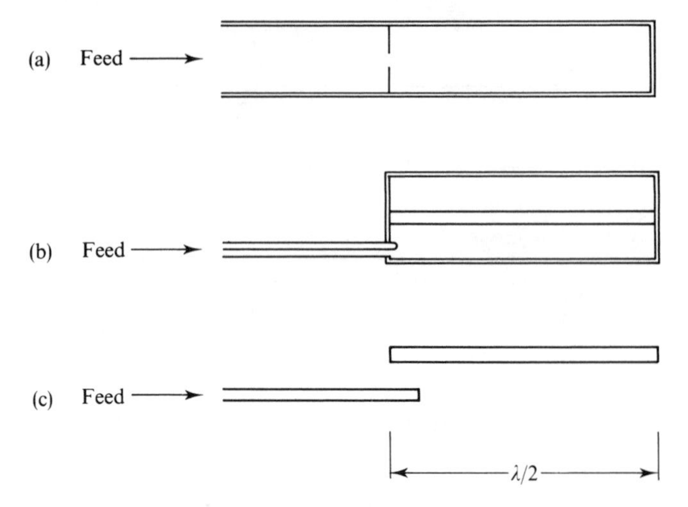

Figure 5.12 Resonant cavities: (a) iris-coupled waveguide; (b) loop-coupled coaxial resonator; (c) capacitatively coupled microstrip resonator.

Figure 5.12(c) shows a microstrip resonator open circuited at both ends, capacitatively coupled to a transmission line. All may be converted into 2-port devices by an additional coupling element at the far end.

Figure 5.13(a) shows a circuit illustrative of these configurations, consisting of a length of transmission line short-circuited at one end and shunted by a low valued inductance at the other. The lowest frequency of resonance will be approximately that which makes l one-half wavelength, although additional higher order resonances exist. If this angular frequency is denoted by ω_0, the input impedance of the line is given by equation (2.29) as

$$Z = \mathrm{j}Z_{oc} \tan(\omega\pi/\omega_0) \tag{5.1}$$

This can be represented in the neighbourhood of ω_0 by the equivalent circuit of Figure 5.13(b). Figure 5.13(c) shows the circuit equivalent to the cavity in association with the coupling loop. A simplification of this circuit can be made by transferring the reference plane for the terminals. Shown in Figure 5.13(d) is a circuit transformation, enabling the shunt element to be replaced by a series element, giving finally the equivalent circuit of Figure 5.13(e). In this last circuit a resistance R has been inserted to represent resonator losses. The impedance presented to the line at resonance is n^2R: the ratio $\beta = Z_0/n^2R$ is termed the **coupling coefficient**. The cavity is said to be over- or under-coupled according as $\beta > 1$ or $\beta < 1$ respectively. Since the impedance at resonance is real and equal to $\beta^{-1}Z_0$, the VSWR is equal to β, or β^{-1}.

The quality factor Q may be defined as for circuits. If L' is replaced by a short circuit the cavity, Q is referred to as the **unloaded** $\boldsymbol{Q}$, Q_u; when the feed line is terminated at the source end in Z_0, the **loaded** $\boldsymbol{Q}$, Q_l, results. The

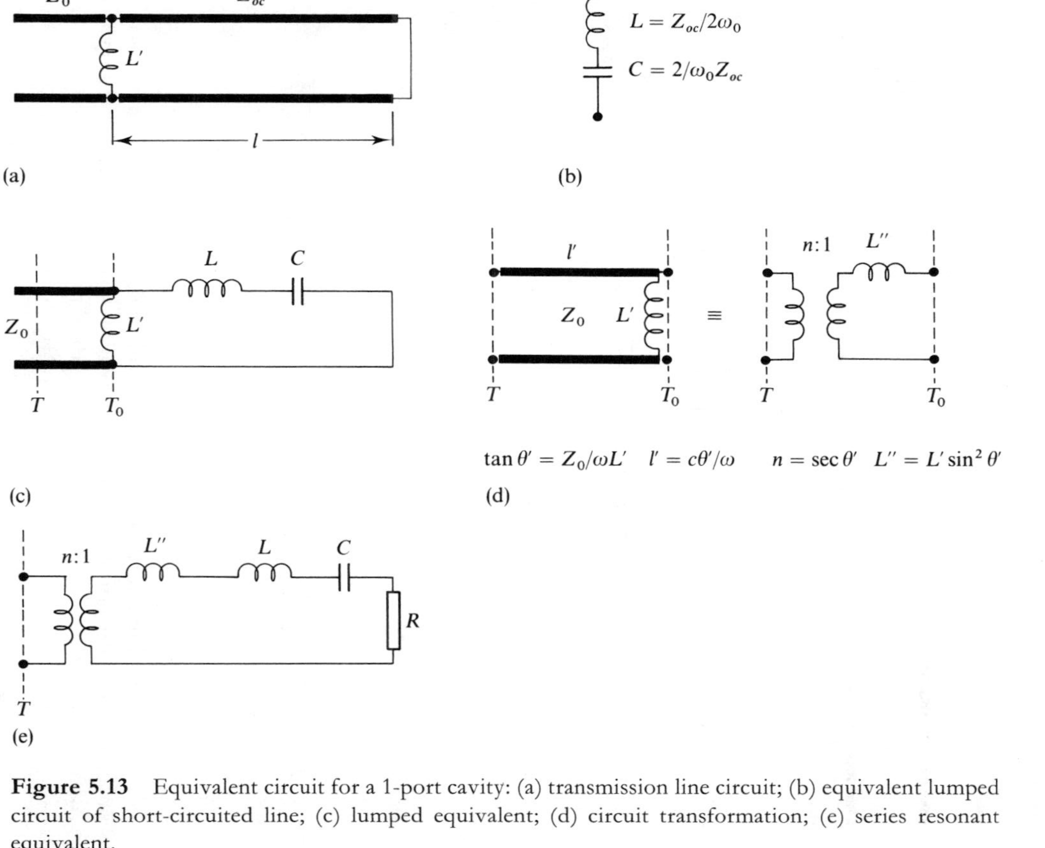

Figure 5.13 Equivalent circuit for a 1-port cavity: (a) transmission line circuit; (b) equivalent lumped circuit of short-circuited line; (c) lumped equivalent; (d) circuit transformation; (e) series resonant equivalent.

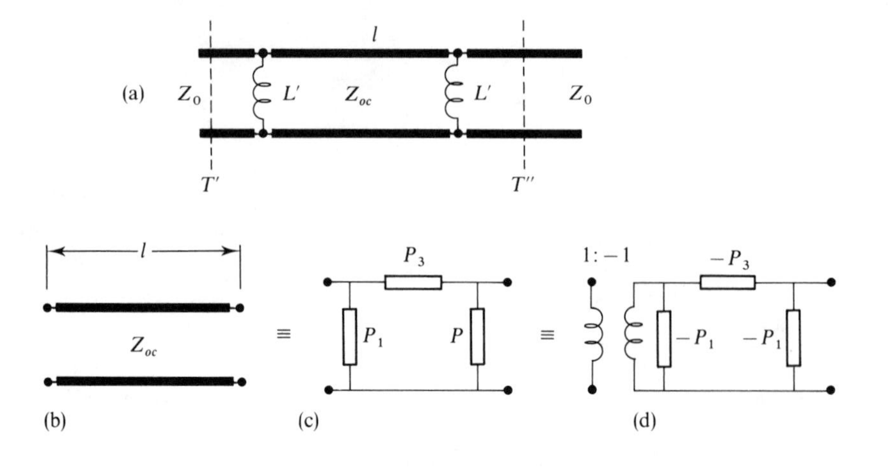

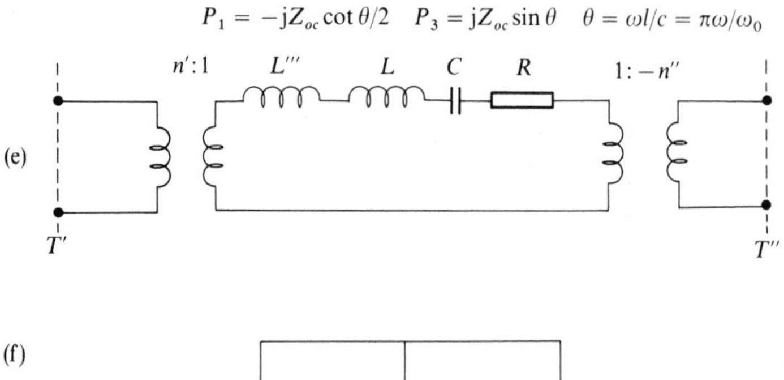

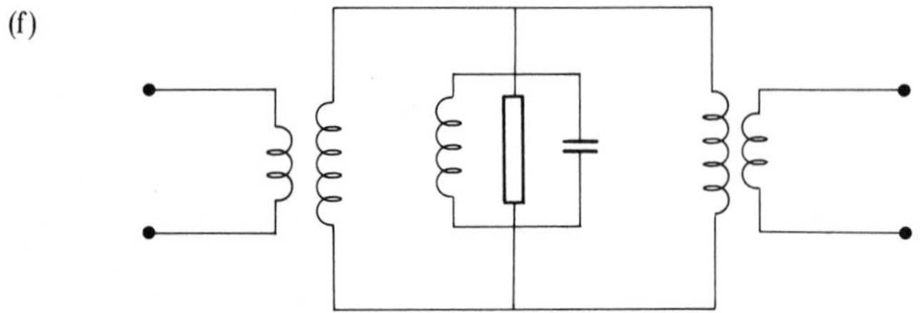

Figure 5.14 Equivalent circuit for a 2-port cavity: (a) transmission line circuit; (b), (c) and (d) circuit transformations; (e) series equivalent circuit; (f) parallel equivalent obtained by variation of reference planes. These equivalent circuits will only be valid over a small range around the mid-frequency.

external, or **radiation** $\boldsymbol{Q}$, Q_e, is defined by

$$\frac{1}{Q_l} = \frac{1}{Q_u} + \frac{1}{Q_e}$$

The basic circuit of a 2-port cavity is shown in Figure 5.14(a). Circuit transformations for a length of line are shown in Figure 5.14(b), (c), (d): the

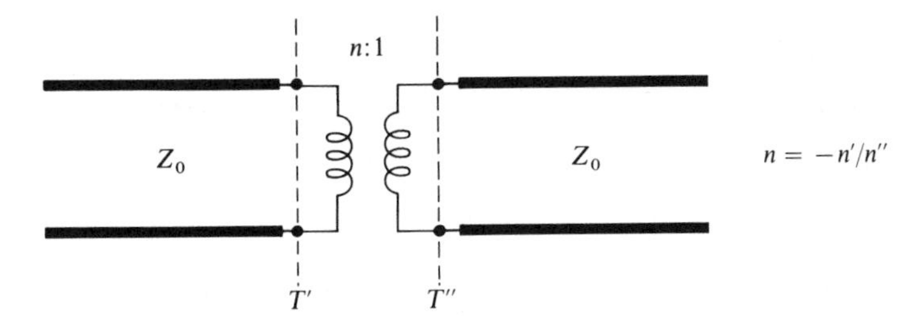

Figure 5.15 Transformer equivalent for 2-port cavity at resonance.

final step is to obtain realizable impedances in the neighbourhood of $\omega = \omega_0$. It will be seen that in this region P_3 is represented by the circuit of Figure 5.13(b); P_1 is large, and may be neglected when in shunt with L', L''. Use of the transformation of Figure 5.13(d) finally results in the circuit of Figure 5.14(e). This circuit can be taken as a basis for describing any resonance well separated from others.

Although any arbitrary 2-port resonator can be described by a circuit of the form of Figure 5.14(e), the appropriate reference planes must be determined. Consider the situation of a lossless cavity at resonant frequency, when the circuit of Figure 5.14(e) reduces to that of Figure 5.15, with $n = n'/n''$. The scattering matrix of this circuit with respect to T′, T″ is given by

$$[S] = \begin{bmatrix} \dfrac{n^2 - 1}{n^2 + 1} & \dfrac{2n}{n^2 + 1} \\ \dfrac{2n}{n^2 + 1} & -\dfrac{n^2 - 1}{n^2 + 1} \end{bmatrix} \tag{5.2}$$

If the resonator is set up with a matched termination on side 2 and is fed from side 1, a standing wave pattern will exist in the feed line. The plane T′ may be taken at a maximum of this pattern, where the reflection coefficient will be real and positive. The VSWR will be equal to n^2. With the feed on side 2 and a matched termination on side 1, T″ must be located at minimum of the standing wave pattern, at which the reflection coefficient will be real and negative. Evidently, this identification is not unique and various alternative positions are possible. To ensure the maximum range of validity for the equivalent circuit, the reference planes should be chosen as close to the resonator as possible.

The scattering matrix of equation (5.2) applies to any lossless 2-port. It was shown in Section 2.11.1 that a lossless 2-port could be described by the scattering matrix.

$$\begin{bmatrix} (1 - K^2)^{\frac{1}{2}} & \mathrm{j}K \\ \mathrm{j}K & (1 - K^2)^{\frac{1}{2}} \end{bmatrix} \tag{5.3}$$

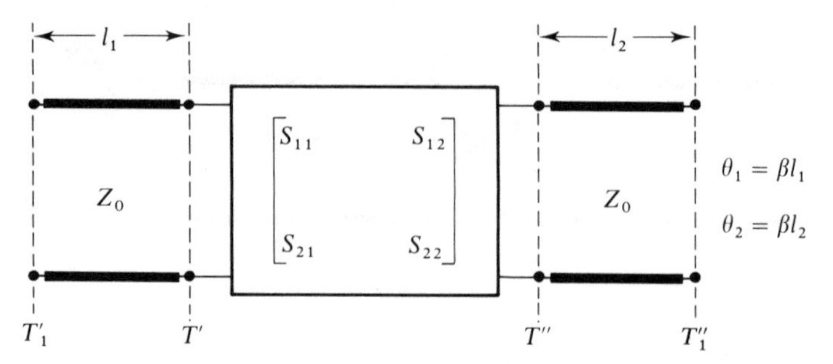

Figure 5.16 An arbitrary reactive 2-port, with shift of reference planes.

provided that the reference planes were taken where S_{11}, S_{22} were real and positive, i.e. both at maxima in the arrangement described above. Displacement of the reference planes to T'_1, T''_1 as shown in Figure 5.16 is easily show to result in the scattering matrix

$$[S'] = \begin{bmatrix} S_{11}\exp(-2j\theta_1) & S_{12}\exp(-j(\theta_1+\theta_2)) \\ S_{21}\exp(-j(\theta_1+\theta_2)) & S_{22}\exp(-2j\theta_2) \end{bmatrix} \tag{5.4}$$

in which $\theta_1 = \beta l_1$, $\theta_2 = \beta l_2$.

It will be seen that a choice of $\theta_1 = 0$, $\theta_2 = \pi/2$ makes equations (5.2) and (5.4) identical.

The above discussion has been carried out for the series resonant circuit. By change of reference planes a parallel connected equivalent circuit may be obtained, as shown in Figure 5.14(f).

The analysis in this section has been carried out for the circuit of Figure 5.13(a). The reduction of the waveguide cavity of Figure 5.12(a) would use an equivalent for the iris coupling hole: an example of this equivalent for a symmetrically placed circular iris in rectangular waveguide is indicated in Figure 5.17.

5.6 3-port junctions

In the previous sections 1-port and 2-port networks have been considered: 3-port junctions are a common occurrence in, for example, mixing two signals together or in power splitting. The concept of placing an admittance in shunt on a coaxial line by using a short-circuited stub, as in Figure 5.1(a), is in fact a junction of three lines, and may be properly called a shunt junction. A series junction is difficult to envisage in coaxial line, when balance-to-unbalance considerations are relevant. In waveguide systems both series and shunt junctions are used, as shown in Figure 5.18(a) and (b) for rectangular waveguide. The series, or E-plane, junction is so called because in the trans-

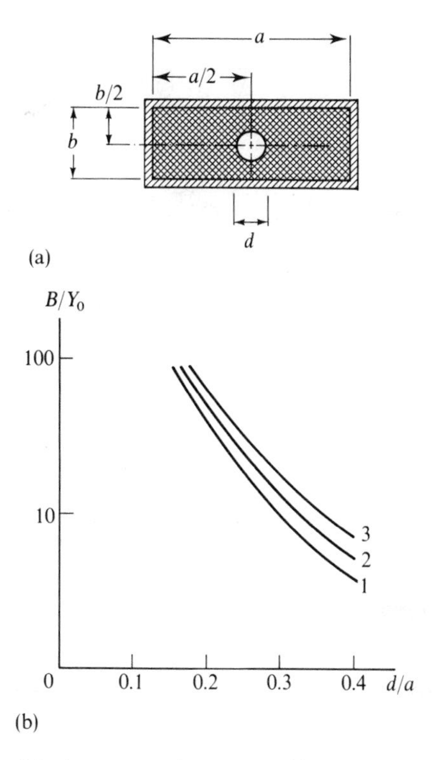

Figure 5.17 Circular iris in rectangular waveguide: (a) geometry; (b) normalized susceptance as a function of d/a at various frequencies, in WG 16 waveguide. Curve 1: $f/f_{10} = 1.67$; curve 2: $f/f_{10} = 1.45$; curve 3: $f/f_{10} = 1.25$.

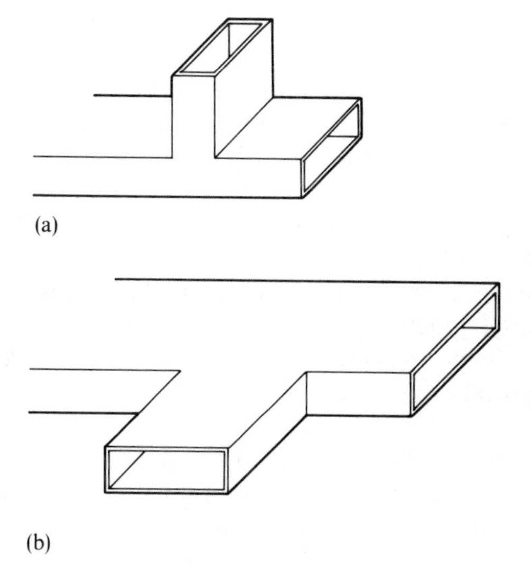

Figure 5.18 Waveguide T-junctions: (a) E-plane; (b) H-plane.

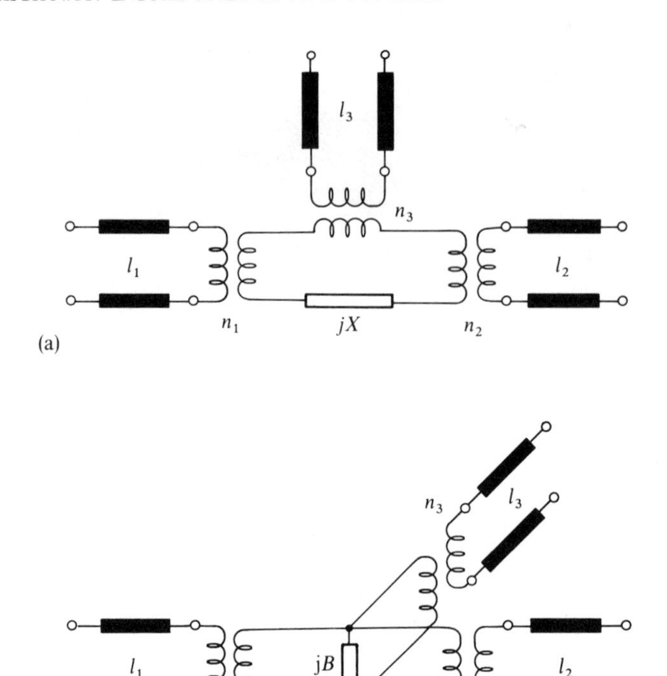

Figure 5.19 Equivalent circuits for T-junctions: (a) series form; (b) shunt form.

mission line analogue the electric field becomes voltage. Similarly in the shunt, or H-plane, junction current is the analogue of magnetic field. Equivalent circuits for such junctions can be found in the form of 3-port circuits of reactances, requiring in general six parameters. Also valid are circuits such as those of Figure 5.19(a) and (b). It must be emphasized that although these have the appearance of series and shunt junctions, they can be transformed into each other by choice of the parameters. One may be more convenient than the other because, for example, the physical configuration resembles the circuit, or the reference planes are close to the physical junction. Similarly, junctions between waveguides and coaxial lines are frequently used. Some general theorems concerning lossless 3-ports are available, which will be quoted without proof:

(1) It is always possible to place a short-circuit in one arm of the T-junction so that there is no transmission of power between the other two arms. This can be seen from the equivalent circuits in Figure 5.19.

(2) It is impossible to match a general T-junction completely, i.e. if any two arms have matched loads, a reflection will take place from the third.

(3) If a T-junction is symmetrical about one arm then a short-circuit may be placed in that arm to give reflectionless transmission between the other two. This also follows from the equivalent circuits.

5.6.1 Junctions containing resistors

Matching to all ports and power division can be obtained by circuits such as that of Figure 5.20, but power is lost in the resistors. A useful circuit is shown in Figure 5.21(a): in this case power into port one is shared equally between

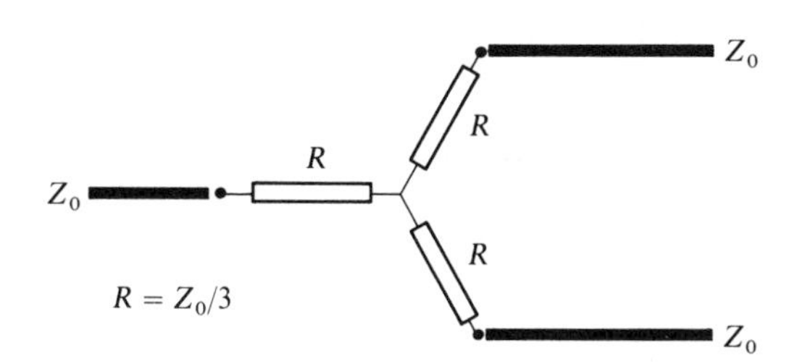

Figure 5.20 A power divider. With matched loads one-half of the input power is lost in the resistors.

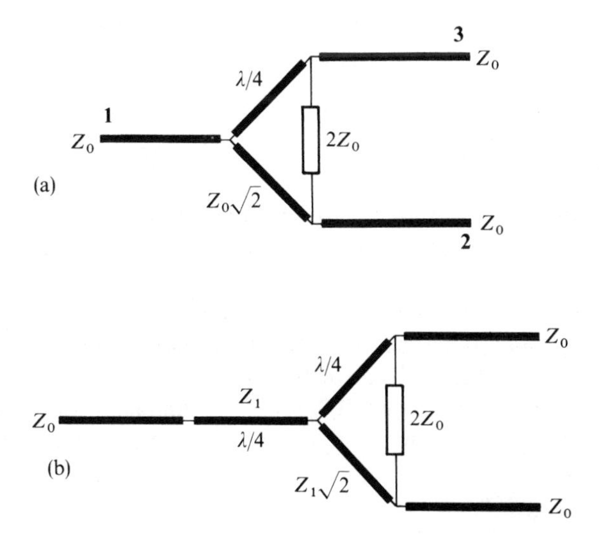

Figure 5.21 The 'Wilkinson' divider: an input to port one is split equally between ports two and three without loss; an input to port two is decoupled from port three, and one-half the input power is lost in the resistor.

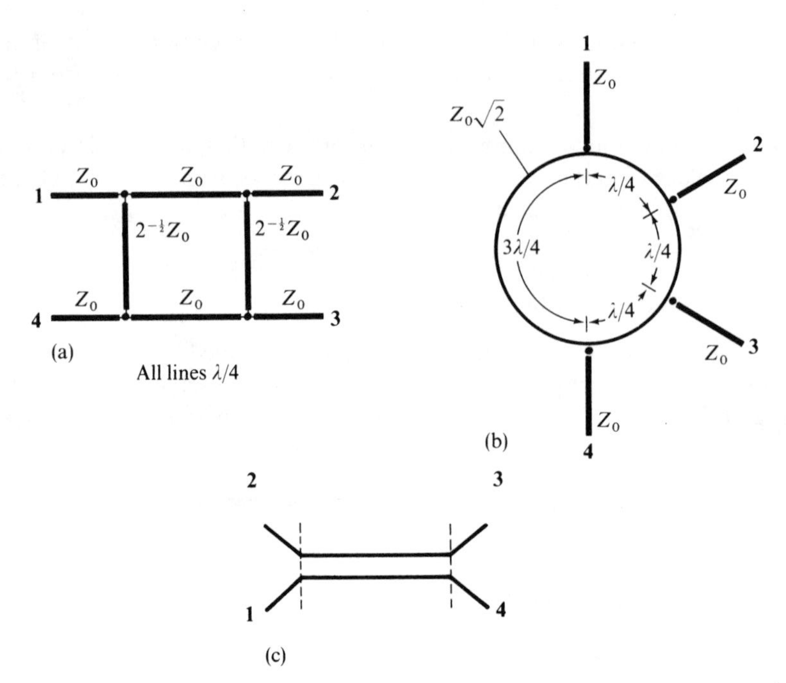

Figure 5.22 Hybrid junctions: (a) transmission line configuration; (b) the 'rat race'; (c) continuously coupled lines.

ports two and three with no loss in the resistor; power into ports two or three with port one matched is also matched, but power is lost in this case. The bandwidth may be increased by circuits such as Figure 5.21(b).

5.7 4-port junctions

Multi-port junctions find little use at low frequencies. At microwave frequencies, however, a number of such junctions have useful properties and are easy to realize. In this section, two lossless 4-port junctions that exhibit the property of complete matching will be described, i.e. if any three ports have matched terminations then the remaining port presents zero reflection coefficient. It was mentioned above that this is not possible for a 3-port system. These circuits are frequently used, for example, in the realization of measuring techniques and when it is necessary to combine signals from different sources.

5.7.1 Hybrid junctions

An ideal hybrid junction is a lossless 4-port device which has the property that with each port matched, the power fed into one port is equally divided

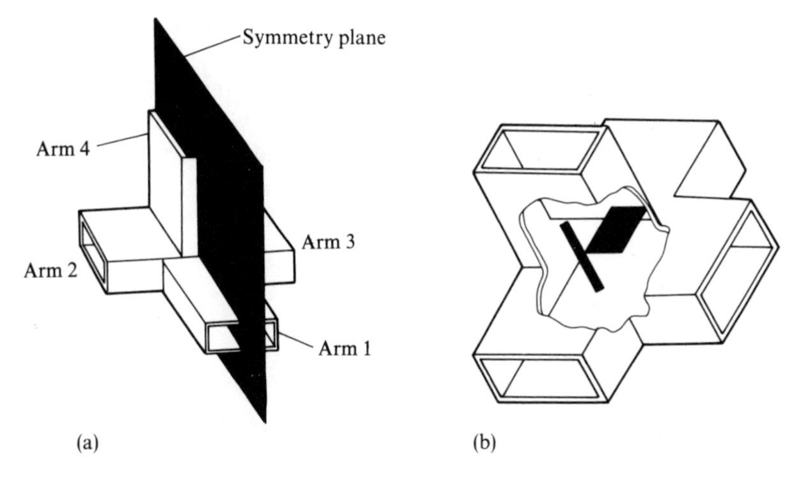

Figure 5.23 The waveguide 'magic-T' junction.

between two ports, with no coupling to the fourth port. Some examples in transmission line are shown in Figure 5.22. The diagrams can perhaps best be visualized as the conductors in a microstrip realization with a ground plane below. Figure 5.23 shows the 'magic-T' waveguide. Analysis shows that the scattering matrix for the configuration of Figure 5.22(a) is

$$S = \frac{1}{2^{\frac{1}{2}}} \cdot \begin{bmatrix} 0 & 0 & -1 & -j \\ 0 & 0 & -j & -1 \\ -1 & -j & 0 & 0 \\ -j & -1 & 0 & 0 \end{bmatrix} \tag{5.5}$$

showing that power into port one is divided equally between ports three and four, with port two decoupled. The two outputs are in phase quadrature. In the 'rat race' hybrid of Figure 5.22(b) the characteristic impedance of the ring must be $2^{\frac{1}{2}}Z_0$, when the scattering matrix is

$$S = -\frac{j}{2^{\frac{1}{2}}} \cdot \begin{bmatrix} 0 & 1 & 0 & -1 \\ 1 & 0 & 1 & 0 \\ 0 & 1 & 0 & 1 \\ -1 & 0 & 1 & 0 \end{bmatrix} \tag{5.6}$$

Thus an input to port one gives antiphase outputs in ports two and four; an input to port two gives in-phase outputs in ports one and three.

If the diagram of Figure 5.22(c) is considered in microstrip there is evidently coupling between the two lines on a continuous basis. Provided the geometrical layout can be designed to give the required coupling a 90°-hybrid can be realized. It may be necessary to split and interleave the two lines to obtain the required coupling.

The analysis of this type of coupler is considered in Section A.8. It may be shown that, for a homogeneous dielectric and with the numbering and reference planes of Figure 5.22(c), the scattering matrix takes the form

$$[S] = \begin{bmatrix} 0 & j\alpha & 0 & \beta \\ j\alpha & 0 & \beta & 0 \\ 0 & \beta & 0 & j\alpha \\ \beta & 0 & j\alpha & 0 \end{bmatrix}$$

in which $\arg(\alpha) = \arg(\beta)$ and $|\alpha|^2 + |\beta|^2 = 1$.

The operation of the waveguide magic-T of Figure 5.23 can be deduced from symmetry, bearing in mind the field configuration of the dominant mode. An input to port one will divide equally into arms two and three, giving cophased outputs; the electric vector is in the wrong direction to excite a wave in arm four. The power into port four will not excite a wave in arm one but will divide equally between arms two and three with antiphase outputs. It can be shown that the scattering matrix to suitable reference planes takes the form

$$S = \frac{j}{2^{\frac{1}{2}}} \cdot \begin{bmatrix} 0 & 1 & 1 & 0 \\ 1 & 0 & 0 & 1 \\ 1 & 0 & 0 & -1 \\ 0 & 1 & -1 & 0 \end{bmatrix} \tag{5.7}$$

5.7.2 Directional couplers

In Figure 5.24 one type of directional coupler is shown schematically: coupling between the two guides is by means of two small holes in the broad faces of the waveguide which are one-quarter of a guide wavelength apart. A wave propagating from port one to port two of the main waveguide will give rise to waves in the auxiliary waveguide from each of the two holes. The phasing is such that they reinforce in the direction of port four and cancel in

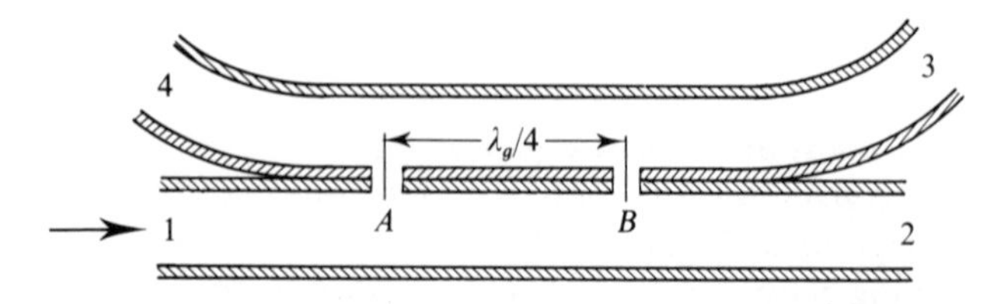

Figure 5.24 A waveguide directional coupler.

the direction of port three. A wave propagating in the direction from port two to port one will produce an output in port three and zero in port four. Thus the outputs from ports four and three will respectively measure forward and backward waves in the main waveguide. Hence the term directional coupler. With suitable reference planes the scattering matrix takes the form

$$[S] = \begin{bmatrix} 0 & \alpha & 0 & \beta \\ \alpha & 0 & \beta & 0 \\ 0 & \beta & 0 & -\alpha \\ \beta & 0 & -\alpha & 0 \end{bmatrix} \tag{5.8}$$

in which $\alpha^2 + \beta^2 = 1$, and β is the coupling factor. A coupler might be described as a '20 dB coupler', meaning that β corresponds to 20 dB attenuation.

Other varieties of directional coupler can be made. It will be seen that the hybrid junctions described earlier may be regarded as 3 dB directional couplers.

5.8 Mode filters and transducers

For the purpose of power transmission in a waveguide it is usual to use a system with a dominant mode. For some purposes, for example rotary joints, it is necessary to use a section of circular waveguide with a circularly symmetrical mode such as TM_{01} (Figure 2.15). Configurations carrying out such mode conversions are mode transducers. It may also be necessary to use a mode which is not dominant: at such a frequency imperfections in the waveguide will cause mode conversion to other propagating modes. Mode filters may be used to suppress unwanted modes.

5.9 Microwave filters

Low frequency systems make extensive use of filters to separate various frequency bands. It is possible to use 2-port cavities (as discussed in Section 2.5) as equivalent to tuned circuits and thus to make band-pass filters at microwave frequencies. Another class of filter may be derived from low frequency circuits such as the Butterworth low-pass circuit of Figure 5.25(a). The inductors may be replaced by lengths of short-circuited transmission line and the capacitors by open circuited lengths, each designed to give the correct reactance at some suitable frequency, as shown in Figure 5.25(b). Obviously the response will not be the same as that of the original, and it will exhibit a series of pass bands because of the periodic nature of the reactances. One possibility leading to a design method is to make the lines of equal length but

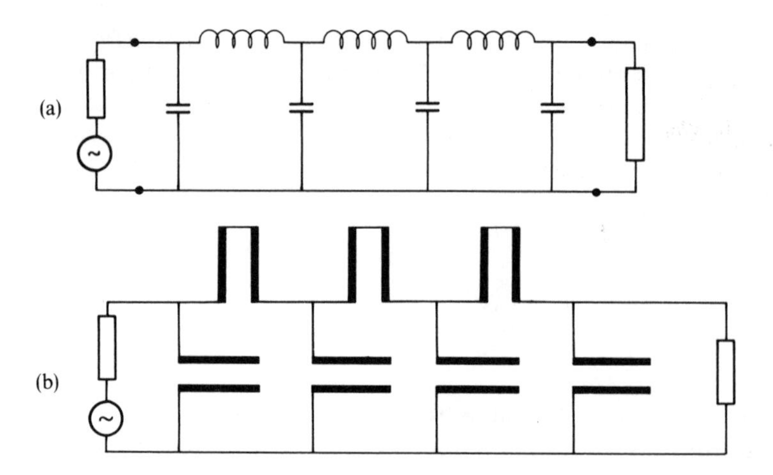

Figure 5.25 Filters: (a) circuit of a three stage Butterworth filter; (b) replacement of components with lengths of transmission line.

different characteristic impedances: this can easily be shown to replace the angular frequency variable in the original by $\tan(\omega l/c)$. Reference must be made to the literature for detailed information. The 'YIG' tunable filter is considered later together with other ferrite devices.

5.10 Ferrite components

Ferrites form a class of material which shows strong magnetic effects by reason of the magnetic dipoles of unpaired spinning electrons. The effective magnetic permeability can be comparable to that of ferromagnetic materials, but ferrites have sufficiently high resistivity to behave like a material with a loss tangent of around 10^{-3}. An external magnetic field will tend to align the individual dipoles, but disturbance from alignment causes the dipoles to precess, as indicated in Figure 5.26. Rotation takes place at a frequency proportional to the local magnetic field (27.9 GHz T^{-1}), with a definite sense rotation with respect to the magnetic field direction. It will be readily appreciated that a circularly polarized wave travelling parallel to the static magnetic field will be treated differently according to whether its polarization is right-handed or left-handed. The two waves travel with different velocities, and suffer different phase delays in traversing a given path length in the ferrite. Further, since the sense of rotation is determined by the static field, the effects on right- and left-handed polarization are interchanged for a wave propagating antiparallel to that field. These effects are illustrated in Figure 5.27. A plane wave travelling parallel to the static field can be regarded as the sum of a right-handed wave and a left-handed wave. The result, as illustrated in Figure 5.28, is a rotation of the plane of polarization. A plane wave travelling antiparallel to the static field suffers a rotation opposite in sense, so

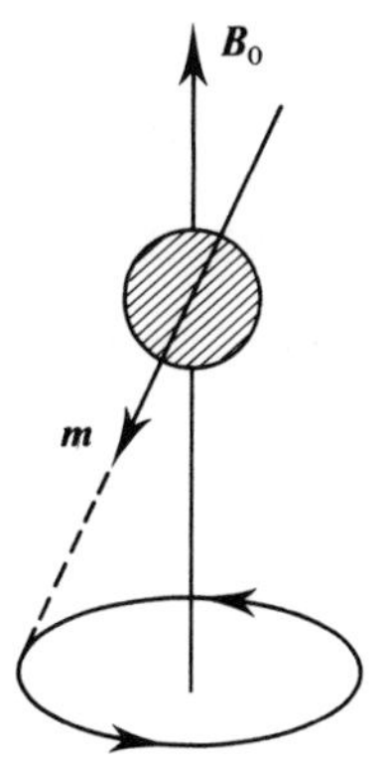

Figure 5.26 Illustrating precession of a spinning electron in a magnetic field.

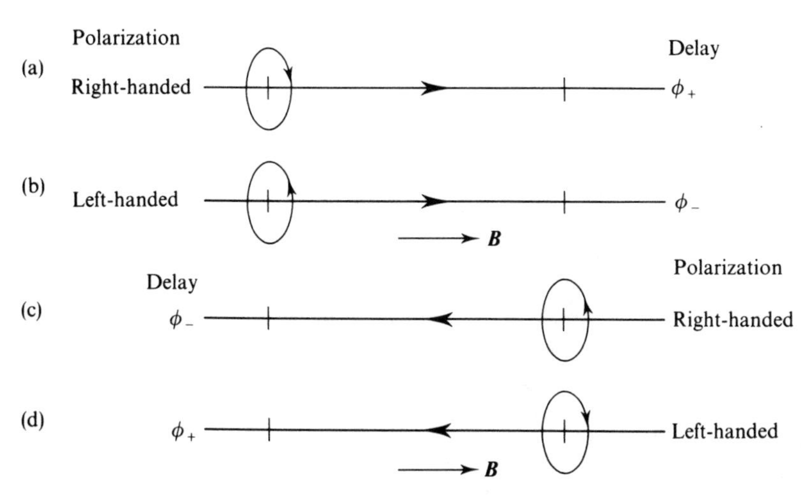

Figure 5.27 Propagation of circularly polarized waves through a ferrite material: (a) wave direction parallel to static magnetic field, right-handed polarization, phase delay ϕ_4; (b) as in (a) but left-handed polarization, phase delay ϕ_-; (c) wave direction antiparallel to static magnetic field, right-handed polarization, phase delay ϕ_-; (d) as in (c) but left-handed polarization, phase delay ϕ_+.

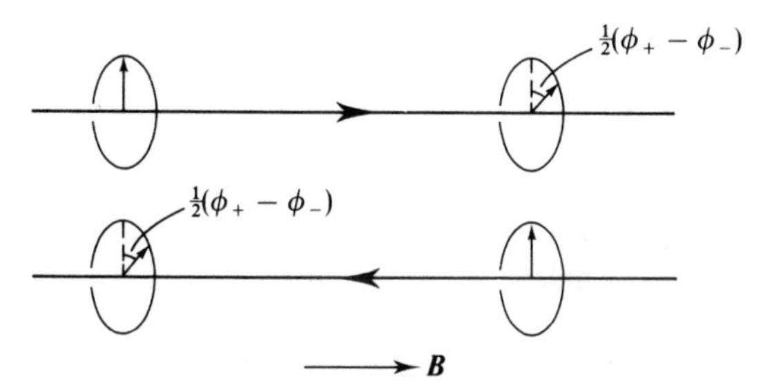

Figure 5.28 The Faraday effect: the plane of polarization rotates on traversing a length of ferrite material; the sense of rotation depends on direction of traverse, leading to non-reciprocity.

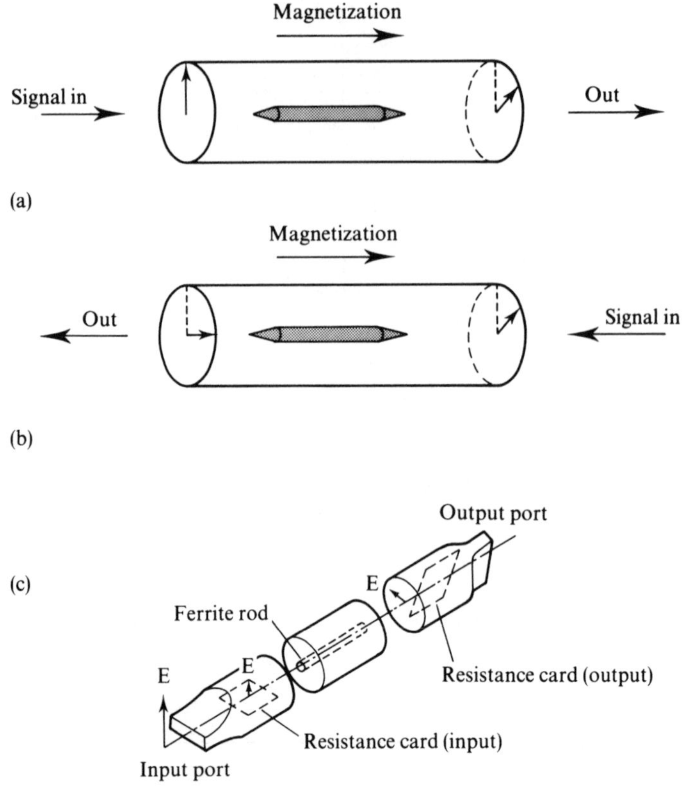

Figure 5.29 An isolator: (a) rotation of plane of polarization in the forward direction; (b) rotation of plane of polarization in reverse direction; (c) schematic form for waveguide isolator.

that a wave reflected back through the same path does not return to its original plane of polarization. Thus, devices depending on propagation through ferrites are non-reciprocal. The same mechanism applies to the Faraday effect in optics, so that term is also used in connection with microwave devices.

5.10.1 Isolators

The non-reciprocity of propagation through ferrite material can be used to make a device that propagates a signal from an input port to an output port in one direction only. Such a device is called an isolator, and is used, for example, after an oscillator, to prevent a reflected wave affecting operation. The basic elements of one type of isolator are depicted schematically in Figure 5.29. A TE_{11} wave can be launched into the circular waveguide with a specific orientation by a rectangular-to-circular transducer. The effect of propagation along the ferrite filled guide will be to rotate the orientation of the pattern in a

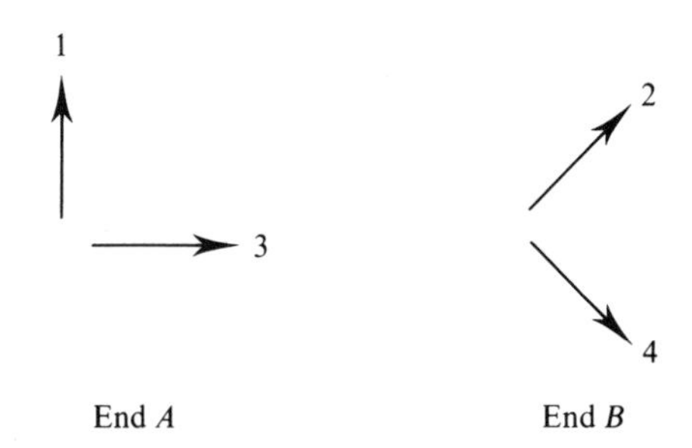

Figure 5.30 Labelling of ports associated with the different planes of polarization in the configuration of Figure 5.29.

way similar to that described in the previous section, and an appropriately oriented transducer will accept the wave. In the case when the angle of rotation is 45°, a wave propagating in the reverse direction will arrive at the input port with the mode pattern oriented at 90° to that of the original input wave and will thus not be accepted by the input transducer. In a practical device lossy material will be arranged to absorb the unwanted wave, as indicated in Figure 5.29(c).

5.10.2 Circulators

The device discussed in the previous section can be extended by separating the orientations at each end by means of mode filters, thus giving a 4-port device. If these orientations are numbered as in Figure 5.30, it will be seen that the input/output relationship, assuming ports are matched, is as follows:

Input port	*Output port*
1	2
2	3
3	4
4	1

Such a device is called a **circulator** and it has the circuit symbol shown in Figure 5.31(a). Other configurations are possible: a 3-port circulator is shown in waveguide in Figure 5.31 (b) and in microstrip in Figure 5.31 (c).

Excitation of the system from port one sets up a field pattern which has a null at port three. This is only possible in the presence of the ferrite: without ferrite, a pattern with a maximum at port one would result in equal signals at ports two and three. The size diameter and thickness of the ferrite and the magnetic field are design parameters and have to be chosen correctly.

5.10.3 YIG filters

The acronym YIG is derived from the Yttrium Iron Garnet ferrite found to be most suitable for these devices. The mode of operation can be seen by

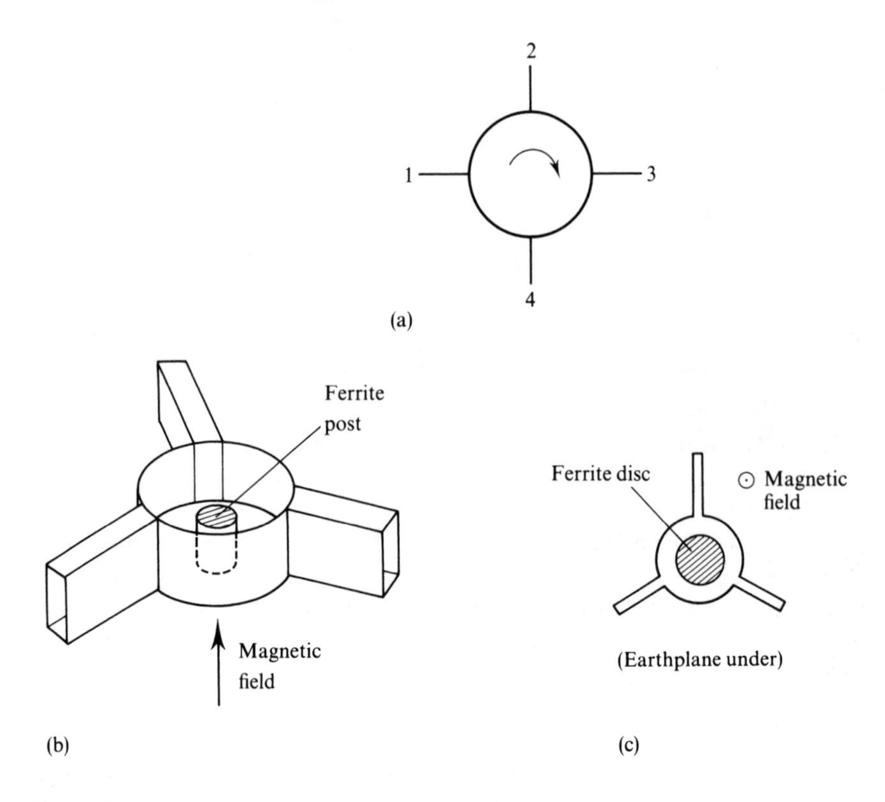

Figure 5.31 Circulators: (a) symbol for 4-port circulator; (b) 3-port waveguide circulator; (c) 3-port circulator in microstrip.

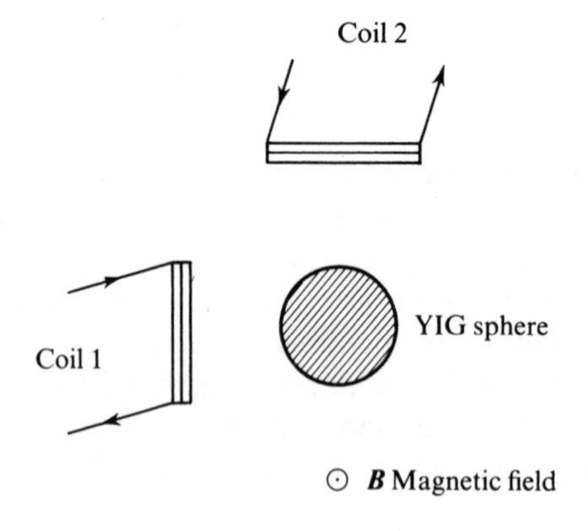

Figure 5.32 YIG filter. In the absence of the sphere, coils one and two are decoupled from each other. With the magnetized sphere coupling occurs near the precession frequency.

considering the two coil system shown in Figure 5.32. (At microwave frequencies the sphere and coils would be very small.) The two coils are arranged so that in the absence of magnetic effects from the YIG sphere there is no magnetic coupling between them. When a uniform magnetic bias field is applied to the YIG sphere as indicated, the internal magnetic field will be uniform. (This is a property of a sphere of magnetic material.) This means that all the dipoles can precess at precisely the same frequency. An RF magnetic field produced by currents in one coil will then induce this precession only if its frequency is close to the frequency of precession. The precessing dipoles will then induce a field and resultant emf in the cross-coupled coil. It is found that bandwidths of the order of 20 MHz are obtainable at microwave frequencies. Since the frequency of precession depends on the static magnetic field, the centre frequency of the filter can be controlled by altering the static field. Thus narrow-band tunable filters can be made, and if necessary the centre frequency can be swept in time by sweeping the bias magnetic field.

5.11 Phase shifters

Variable phase shifters are required in various contexts, in measurement systems, for example, or as a component in a phased array antenna. For some purposes variation by means of a mechanical control suffices; for others electrical control is necessary. A further point is whether or not calibration suffices, or is a calculable phase change necessary. A number of phase shifters will now be described.

5.11.1 Rotary phase shifter

Before describing this device, it is necessary to introduce the quarter-wave pipe, the microwave analogue of the optical quarter-wave plate. (This device transforms plane polarization into circular.) A circular waveguide will support TE_{11} (Figure 2.15) with any orientation: if the middle of a length of such a waveguide is deformed about a diametrical plane, two specific

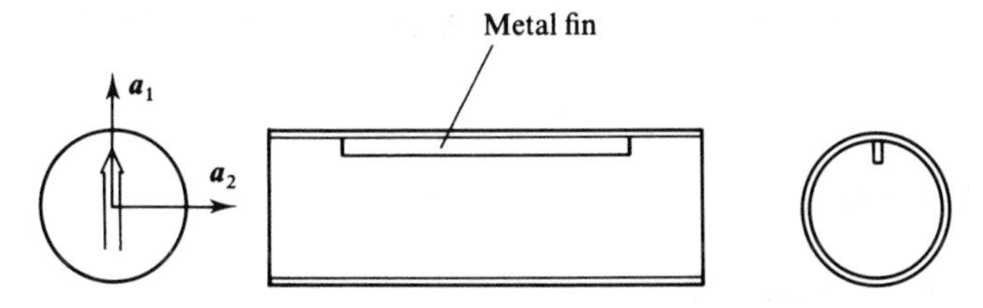

Figure 5.33 The quarter-wave pipe. The fin causes waves with the two orientations to travel with different phase velocities. A phase difference of $\pi/2$ gives a waveguide equivalent of the optical quarter-wave plate.

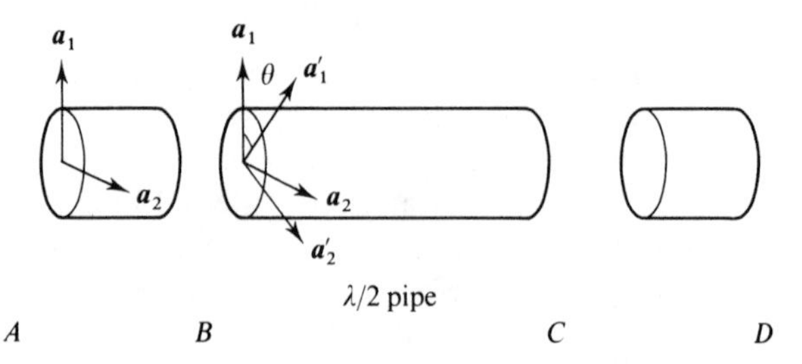

Figure 5.34 The rotary phase shifter. The end components are quarter-wave pipes, the centre element a half-wave pipe. Rotation of the centre element about its axis by an angle θ produces a change of 2θ in the phase of the output.

orientations come into being, so that the delay in one orientation is different from that in the other. Such a situation is shown in Figure 5.33. A quarter-wave pipe results if the phase delay to a wave oriented along $\boldsymbol{a}_1$ exceeds that of one oriented along $\boldsymbol{a}_2$ by $\pi/2$; a half-wave pipe has a differential delay of π.

An electric field $E_1\boldsymbol{a}_1 + E_2\boldsymbol{a}_2$ at the entrance to a quarter-wave pipe emerges as $-\mathrm{j}E_1\boldsymbol{a}_1 + E_2\boldsymbol{a}_2$, apart from a common delay. Thus a plane polarized wave will be transformed into one circularly polarized. The rotary phase shifter is made of two aligned quarter-wave pipes with a half-wave pipe in between, oriented at an angle θ to the quarter-wave pipes, as indicated in Figure 5.34.

An input at plane A given by

$$\boldsymbol{E}_A = E_1\boldsymbol{a}_1 + E_2\boldsymbol{a}_2 \tag{5.9}$$

emerges at plane B as

$$\boldsymbol{E}_B = -\mathrm{j}E_1\boldsymbol{a}_1 + E_2\boldsymbol{a}_2 \tag{5.10}$$

By transforming to unit vectors $\boldsymbol{a}'_1$, $\boldsymbol{a}'_2$ and delaying the component along $\boldsymbol{a}'_1$ by π it will be found that at plane C, $\boldsymbol{E}_C$ can be written in the form

$$\boldsymbol{E}_C = \boldsymbol{a}_1(\mathrm{j}E_1 \cos 2\theta - E_2 \sin 2\theta) + \boldsymbol{a}_2(\mathrm{j}E_1 \sin 2\theta + E_2 \cos 2\theta) \tag{5.11}$$

Finally after the second quarter-wave pipe

$$\boldsymbol{E}_D = \boldsymbol{a}_1(E_1 \cos 2\theta + \mathrm{j}E_2 \sin \theta) + \boldsymbol{a}_2(\mathrm{j}E_1 \sin 2\theta + E_2 \cos 2\theta) \tag{5.12}$$

Thus a wave oriented so that

$$\boldsymbol{E}'_A = \frac{E}{2^{\frac{1}{2}}}(\boldsymbol{a}_1 + \boldsymbol{a}_2)$$

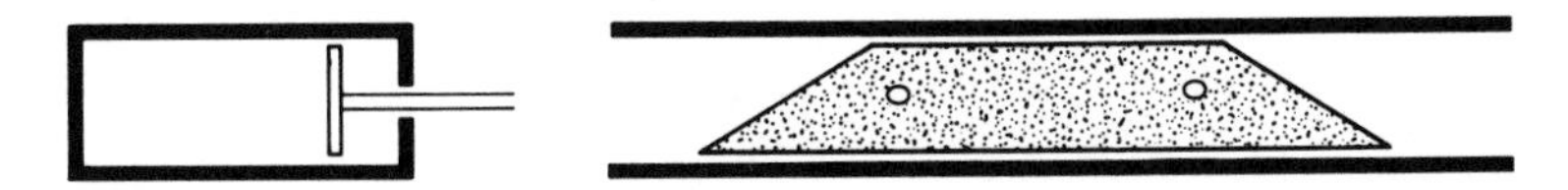

Figure 5.35 A dielectric vane phase shifter. The phase delay increases as the vane is moved towards the centre of the waveguide.

emerges as

$$\boldsymbol{E}'_D = \exp(2j\theta)\boldsymbol{E}'_A \tag{5.13}$$

A wave

$$\boldsymbol{E}''_A = \frac{E}{2^{\frac{1}{2}}}(\boldsymbol{a}_1 - \boldsymbol{a}_2)$$

emerges as

$$\boldsymbol{E}''_D = \exp(-2j\theta)\boldsymbol{E}''_A \tag{5.14}$$

Thus, altering the angle θ alters the phase delay or advance, depending on polarization.

5.11.2 Dielectric vane phase shifters

The phase delay in propagation along a waveguide depends on the dielectric filling within the guide. The arrangement shown in Figure 5.35 in which a vane of dielectric can be moved across the guide from a region of weak field near the wall to the maximum in the centre forms a variable phase shifter. To assist matching, the vane is tapered at the ends. This device needs calibration. Similarly, a dielectric vane extending across a diametral plane in a circular waveguide carrying the TE_{11} mode will give a phase shift that will depend on the angle between the vane and the mode pattern. Rotation of the vane provides variable phase.

5.11.3 Ferrite phase shifters

The effective permeability of a ferrite depends on the magnetic bias field, so that phase delay through a system containing ferrite can be controlled by controlling the magnetic field. The ferrite may be used in a 2-state mode, either unmagnetized or saturated, or it may be used partially magnetized to give continuous variation. Three types may be distinguished. Firstly, a rotary device may be made by realizing the asymmetry required to make the half-wave pipe in the rotary phase shifter by asymmetrically magnetizing a

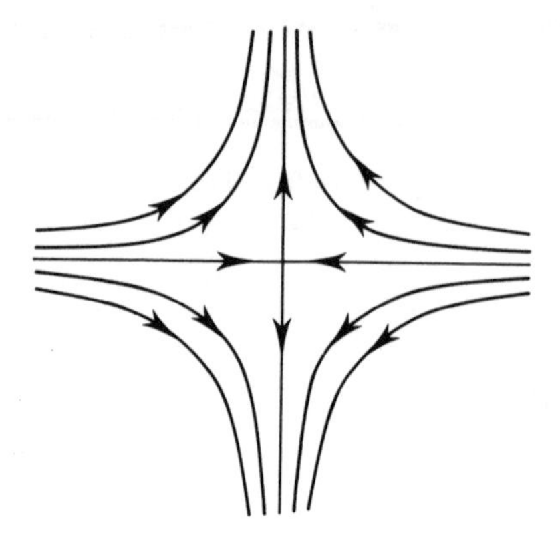

Figure 5.36 Quadripole magnetic field. A waveguide filled with ferrite and magnetized in this form can act as a half-wave pipe.

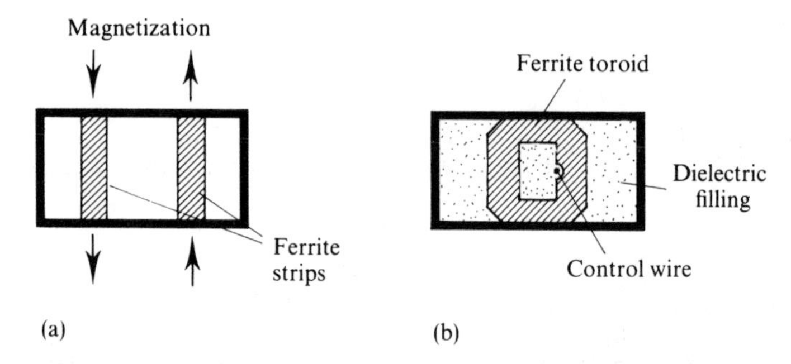

Figure 5.37 Ferrite phase shifters: (a) twin slab; (b) toroidal.

ferrite filled circular waveguide, as indicated in Figure 5.36. Rotation of the magnetic field can be done electrically. Secondly, the 'twin slab' configuration of Figure 5.37(a) may be used. This is realized by the toroidal form of Figure 5.37(b). A third way is the 'dual mode' type, in which propagation takes place through a ferrite filled waveguide, indicated in Figure 5.38. The 'twin slab' is non-reciprocal, which may be a disadvantage in some systems.

5.11.4 Diode phase shifters

Semiconductor diodes of the *p–i–n* type can be made to have high reverse impedance with low capacitance and very low forward impedance at microwave frequencies. The switching of different elements into (and out of) microwave circuits becomes possible. Such a facility might be of use, for example, in loading a transmission line with extra reactance to alter the phase, or switching in extra lengths of transmission line to give extra delay. These forms are indicated in Figure 5.39. Control currents will necessitate

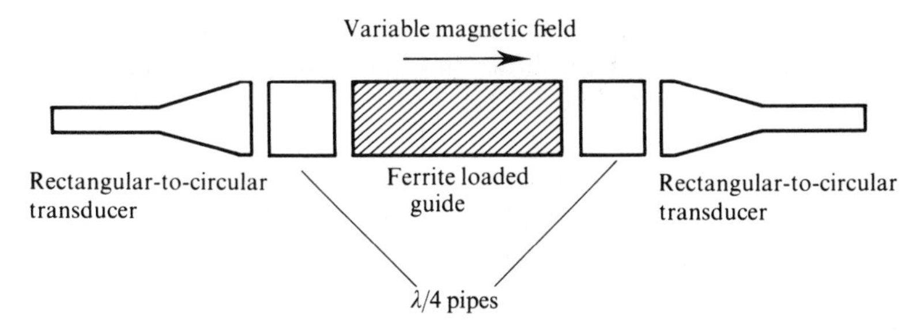

Figure 5.38 Phase shifter utilizing change of phase delay with the magnetization of a ferrite material.

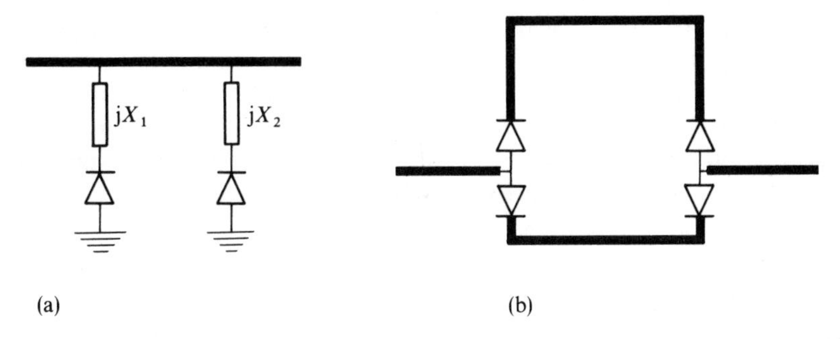

Figure 5.39 Diode phase shifters: (a) switching additional reactances into circuit; (b) switching between different lengths of transmission line. DC connections for switching currents are not shown.

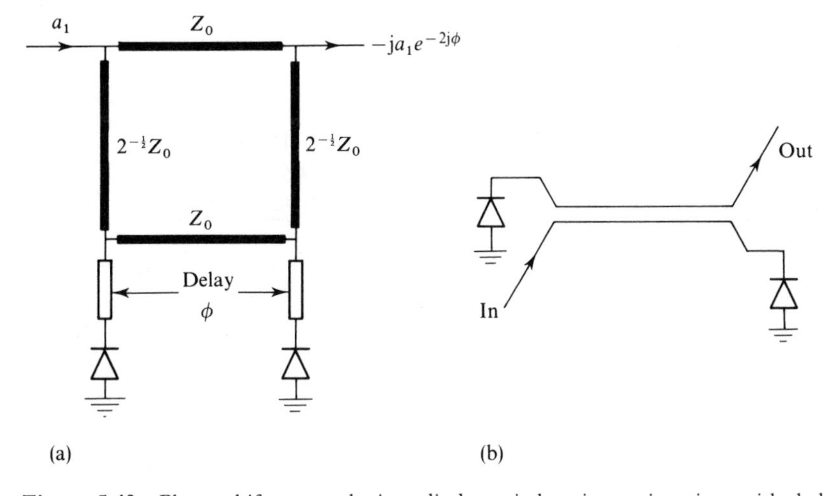

Figure 5.40 Phase shifters employing diode switches in conjunction with hybrid couplers.

chokes and DC blocks which are not shown. Two forms of phase shifters using 90°-hybrid junctions are shown in Figure 5.40. The first figure utilizes the hybrid junction of Figure 5.22(a), and use of the scattering matrix of equation (5.5) will verify that a delayed output appears at port two. An alternative realization using coupled lines is shown in Figure 5.40(b). Such circuits may be cascaded, giving overall a switched variable phase.

5.12 Attenuators

5.12.1 Resistive attenuators

The standard low frequency type of attenuator using a *T*-pad resistor assembly can be used in transmission line form at microwave frequencies. Resistors are made by evaporating a thin film of material such as nichrome onto ceramic; mounting and housing are conformable with a transmission line configuration. Examples are shown in Figure 5.41. Switching can be satisfactorily achieved in several forms. The use of low capacitance semi-conductor *p–i–n* diodes can provide electrically controllable attenuators.

5.12.2 Waveguide attenuators

Attenuation in a waveguide circuit is achieved by inserting a resistive material into the field. Thus vanes of thin film can be placed in rectangular waveguide as indicated in Figure 5.42. Figure 5.42(a) shows one configuration. A variable attenuator can be made by moving a resistive sheet across the waveguide from a position of zero electric field by the wall to a maximum in the middle, as in Figure 5.42(b). Such attenuators need calibration. A non-reflective termination can be made by using a tapered vane of sufficient

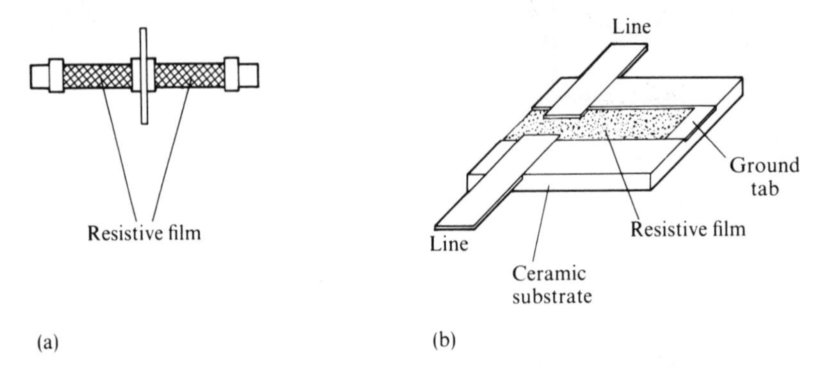

Figure 5.41 Microwave resistive attenuators: (a) coaxial; (b) planar.

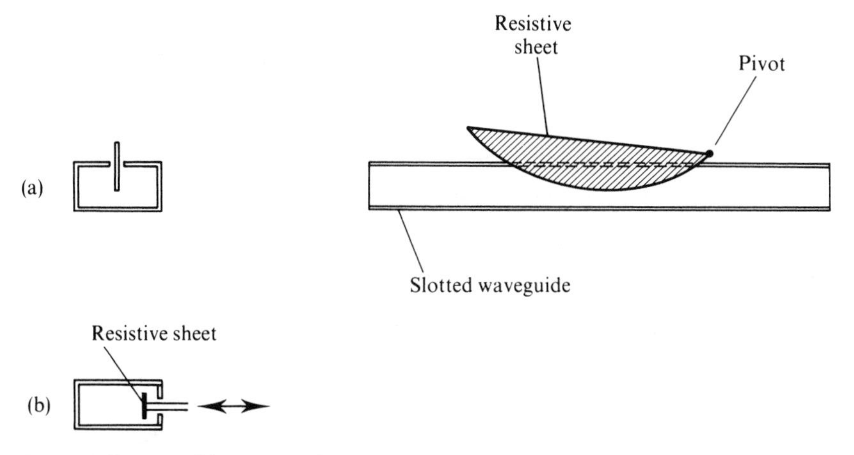

Figure 5.42 Variable waveguide attenuators: (a) resistive sheet on plane of maximum electric field; (b) resistive sheet movable across the waveguide.

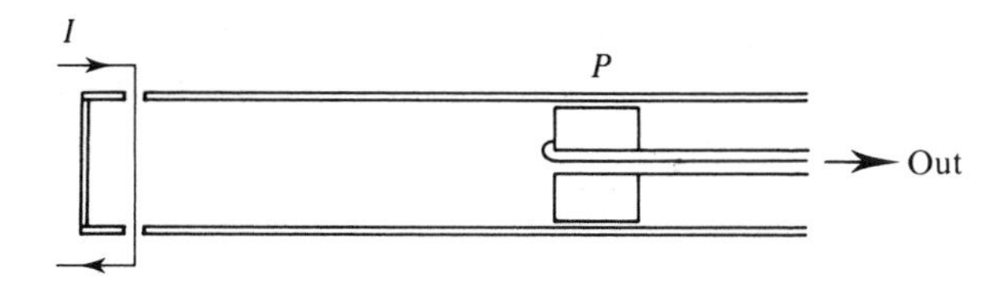

Figure 5.43 The piston attenuator. The diagram shows a circular waveguide below cut-off fitted with a movable pickup loop.

length. A lossy wedge will act in the same way, and a fluid filled wedge is used in measurement of high power.

5.12.3 The piston attenuator

Figure 5.43 shows a circular waveguide excited at one end and containing a movable plunger with a pickup loop. If the frequency is below the cut-off for the waveguide the field is attenuated exponentially along the guide and thus the output from the pickup loop gives an accurate decibel scale with distance. Such a device is known as a piston attenuator, and can be used up to lower microwave frequencies.

5.13 Switches

The use of *p–i–n* diodes for switching has already been mentioned when discussing phase shifters in Section 5.11.4. Such switches are not suitable at high power levels. Instead, ferrites may be used as switches. For example, the

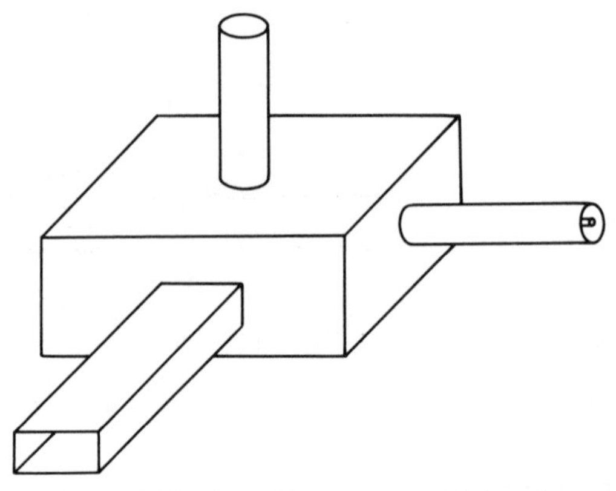

Figure 5.44 An arbitrary waveguide junction.

length of a ferrite filled stub short-circuited at one end may be chosen to be one-half wavelength with the ferrite unmagnetized, thus presenting a short at the open end. The correct bias will increase the permeability to make the length three-quarters of a wavelength, thus presenting an open circuit at the open end. Such devices can be used to switch power from one line to another or to reroute transmit and receive signals on a single antenna.

5.14 General circuit theorems

There are a number of general results comparable with low frequency circuit theory which can be derived from Maxwell's equations, and which are of use in obtaining a general understanding. Some of these are presented below with outline derivations.

5.14.1 Impedance and admittance matrices

The theorems concern an arbitrary waveguide junction, but it is assumed that the 'terminal' planes are in regions where only one mode of each waveguide propagates. This enables a voltage and current to be assigned at each port as discussed in Section 2.4.

Consider the junction shown schematically in Figure 5.44 in which the enclosing walls are considered to be lossless and over which therefore the tangential electric field must be zero. The field configuration throughout the volume is defined once either $\boldsymbol{E}$ or $\boldsymbol{H}$ in the terminal planes is specified, and hence, since the system is linear, the voltages and currents are related by an impedance matrix Z_{ij} such that

$$V_i = \sum_{i=1}^{N} Z_{ij} I_j \tag{5.15}$$

In general this may be inverted to define an admittance matrix

$$Y = Z^{-1} \tag{5.16}$$

Within the volume of the junction, the electric and magnetic fields satisfy Maxwell's equations:

$$\begin{aligned} \operatorname{curl} \boldsymbol{E} &= -\mathrm{j}\omega\mu\boldsymbol{H} \\ \operatorname{curl} \boldsymbol{H} &= \mathrm{j}\omega\varepsilon\boldsymbol{E} \end{aligned} \tag{5.17}$$

By using the identity

$$\operatorname{div}(\boldsymbol{A} \times \boldsymbol{B}) = \boldsymbol{B}\cdot\operatorname{curl}\boldsymbol{A} - \boldsymbol{A}\cdot\operatorname{curl}\boldsymbol{B} \tag{5.18}$$

coupled with Gauss' theorem, a number of useful results may be obtained.

5.14.2 Reciprocity

If $\boldsymbol{E}'$, $\boldsymbol{H}'$ and $\boldsymbol{E}''$, $\boldsymbol{H}''$ are two electromagnetic fields arising from different source distributions external to a volume bounded by a surface S it may be shown that

$$\oint_S \boldsymbol{E}' \times \boldsymbol{H}''\cdot \mathrm{d}\boldsymbol{S} = \oint_S \boldsymbol{E}'' \times \boldsymbol{H}'\cdot \mathrm{d}\boldsymbol{S} \tag{5.19}$$

In this case the integrals reduce to a summation of integrals over each terminal plane since the electric field is zero over the conducting surface. Hence equation (5.19) reduces to

$$\sum_{i=1}^{N} V_i' I_i'' = \sum_{i=1}^{N} V_i'' I_i' \tag{5.20}$$

This is the reciprocity theorem of circuit theory, and substitution of equation (5.15) shows that the impedance matrix is symmetrical

$$Z_{ij} = Z_{ji}$$

By expressing $\boldsymbol{E}$ and $\boldsymbol{H}$ in terms of complex wave amplitudes the result

$$S_{ij} = S_{ji}$$

also follows.

5.14.3 Power flow

Another result which is simply derived is

$$\oint \boldsymbol{E} \times \boldsymbol{H}^* \cdot \mathrm{d}\boldsymbol{S} = -\int [\sigma \boldsymbol{E} \cdot \boldsymbol{E}^* + \mathrm{j}\omega(\mu \boldsymbol{H} \cdot \boldsymbol{H}^* - \varepsilon \boldsymbol{E} \cdot \boldsymbol{E}^*)]\,\mathrm{d}v \tag{5.21}$$

The left-hand side reduces to

$$-\sum_{i=1}^{N} V_i I_i^*$$

The negative sign arises because the integral represents power flowing outwards. The right-hand side may be written in the form

$$-2(P + 2\mathrm{j}\omega(W_m - W_e))$$

where P is the power loss, and W_m and W_e are the average stored magnetic and electric energies respectively. Thus

$$\frac{1}{2}\sum_{i=1}^{N} V_i I_i^* = P + 2\mathrm{j}\omega(W_m - W_e) \tag{5.22}$$

A single port junction is specified by an impedance Z, given by

$$\tfrac{1}{2}|I|^2 Z = P + 2\mathrm{j}\omega(W_m - W_e) \tag{5.23}$$

The corresponding admittance is given by

$$\tfrac{1}{2}|V|^2 Y = P - 2\mathrm{j}\omega(W_m - W_e) \tag{5.24}$$

At resonance the reactive power flow is zero and $W_e = W_m$, so that for any system at resonance the average stored magnetic and electric energies are equal.

5.14.4 A lossless junction

For a lossless junction the power P is zero, requiring

$$Re \sum_{i=1}^{N} \sum_{j=1}^{N} (I_i^* Z_{ij} I_j) = 0$$

By considering single ports and then pairs of ports this shows all Z_{ij} are purely imaginary. Consider the case when all ports except one are short circuited. For given excitation the fields may be separated into real and imaginary parts:

$$\boldsymbol{E} = \boldsymbol{E}_r + \boldsymbol{E}_i \quad \boldsymbol{H} = \boldsymbol{H}_r + \boldsymbol{H}_i$$

Substitution into Maxwell's equations (5.17) shows that

$$\begin{aligned} \text{curl}\, \boldsymbol{E}_r &= -\mathrm{j}\omega\mu \boldsymbol{H}_i \\ \text{curl}\, \boldsymbol{H}_i &= \mathrm{j}\omega\varepsilon \boldsymbol{E}_r \end{aligned} \tag{5.25}$$

Over the single terminal plane, $\boldsymbol{E}$ may be specified as real and hence everywhere $\boldsymbol{E}$ will be real and $\boldsymbol{H}$ in phase quadrature.

5.14.5 Frequency dependence of a lossless junction

Consider a change of frequency, $\delta\omega$, which alters the electric and magnetic fields to $\boldsymbol{E} + \delta\boldsymbol{E}$, $\boldsymbol{H} + \delta\boldsymbol{H}$. Substitution in Maxwell's equations shows

$$\begin{aligned} \text{curl}\, \delta\boldsymbol{E} &= -\mathrm{j}\omega\mu\, \delta\boldsymbol{H} - \mathrm{j}\, \delta\omega\, \mu\boldsymbol{H} \\ \text{curl}\, \delta\boldsymbol{H} &= \mathrm{j}\omega\varepsilon\, \delta\boldsymbol{E} + \mathrm{j}\, \delta\omega\, \varepsilon\boldsymbol{E} \end{aligned} \tag{5.26}$$

It may then be shown that

$$\oint (\boldsymbol{E} \times \delta\boldsymbol{H} - \delta\boldsymbol{E} \times \boldsymbol{H}) \cdot \mathrm{d}\boldsymbol{S} = -\mathrm{j}\, \delta\omega \int (\varepsilon \boldsymbol{E} \cdot \boldsymbol{E} - \mu \boldsymbol{H} \cdot \boldsymbol{H})\, \mathrm{d}v$$

Applying this to the case of single-port excitation with shorts on the remaining ports, when $\boldsymbol{E}$ may be taken as real and $\boldsymbol{H}$ as imaginary, this equation becomes

$$I\, \delta V - V\, \delta I = -4\mathrm{j}\, \delta\omega (W_e + W_m)$$

For the single port junction

$$V = \mathrm{j} X I$$

with V real and I imaginary. Hence

$$\frac{\mathrm{d}X}{\mathrm{d}\omega} = \frac{4(W_e + W_m)}{|I|^2} \tag{5.27}$$

Stabilizing the frequency of any oscillator depends on having a large change of reactance with frequency. This theorem, sometimes known as Foster's reactance theorem, shows that for such stabilization it is necessary to increase the stored energy. This simple result holds at all frequencies of the electromagnetic spectrum and implies that it is difficult to produce highly stable miniature power sources unless clever techniques are employed to increase the stored energy.

5.14.6 Slater's theorem

This theorem concerns the change of resonant frequency of a cavity when a small lossless perturbation is made, such as a small dielectric insert. At the new resonance $\delta\omega$ above the old, the reactance must vanish. The reactance has changed by an amount δX because of the frequency change, given by

$$|I|^2\delta X = 4\delta\omega(W_e + W_m)$$

A change ΔX also occurs because of the change in stored energy on perturbation. Since

$$|I|^2 X = 4\omega(W_m - W_e)$$
$$|I|^2\,\Delta X = 4\omega(\Delta W_m - \Delta W_e)$$

Hence for zero change in X, $\delta X + \Delta X = 0$,

$$\frac{\delta\omega}{\omega} = \frac{\Delta W_e - \Delta W_m}{W_e + W_m} \tag{5.28}$$

Since both perturbation and frequency change are small, we may estimate ΔW_e and ΔW_m from the unperturbed fields. This theorem is invaluable in applying certain measurement techniques. It also shows that the effect of increasing electrical stored energy is to decrease the resonant frequency, and increasing magnetic stored energy raises the resonant frequency.

5.14.7 Transmission line terminated in reactance

Consider a transmission line terminated in a reactance X. The reflection coefficient may be written

$$\Gamma = \frac{\mathrm{j}X - Z_0}{\mathrm{j}X + Z_0} = -\exp(-2\mathrm{j}\phi)$$

where

$$\tan\phi = X/Z_0$$

Hence, using equations (5.27), we find

$$\sec^2\phi \frac{d\phi}{d\omega} = \frac{4W}{Z_0|I|^2}$$

where W is the stored energy. I is the current in X, and is related to the incident complex wave amplitude a by

$$IZ_0^{\frac{1}{2}} = a(1 - \Gamma) = 2a\exp(-j\phi)\cos\phi$$

Hence

$$2\frac{d\phi}{d\omega} = \frac{W}{P} \tag{5.29}$$

where $P = \frac{1}{2}|a|^2$ is the incident power.

For a small range of frequencies we may write

$$\Gamma \cong -\exp(2j\omega(d\phi/d\omega))$$

corresponding to a delay time on reflection of τ, given by

$$\tau = 2\frac{d\phi}{d\omega} = \frac{W}{P} \tag{5.30}$$

Thus the delay time is that required to build up the stored energy.

EXERCISES

5.1 Investigate the matching properties of transformers made from

(a) a single quarter-wave section,

(b) two sections as shown in Figure 5.8(b).

Show that as the frequency is varied about the design value, the VSWR on a line terminated by the transformer working into its design load has a flat minimum in the second case but not in the first. Such transformers are designed for matching a load $R_L = 4Z_0$. Calculate VSWR on the input line as a function of l/λ, and find the fractional bandwidths in the two cases over which VSWR < 1.2.

(The calculations may be performed with the aid of a Smith Chart or, more easily, using a programmable calculator.)

5.2 A microwave resonant cavity is connected through a length of waveguide. Using the equivalent circuit of Figure 5.13(e) obtain expressions for the unloaded and loaded Q-factors of the system. A termination on the waveguide effectively places an impedance $R_1 + jX_1$ in parallel with the cavity at the reference plane used for Figure 5.13(e). Show that the system of cavity plus waveguide has a resonant frequency differing from that of the unloaded cavity by

$$\frac{\Delta\omega}{\omega_0} \cong -\frac{X_1}{2Z_0}\frac{1}{Q_e}$$

assuming $\Delta\omega \ll \omega_0$. In this expression Q_e is the radiation Q of the cavity. Use this expression to discuss frequency pulling of an oscillator by its load.

5.3 Arrangements are made to measure the standing wave pattern in the feed to the cavity of question 2. Show that:

(a) the VSWR nearest unity, S_0, occurs at resonance and is equal to β or β^{-1} according to whether $\beta > 1$ or $\beta < 1$;

(b) if $\beta < 1$, change of frequency causes the position of a minimum to move and eventually return to the same place;

(c) if $\beta > 1$, the position of a minimum moves by $\lambda/4$ as frequency is changed from resonance to well off resonance.

Show also that when $Q_L \gg 1$, at the frequencies for which

$$\omega = \omega_0\left(1 \pm \frac{1}{2Q_L}\right)$$

the VSWR is given by

$$S_1 = \frac{1 + S_0 + (1 + S_0^2)^{\frac{1}{2}}}{1 + S_0 - (1 + S_0^2)^{\frac{1}{2}}}$$

5.4 With the system of question 3, the resonant frequency of the cavity is found to be 9.370 GHz, when $S_0 = 1.54$ and it is found that the position of the minimum shifts by $\lambda_g/4$ going off resonance. The frequency separation between values for which $S = 6.25$ is 1.2 MHz. Determine β, Q_u and Q_l for the cavity.

5.5 A system has the equivalent circuit shown in Figure 5.14(e). New reference planes are taken, each displaced from T′, T″ a distance $\lambda/4$ away from the cavity. Show that the equivalent circuit referred to these new planes consists of coupling transformers with a parallel LCR circuit in shunt connection, and obtain expressions for the new parameters.

5.6 One port of a 2-port cavity is connected to a source matched to the line. The other is terminated in a matched load. Using the equivalent circuit of Figure 5.14(e) show that the power in the load is given by

$$P_a \frac{4\beta'\beta''}{(1+\beta'+\beta'')^2 + 4Q_u^2(\delta f/f_0)^2}$$

in which P_a is the available power of the source and β', β'' are the coupling coefficients Z_0/n'^2R, Z_0/n''^2R respectively.

5.7 (a) Obtain the scattering matrix for the 3-port network shown in Figure 5.20.

(b) Show that the scattering matrix for the divider of Figure 5.21(a) is

$$\begin{bmatrix} 0 & \alpha & \alpha \\ \alpha & 0 & 0 \\ \alpha & 0 & 0 \end{bmatrix}$$

where $\alpha = -j/2^{\frac{1}{2}}$.

Investigate the behaviour of the circuit as frequency changes from the design value. (N.B. If the transmission lines are replaced by Π-equivalent networks with common earth the nodal admittance matrix can be written down by inspection.)

5.8 Discuss fully any *three* of the following microwave components:

(a) three-port circulator in waveguide or microstrip line;

(b) Faraday rotation isolator;

(c) 3 dB attenuator for a 50 Ω coaxial system;

(d) three-stub tuner for a coaxial system;

(e) 10 dB directive coupler for a TE_{10} rectangular waveguide system.

The discussion should include (i) an account of the ideal simple theory, (ii) practical considerations, (iii) a schematic sketch of the construction.

5.9 Using the fact that the scattering matrix for a lossless junction has the property $SS^* = I$ (Section 2.11.1), prove the second theorem of Section 5.6. (The junction will be completely matched if all the diagonal terms are zero.)

5.10 Verify equations (5.5) and (5.6) referring to Figure 5.22(a) and (b). Take as reference planes the junctions of the lines.

5.11 A lossless 4-port waveguide junction has the form of the 'magic-*T*' of Figure 5.23 with complete symmetry about the plane shown. Explain why symmetry and the field configuration require

$$S_{31} = S_{21} \quad S_{41} = 0 \quad S_{34} = -S_{24} \quad S_{33} = S_{22}$$

Show that additional matching elements can be added so as to make $S_{11} = S_{44} = 0$ without disturbing the above conclusions. From these conditions and the requirement $SS^* = I$ show that the scattering matrix must take the form

$$\begin{bmatrix} 0 & \alpha & \alpha & 0 \\ \alpha & 0 & 0 & \beta \\ \alpha & 0 & 0 & -\beta \\ 0 & \beta & -\beta & 0 \end{bmatrix}$$

in which $|\alpha|^2 = |\beta|^2 = \frac{1}{2}$. (N.B. Notice that the decoupling of arms two and three is the consequence of matching arms one and four.)

5.12 A circular waveguide of inner diameter 15 mm is used as a piston attenuator. Calculate the attenuation in dB mm^{-1} at a frequency of

(a) 3 GHz,

(b) 4.5 GHz.

5.13 A length d of rectangular waveguide of cross-section $a \times b$ where $b < a$, is closed at each end to make a resonant cavity with its lowest resonance in the TE_{10} mode. Show that the fields in the cavity may be expressed in the form

$$E_y = -E_0 \sin\frac{\pi x}{a} \sin\frac{\pi z}{d}$$

$$H_x = E_0 \frac{j\beta}{\omega\mu} \sin\frac{\pi x}{a} \cos\frac{\pi z}{d}$$

$$H_z = -E_0 \frac{j\pi}{\omega\mu a} \cos\frac{\pi x}{a} \sin\frac{\pi z}{d}$$

in which $\beta d = \pi$.

Show that the total stored energy at resonance is

$$W = \tfrac{1}{8}\varepsilon_0 E_0^2 V$$

in which V is the volume of the cavity.

A small piece of dielectric material, relative permittivity ε_r and of volume v, is placed at the centre of a broad face within the cavity. Show that the fractional increase in frequency is approximately $2(\varepsilon_r - 1)v/V$. Determine the change in frequency for a cavity for which $a = d = 25$ mm, $b = 12$ mm, and the dielectric is a wafer $5 \times 5 \times 2$ mm with $\varepsilon_r = 2.3$.

References

Aitcheson, C. L. (1971). 'Lumped circuit elements at microwave frequencies'. *IEEE Trans.* MTT-**19**, 928–37.

Edwards, T. C. (1981). *Foundations for Microstrip Circuit Design.* New York: Wiley.

Ginzton, E. L. (1957). *Microwave Measurements.* New York: McGraw-Hill.

Harvey, A. F. (1963). *Microwave Engineering.* New York and London: Academic Press.

Helszajn, J. (1978). *Passive and Active Microwave Circuits.* New York: Wiley.

Howe, H. (1974). *Stripline Circuit Design.* Norwood, Mass.: Artech House.

Kurokawa, K. (1969). *An Introduction to the Theory of Microwave Circuits.* New York and London: Academic Press.

Marcovitz, N. (1965). *Waveguide Handbook.* Radiation Lab. Series **10**. New York: Dover reprint.

Matthei, G. L. and Young, L. (1980). *Microwave Filters, Impedance Matching Networks and Coupling Structures.* Norwood, Mass.: Artech House.

Montgomery, C. G., Dicke, R. H. and Purcell, E. M. (1948). *Principles of Microwave Circuits.* Radiation Lab. Series **8**. New York: McGraw-Hill.

Ragan, G. L. (1948). *Microwave Transmission Circuits.* Radiation Lab. Series **9**, New York: McGraw-Hill.

CHAPTER 6

MICROWAVE TRANSISTORS AND AMPLIFIERS

OBJECTIVES

The purpose of this chapter is twofold: firstly it gives a brief background to the construction and properties of microwave transistors, and secondly it considers the system parameters of nominally linear amplifier modules, whether or not the amplification is by means of transistors. The system parameters adequately describe the modules for the purposes of interconnection.

6.1 Introduction

Both bipolar and field-effect transistors are important in the microwave region, the former up to about 6 GHz, the latter as high as 20 GHz, and the ranges are extending as processing techniques develop. The subject is much too intricate to attempt to cover in this text, and for details reference must be made to more specific literature, some of which is listed at the end of the chapter.

This chapter describes the salient points about the construction of microwave transistors and outlines the various ways in which they may be used to form amplifier modules. This is followed by consideration of the system parameters of such modules. Such parameters are not restricted to transistor amplifiers but are relevant whatever the active device used.

6.2 Construction of microwave transistors

The principles of transistor action are independent of frequency, but their physical construction of necessity builds in various unwanted parasitic reactances and resistances. In the case of the bipolars the collector base capacitance and base spreading resistance may be instanced. The values of such parasitic components depend on size and material properties. It is possible to set up equivalent circuits with components closely related to geometry and material, which can then be used to investigate the gain available in various circuit configurations. Optimization of the various parameters can then be attempted in terms of geometry and, for example, doping levels. As might be expected, it proves desirable to reduce the size, a process which is limited by the available technology. At present, strips of the order of 1 μm can be satisfactorily defined, and doping profiles may be controlled to an order of magnitude better.

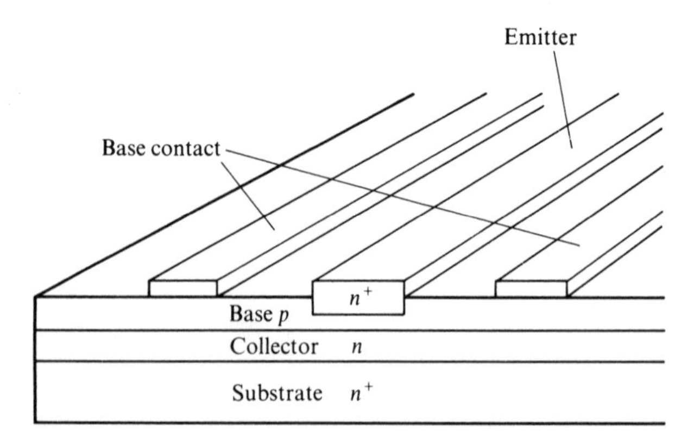

Figure 6.1 The construction of bipolar transistors for use at microwave frequencies.

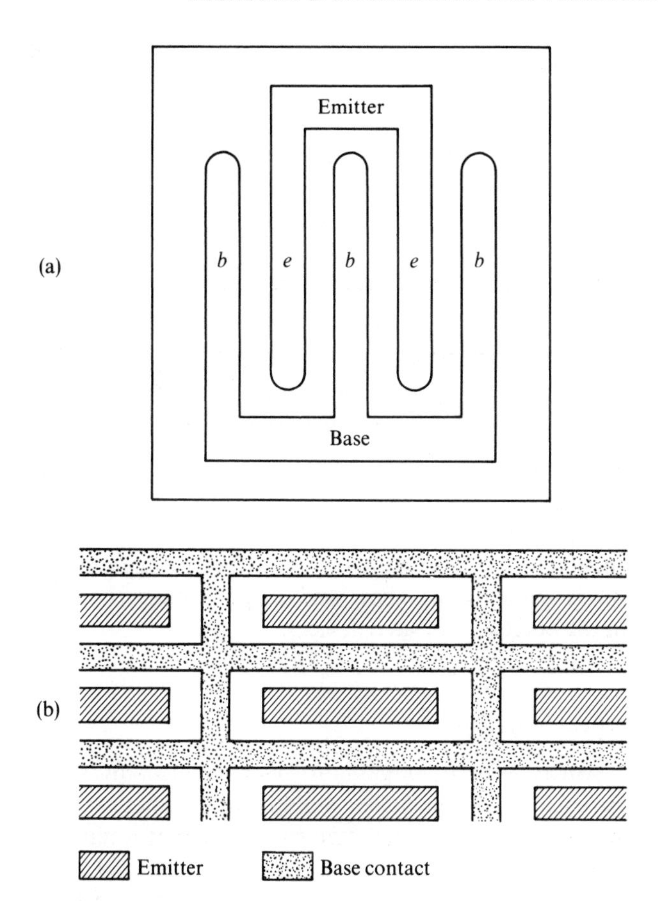

Figure 6.2 Two methods of paralleling bipolar transistors to increase power handling capacity: (a) interdigital; (b) overlay.

6.2.1 Bipolar transistors

The standard construction of bipolar transistors is the planar form shown in Figure 6.1, using silicon. It proves necessary to keep both emitter width and the spacing between base contact and emitter small, and in general to make the ratio of emitter periphery to emitter area as large as possible. These widths are typically of the order of 1 μm. The base thickness has to be minimized; this is controlled by the diffusion process and may be of the order of 0.1 μm. Detailed theory will also dictate the way in which doping levels and profiles can be optimized. An obvious result of reducing emitter size is reduction of collector current and hence of power handling capacity. In order to increase collector current transistors may be paralleled, using such forms as those indicated in Figure 6.2(a) and (b). The latter is known as an overlay pattern since the grid connected to the emitters must overlay and be insulated

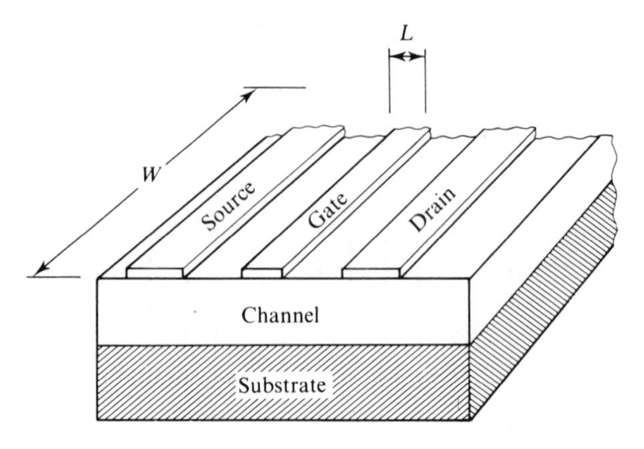

Figure 6.3 The construction of a microwave field-effect transistor.

from that connecting the base contacts. A further problem associated with size reduction is removal of heat, and heat sinking is of the greatest importance in the operation of such devices.

6.2.2 Field effect transistors

The type of structure used in gallium arsenide field-effect transistors is shown in Figure 6.3. GaAs has higher mobility and higher saturated electron velocity than silicon resulting in better frequency performance. Frequency limitations are determined primarily by the gate length, L, which can be reduced to less than 1 μm. The channel is made by doping a semi-insulating GaAs substrate, giving channel depths of the order of 0.1 μm. The gate-width, W, is typically of the order of 250 μm.

As with bipolar transistors, power output can be increased by paralleling individual transistors. Because the structure of the FET is horizontal rather than vertical the problem is different from that encountered in paralleling bipolar transistors. As Figure 6.4 makes clear, drain and gate connections can be made without difficulty, but the interconnection of the source electrodes requires bridging: this can be done by dielectric or air bridges on the top surface, or by using plated holes (**vias**) connecting each source electrode through the substrate to a common source connection. Gate widths of the order of 2.5 mm can be achieved.

6.2.3 Parasitics

As indicated earlier, some unwanted components are intrinsic to the device and can never be avoided. In a slightly different class are the parasitic components introduced by packaging: for example, a connecting lead introduces

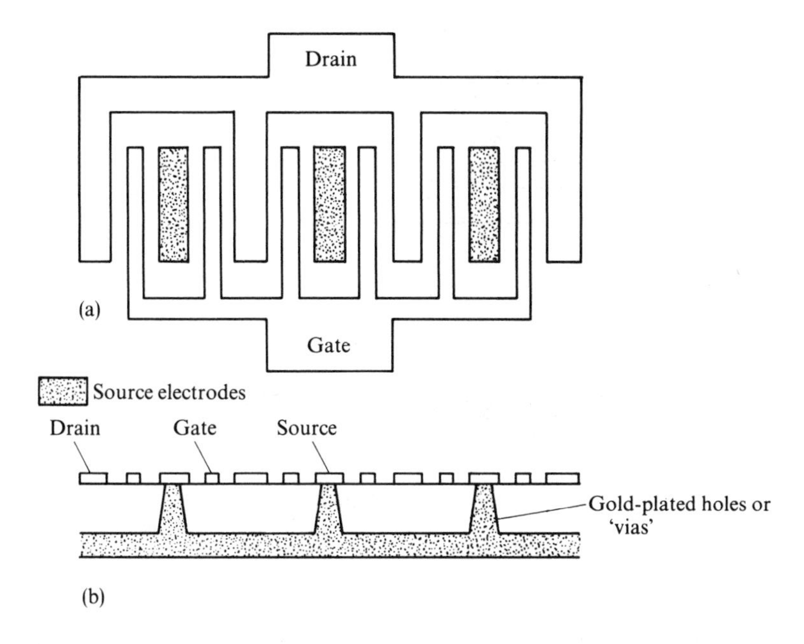

Figure 6.4 Paralleling field-effect transistors: (a) illustrates the necessity for bridging connections for the source electrode; (b) illustrates the use of 'vias' for source connection.

a bonding resistance, it will have inductance, and will introduce extra stray capacities. These can be reduced to some extent and in some cases incorporated into the external circuit: the inductance associated with a lead can be absorbed by making the lead part of a transmission line. This approach extends to the type of packaged device shown schematically in Figure 6.5, which is designed to be used on a ground plane when the leads become 50 Ω microstrip transmission lines. In any event parasitics must be included in the analysis of a transistor circuit.

6.3 Figures of merit

In use a transistor must be embedded into a circuit, and very frequently it will be required to operate into 50 Ω transmission lines at input and output. In making comparisons one is looking for overall figures of merit that characterize the transistor. Some of these are listed here.

- *Frequency of unity current gain, f_T* This is as at low frequencies, based on short-circuit current gain.
- *Maximum frequency of oscillation, f_{max}* The power gain of the transistor depends on input and output impedances as well as current gain,

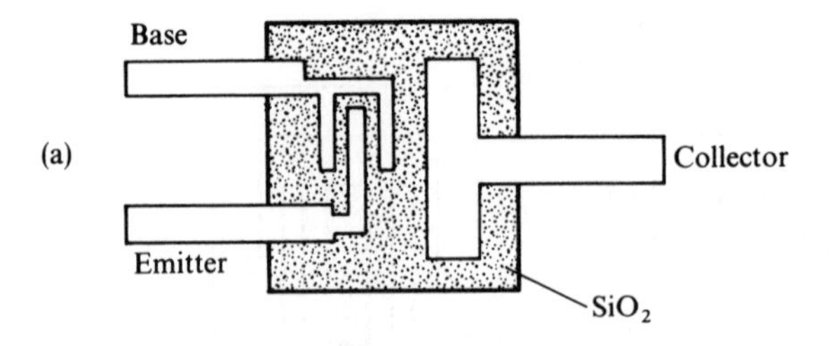

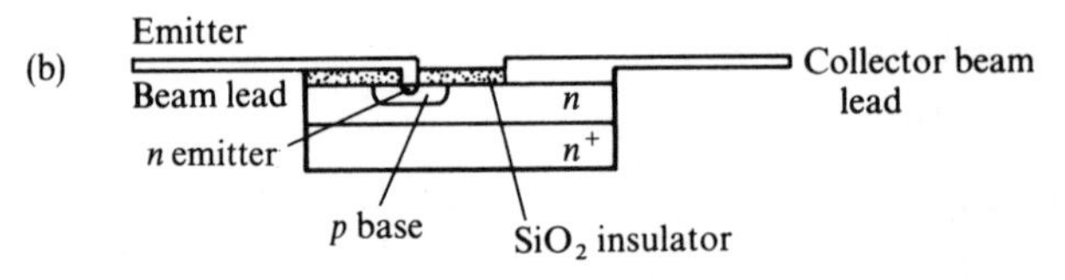

Figure 6.5 Microwave transistor in beam-lead package.

so that the maximum frequency at which oscillation may be obtained will be greater than f_T. At present, values of 50 GHz are attainable with GaAs FETs.

- *Power gain* The power gain which may be obtained from a transistor depends on the network in which it is embedded. A convenient approximation in the frequency range involved is:

$$G \cong (f_{max}/f)^2$$

- *Power output* The power an amplifier will deliver depends on operating voltage and current, as mentioned earlier. The optimum design of a device for operation at a given frequency imposes restrictions on size, which in turn influences parameters such as current density and breakdown voltage. Permissible values of these variables are also influenced by material properties. It is found that the maximum power output decreases inversely as frequency squared. Powers of 60 W have been attained at 3 GHz.

6.4 Equivalent circuits

Equivalent circuits for both bipolar and field-effect transistors can be derived, with special forms dictated by the high frequency. The circuits given in the following sections are of use in general terms, although they are not adequate for a detailed analysis over a wide frequency range. This requires the use of measured scattering parameters. It should also be noted that the circuits do

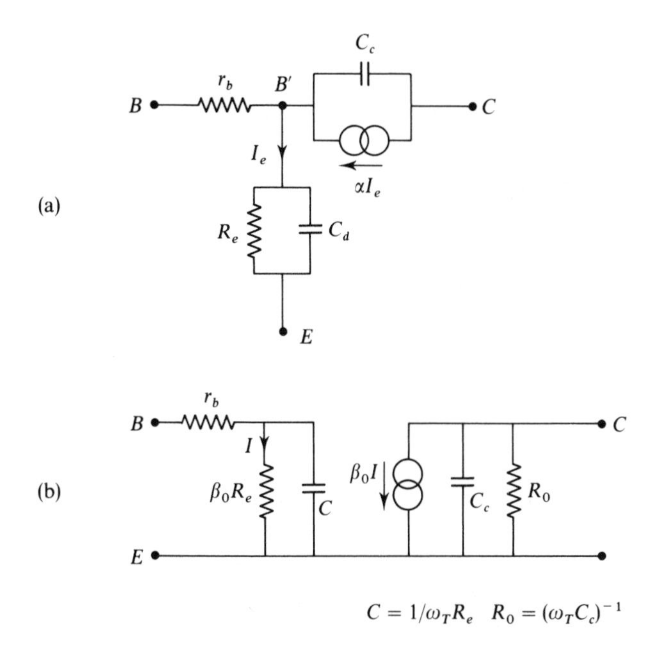

Figure 6.6 Equivalent circuit for a bipolar transistor: (a) a circuit describing the inherent transistor operation; (b) a unilateral approximation. Parasitic components arising from packaging are not included.

not take account of the packaging parasitics referred to in Section 6.2.3. This must be done by introducing other components. The correspondence of the equivalent circuit with the actual measured performance can be improved by allowing for the distribution in space of the transistor, and for further detailed effects.

6.4.1 Bipolar transistors

A circuit that reflects the principal high frequency effects intrinsic to the transistor is given in Figure 6.6(a), in which, in terms of the base transit time, τ_b,

$$\alpha = \alpha_0(1 + j\omega/\omega_b)^{-1} \qquad \omega_b = \tau_b^{-1}$$
$$R_e = kT/e$$
$$C_d = \omega_b/R_e$$

where $e = 1.6 \times 10^{-19}$C, $k = 1.38 \times 10^{-23}$ J K^{-1} and T is in Kelvin.

Straightforward circuit analysis for the h-parameters yields, assuming only $\alpha_0 \cong 1$,

$$h_{ie} = r_b + \beta_0 R_e(1 + j\beta_0\,\omega/\omega_T)^{-1}$$

$$h_{fe} = \beta_0 \frac{1 - j\omega C_c R_e}{1 + j\beta_0\omega/\omega_T}$$

$$h_{re} = j\omega C_c R_e \beta_0 (1 + j\omega\beta_0/\omega_T)^{-1}$$

$$h_{oe} = j\omega C_c \beta_0 \frac{1 + j\omega/\omega_b}{1 + j\beta_0\omega/\omega_T}$$

where

$$\omega_T^{-1} = \omega_b^{-1} + R_e C_c \qquad \beta_0 = \alpha_0/(1 - \alpha_0).$$

At operating frequencies $\omega/\omega_T > 1$, $\omega R_e C_c \ll 1$. With this approximation

$$h_{fe} \cong \beta_0(1 + j\omega\beta_0/\omega_T)^{-1}$$
$$h_{re} \cong \omega_T C_c R_e \cong 0$$
$$h_{oe} \cong \omega_T C_c + j\omega C_c \omega_T/\omega_b$$

These equations may be represented by the circuit of Figure 6.6(b), in which the complex form of h_{fe} has been accounted for by a change of reference current. In any circuit using transistors it is necessary to add stray capacities between all terminals, as well as lead resistance and inductance. The maximum power gain of the transistor is when external capacities are neutralized. Ignoring possible extra lead resistance, the available power into a matched load is

$$P_o = |\beta_0 I|^2/4\omega_T C_c$$
$$P_i \cong |\beta_0 I|^2 r_b \omega^2/\omega_T^2$$

Hence

$$G \cong \omega_T/4r_b C_c \omega^2$$

In this case

$$f_{max} = (\omega_T/r_b C_c)^{\frac{1}{2}}/4\pi$$

6.4.2 Field-effect transistors

Figure 6.7(a) shows an equivalent circuit describing the high-frequency operation of an FET. The control voltage V_g affecting the channel appears across the gate-channel capacity C_g, and the conductance g_{mo} is that valid at

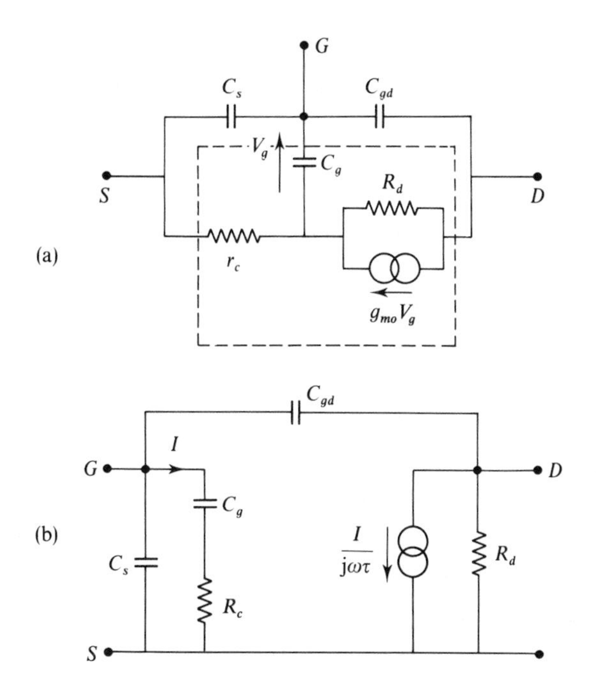

Figure 6.7 Equivalent circuit for a field-effect transistor: (a) circuit describing inherent operation; (b) equivalent form.

low frequencies. The resistance r_c represents the channel resistance. The resistance R_d is assumed to be associated with a depleted region and is therefore high. Whilst the low frequency expressions for g_{mo} may be used, a simple argument relates this directly to transit time under the gate: a change of potential, δV, across C_g must be associated with a change in stored charge of $\delta Q = C_g \delta V$; this extra charge is provided by an increase in drain current δI flowing over the transit time τ. Thus

$$C_g \delta V \cong \tau \delta I$$

or

$$g_{mo} \cong C_g/\tau$$

Analysis of that part of the circuit within the dotted rectangle yields

$$h_{ie} \cong r_c + \frac{1}{j\omega C_g} + \frac{r_c}{j\omega\tau}$$

$$h_{fe} \cong 1/j\omega\tau$$

$$h_{re} \cong r_c/R_d$$

$$h_{oe} \cong 1/R_d$$

assuming $r_c \ll R_d$. Thus, the equivalent circuit shown in Figure 6.7(b) is obtained, neglecting the internal feedback. The angular frequency ω_T for which $|h_{fe}| = 1$ is given by

$$\omega_T = \tau^{-1}$$

Using the same arguments as with the bipolar transistor, the available output power is given by

$$P_o = \frac{1}{4} R_d \left| \frac{I}{j\omega\tau} \right|^2$$

The input power is given by

$$P_i = r_c |I|^2$$

Thus

$$G = (R_d/4r_c\omega^2\tau^2)$$

Hence

$$f_{max} = \tfrac{1}{2} f_T (R_d/r_c)^{\frac{1}{2}}$$

6.5 Types of amplifier

In the previous sections equivalent circuits for microwave transistors were derived. Although these may be used for approximate designs, for a good design it is essential that the device be characterized fully over a wide frequency range, both to obtain the required gain-frequency characteristic and to assess the stability of the circuit. Characterization is usually done by measurement of scattering parameters using automatic test equipment. The direct measurement of these parameters, rather than impedance or admittance parameters, avoids the necessity of realizing short or open circuits and possible instability. Since the transistor is an active device with internal feedback, certain loads on input and output may cause instability, and the choice of input and output networks must avoid this situation. Thus investigation is needed at many frequencies, and the arithmetic of complex quantities is best handled in a computer.

6.5.1 Narrow-band amplifiers

This is the simplest type since it is virtually a single frequency design. A great simplification is introduced if the devices may be taken as unilateral (i.e. $S_{12} \cong 0$). In that case the device presents impedances to input and output

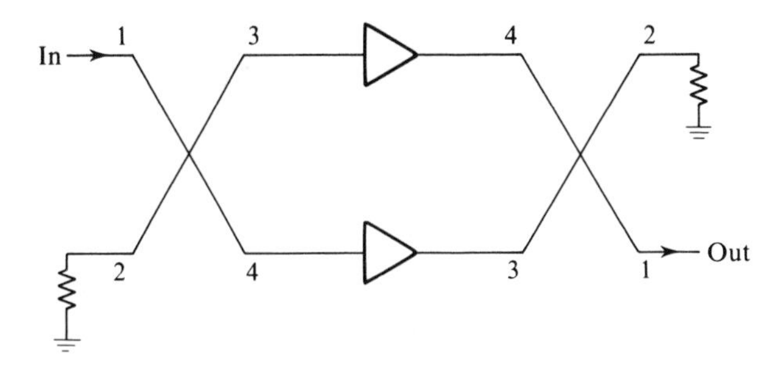

Figure 6.8 Balanced amplifier configuration using hybrid couplers.

circuits which are defined, and the design process is that of deriving suitable matching networks. The validity of the unilateral assumption has to be checked, and possible instability investigated.

6.5.2 Broad-band amplifiers

Single-stage broad-band amplifiers can be designed on the unilateral basis as described for the narrow band case. The complication is that the transistor gain will alter over the band, so that input and output networks must produce a compensating mismatch in order to provide uniform gain over the pass band. This makes it impossible to produce uniformly low reflection coefficients in the input and output circuits at the same time as a flat gain-frequency response, causing problems if such amplifiers are cascaded.

6.5.3 Balanced amplifiers

Cascading whilst maintaining low reflection input and output can be achieved in the balanced structure shown in Figure 6.8. The coupler has been numbered in accordance with Equation (5.5) and Figure 5.22(a). Analysis shows that if the two amplifiers have identical reflection coefficients the reflected waves cancel at the inputs and are absorbed in the load on the fourth port. Similarly, the output will present a matched source to the following stage. Such stages are usually designed using microstrip wide-band couplers (Lange couplers). Other power combiners or dividers may be used.

The penalties for this approach are doubling the hardware and power consumption. Unbalanced amplifiers may be cascaded without interaction if suitable isolators are available.

6.5.4 Distributed amplifiers

The amplifiers described in the previous sections can be cascaded to make one of higher gain, provided that the stage gain is greater than unity. If at

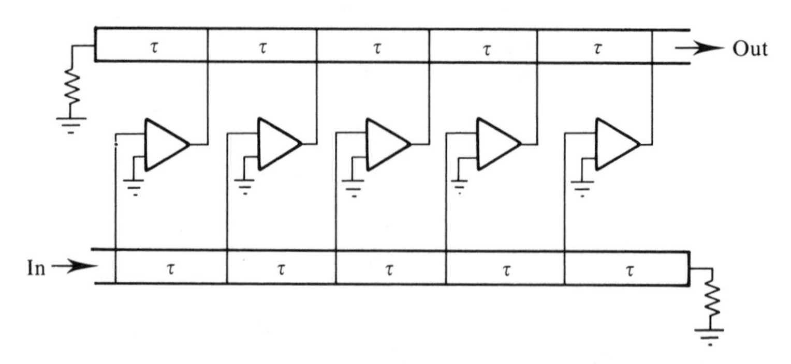

Figure 6.9 Schematic circuit diagram of a 'distributed' amplifier.

some frequency the stage gain is low, an alternative is the distributed amplifier, shown schematically in Figure 6.9. The signal input is fed to several stages in succession with successive time delays. The outputs of those stages are then combined with the appropriate time delays: in Figure 6.9 the time delays are illustrated as transmission lines. The final output is thus the sum of the individual outputs and not the product, so that gain is obtained even if the stage gain is less than unity.

6.6 Noise in transistors

Unwanted random signals are generated in amplifiers, arising from thermal noise in resistors and noise produced by current flow in transistors. The overall characterization of the goodness of an amplifier is treated in Section 6.7. This section merely comments that good circuit design must take into account the noise sources inherent in transistors in order to minimize added noise. This is done by correct choice of transistor and operating conditions. References are given at the end of the chapter.

6.7 System parameters

The previous sections have been concerned with the use of transistors to construct amplifiers. The system designer has a somewhat different outlook: he is concerned with external characteristics and not internal construction. These external characteristics principally concern impedance, gain and noise. They are also relevant to any amplifier whatever active device is used, and the present section is concerned with this wider class of amplifier.

6.7.1 Impedance and gain

In the circuit sense an amplifier operating at low signal level is a linear 2-port, for which well developed theory is available. For the present discussion it is

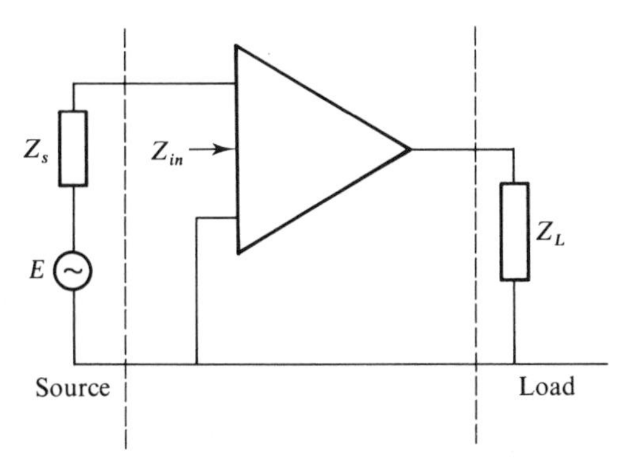

Figure 6.10 Notation for an arbitrary amplifier.

assumed that the amplifier is operating into a specific load and is driven by a specified source, as indicated in Figure 6.10. Very frequently source and load impedance will be determined as the characteristic impedance of a transmission line, often 50 Ω. Because in general at microwave frequencies the amplifier is not unilateral, problems may arise on interconnection. Overcoming these may be regarded as part of the circuit design. Once the load is specified at any one frequency an input impedance is determined: this may be specified in the form of a reflection coefficient with respect to the input transmission line.

Alternative definitions of power gain are of use. The **actual power gain** may be defined as

$$G = \frac{\textit{power into load}}{\textit{power from source}}$$

the **transducer power gain** is defined as

$$G_T = \frac{\textit{power into load}}{\textit{power available from source}}$$

A third definition is the **available power gain**, given by

$$G_a = \frac{\textit{power available from amplifier}}{\textit{power available from source}}$$

Each of these quantities finds use in a different context.

The power gain will be a function of frequency, perhaps of the form shown in Figure 6.11. It is customary to define cut-off frequencies at the half-power points (3 dB points) as shown.

The above definitions refer to single frequency operation. Complete specification to allow calculation of transient waveforms must encompass

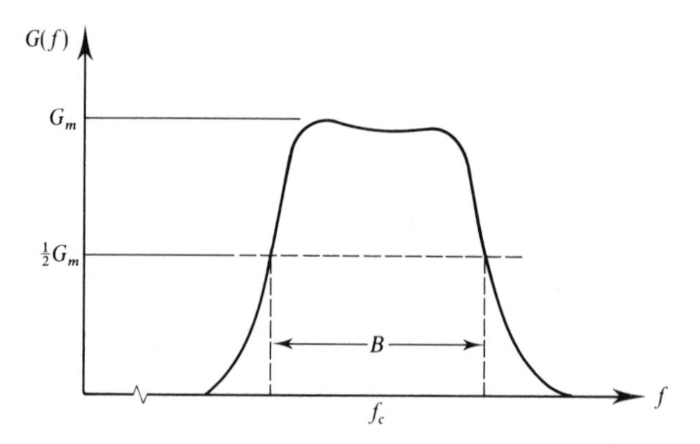

Figure 6.11 Variation of gain with frequency: a typical band-pass form.

phase information. This can be given by defining the 2-port in terms of its scattering matrix.

6.7.2 Noise figure

Noise signals can properly be described only in statistical terms, and calculation of the effect of noise on a detection system must be handled this way. However, the magnitude of the noise signal can be described in terms of average power (or mean square voltage or current). The average power in a small frequency interval δf may be written

$$\delta P = p_n(f)\delta f \tag{6.1}$$

For most instances $p_n(f)$, the **power spectral density**, will in fact be independent of f (**white noise**) although in some instances **flicker noise** with $p_n \propto f^{-1}$ may be encountered. The mean power over a wider frequency interval is given by

$$P = \int_{f_1}^{f_2} p_n(f)\,\mathrm{d}f \tag{6.2}$$

The nuisance value of noise depends on the signal-to-noise ratio, rather than noise power alone. This quantity is conveniently defined in the form

$$S/N = \frac{\textit{signal power}}{\textit{average noise power}}$$

As an example, the performance of a data transmission system might be characterized by a statement such as 'to ensure an error rate better than 1 in

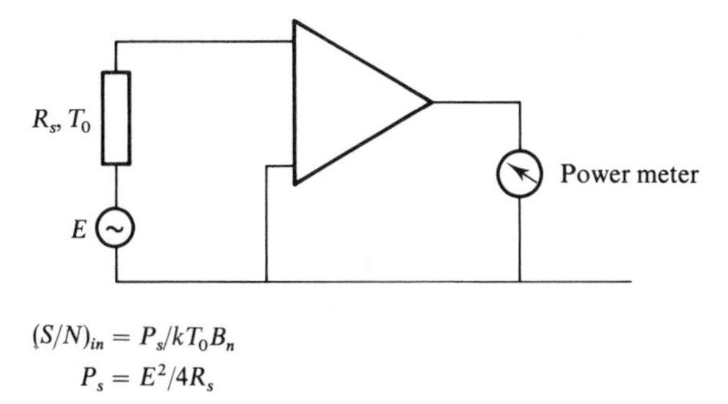

$(S/N)_{in} = P_s/kT_0B_n$
$P_s = E^2/4R_s$

Figure 6.12 Illustrating the definition of noise figure.

10^9 the signal-to-noise ratio must be greater than 22 dB'. Thus the important matter for an amplifier is to characterize its effect on the signal-to-noise ratio. A numerical expression of this effect is gained by considering the ratio

$$\frac{S/N \text{ at input}}{S/N \text{ at output}} \tag{6.3}$$

The numerator of this expression depends on the source. In many cases we are concerned with a source of resistive internal impedance at room temperature as shown in Figure 6.12. In this case the ratio of equation (6.3) is called the **noise figure** of the amplifier. It may be noted that the signal-to-noise ratio is independent of the impedance across source or load, and may be taken as the ratio of available powers. The noise figure is denoted by the symbol F, and is commonly expressed in decibels as a power ratio.

The available noise power from a resistor in a frequency interval δf is given by $kT\delta f$. When $T = 290\,\text{K}$, this gives $4 \times 10^{-15}\,\text{W MHz}^{-1}$. Followed by a noiseless amplifier of transducer gain $G_T(f)$, the noise output will be given by

$$P = \int_0^\infty kTG_T(f)\,\mathrm{d}f \tag{6.4}$$

If, as is usual, $G_T(f)$ approximates to a constant G_m over most of the operational frequency range we can define the **noise bandwidth**, B_n, by

$$\int_0^\infty G_T(f)\,\mathrm{d}f = B_nG_m$$

when equation (6.4) becomes

$$P = kTB_nG_m$$

If the source delivers available power P_s, the noiseless amplifier would have an output signal-to-noise ratio

$$(S/N)_{out} = G_m P_s / kTB_n G_m = P_s / kTB_n$$

In general $(S/N)_{out}$ will be measured, as indicated in Figure 6.12, by measuring noise in the absence of signal followed by signal plus noise. This measurement must be done using a true power meter. The source power P_s must be known using a calibrated source. We then have

$$F = \left(\frac{P_s}{kT_0 B_n}\right) \Big/ (S/N)_{out}$$

in which T_0 is room temperature, usually taken as 290 K.

Measurement of noise figure is considered in Section 11.4.

6.7.3 Noise temperature

The noise figure as defined refers directly to a source at room temperature. It is very likely that although the source may have a resistive impedance, it will not be at room temperature, so other methods of specifying noise performance are more directly informative. It is evident that in the configuration of Figure 6.12 the same $(S/N)_{out}$ would be obtained with a source giving available noise FkT_0B_n followed by a noiseless amplifier, or equivalently, the amplifier has added noise $(F - 1)kT_0B_n$. The **noise temperature** of the amplifier is defined by

$$T_a = (F - 1)T_0$$

6.7.4 Amplifiers in cascade

Frequently, separate amplifiers are cascaded, as in Figure 6.13, and it is the overall noise figure which is of importance. In this instance it is convenient to characterize the individual amplifiers by noise figure and *available* power *gain*, bearing in mind that for each amplifier these quantities must be specified for a source impedance equal to the output impedance of the preceding

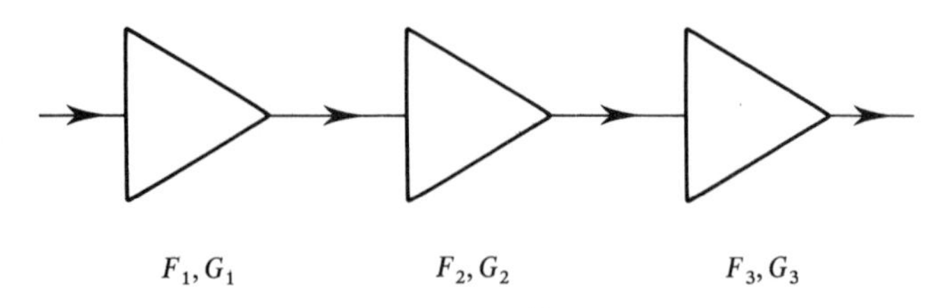

Figure 6.13 Parameters for several amplifiers in cascade.

amplifier. The available signal power at the output of the first amplifier is $P_s G_{a1}$, the available noise output $F_1 k T_0 B G_{a1}$. The noise introduced by the second amplifier is, as discussed in Section 6.7.3, $(F_2 - 1)kT_0B$. Thus the noise figure of the combination is given by

$$F = \frac{P_s}{kT_0B} \frac{F_1 kT_0BG_{a1} + (F_2 - 1)kT_0B}{P_sG_{a1}}$$

$$= F_1 + \frac{F_2 - 1}{G_{a1}}$$

For subsequent amplifiers we find the overall noise figure F_t to be given by

$$F_t = F_1 + \frac{F_2 - 1}{G_{a1}} + \frac{F_3 - 1}{G_{a1}G_{a2}} + \dots \tag{6.5}$$

When considering a chain of identical amplifiers, a useful parameter is the **noise measure**. The overall noise figure of a chain of identical amplifiers is given from equation (6.5) in the form

$$F_t = 1 + (F - 1)\left(1 + \frac{1}{G} + \frac{1}{G^2} + \dots\right)$$

When the number of amplifiers is sufficiently large, or the gain high enough, this expression is given approximately by

$$F_t = 1 + (F - 1)(1 - 1/G)^{-1}$$

The noise measure is then defined as

$$M = F_t - 1 = (F - 1)(1 - 1/G)^{-1}$$

6.7.5 The effect of source impedance

In the previous sections it has been implied that noise figures can only be given a value for a specific source impedance. The noise sources within an amplifier, or individual transistor, are fixed by the device and DC operating conditions; the impedance across the input alters the way in which these sources combine with each other and the signal. It can be shown (see Section A.9) that the dependence can be expressed in terms of source admittance by

$$F = F_0 + \frac{R_n}{G_s}((G_0 + G_s)^2 + (B_0 + B_s)^2)$$

in which F_0, G_0, B_0 and R_n are characteristic of the amplifier and G_s and B_s are respectively the conductance and susceptance of the signal source feeding the amplifier. In particular, F_0 may be identified as the best noise figure that can be achieved. It will be realized that since input matching is used to determine frequency response a compromise with noise performance may have to be made. It must be added that G_0, B_0 are not directly related to the input impedance of the amplifier.

6.8 Non-linearity

Amplifiers are linear for small signals, but at higher levels the active devices become non-linear, and eventually the output will saturate. The non-linearity may have undesirable effects.

6.8.1 Dynamic range

The dynamic range specifies the range of input signal over which the amplifier gives a satisfactory output. The form of output versus input of a typical amplifier is shown in Figure 6.14: at the lower end signals are lost in noise, at the upper the output ceases to respond to change of input. A simple measure at the upper end is the **1 dB compression point**. This is indicated in Figure 6.14, and occurs for the input at which the actual output is 1 dB less than that extrapolated from the linear response.

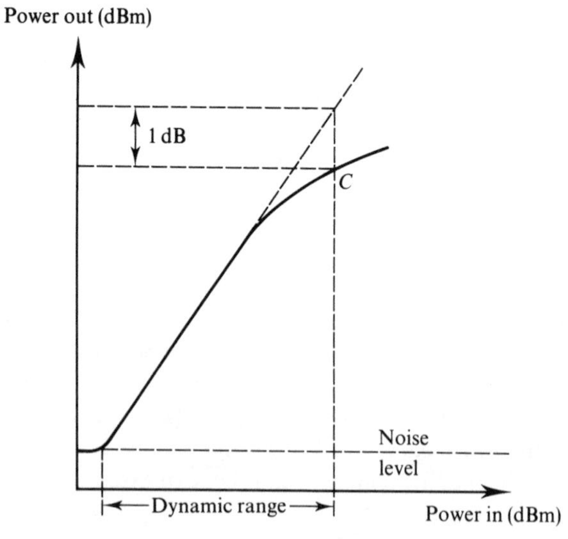

Figure 6.14 The output–input relationship in a real amplifier, showing non-linear behaviour. *C* marks the 1 dB compression point. The dynamic range extends between noise level and *C*.

Table 6.1 Intermodulation products.

Angular frequency	Amplitude
DC	$\frac{1}{2}A_2a_1^2 + \frac{1}{2}A_2a_2^2$
ω_1	$A_1a_1 + \frac{3}{4}A_3(a_1^3 + 2a_1a_2^2)$
ω_2	$A_1a_2 + \frac{3}{4}A_3(a_2^3 + 2a_1^2a_2)$
$2\omega_1$	$\frac{1}{2}A_2a_1^2$
$2\omega_2$	$\frac{1}{2}A_2a_2^2$
$\omega_1 - \omega_2$, $\omega_1 + \omega_2$	$A_2a_1a_2$
$3\omega_1$	$\frac{1}{4}A_3a_1^3$
$3\omega_2$	$\frac{1}{4}A_3a_2^3$
$2\omega_1 - \omega_2$, $2\omega_1 + \omega_2$	$\frac{3}{4}A_3a_1^2a_2$
$2\omega_2 - \omega_1$, $2\omega_2 + \omega_1$	$\frac{3}{4}A_3a_1a_2^2$

6.8.2 Intermodulation products

Non-linearity in the amplifier can give rise to spurious signals which arise from the 'mixing' of two or more input signals. In general, input signals of frequencies f_1 and f_2 will give rise to all frequencies of the form $mf_1 \pm nf_2$, m and n being integers. For the normal band-pass amplifier most of these will be outside the pass-band except for the third order intermodulation products $2f_1 - f_2$, $2f_2 - f_1$. A simple theory predicting magnitudes of those spurious signals can be developed as follows: assume the output voltage v_2 is instantaneously related to the input voltage v_1 by a power series of the form

$$v_2 = A_1v_1 + A_2v_1^2 + A_3v_1^3 + \dots \tag{6.6}$$

Taking only these three terms into account, a two frequency input of the form

$$v_1 = a_1 \cos\omega_1 t + a_2 \cos\omega_2 t \tag{6.7}$$

will give rise to the terms shown in Table 6.1.

Consider the case with input signals of equal strength. The relation of fundamental and spurious third order output signals to the input strength is displayed in Figure 6.15 in the form of a log–log plot, expressing input and output in dBm (decibels above 1 mW). It will be seen that input signals greater than P_2 give rise to spurious signals greater than the minimum detectable signal; below P_1 the signal is lost in noise, so that the spurious free dynamic range is $P_2 - P_1$. The performance of the amplifier in this regard

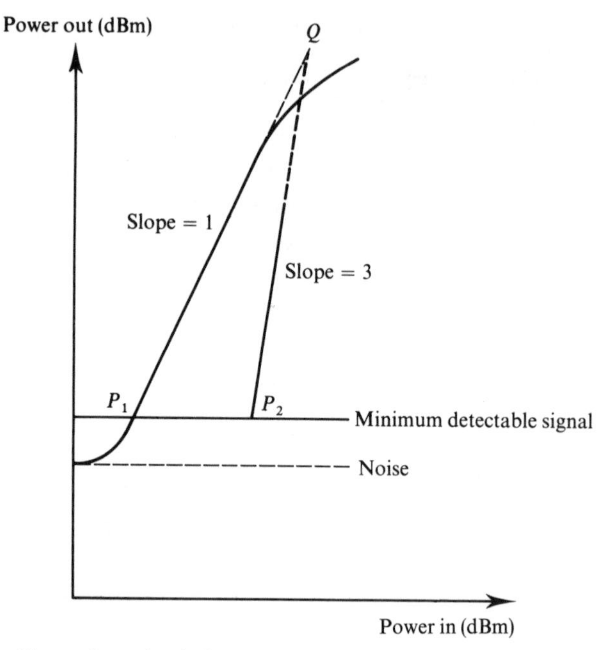

Figure 6.15 Illustrating the behaviour of fundamental and third-order harmonic components in the amplifier output. Q marks the intercept point used in calculation of spurious signals.

can be specified by the **intercept point**, Q, in Figure 6.15: since the slopes of the fundamental and third order product lines are one and three respectively, the rest of the graph may then be constructed. It will be noticed from Table 6.1 that the fundamental varies by an amount equal to the spurious signal; in the region of interest this variation will be negligible.

6.8.3 AM–PM conversion

The treatment of non-linearity in the previous section does not show any variation of phase on account of non-linearity. This follows from the power series representation of equation (6.6). In many cases the non-linearity will occur in a stage containing reactive components in the load, when it is found that positions of zero crossings can alter with input signal. In such a situation, amplitude modulation on a signal can give rise to phase modulation, which could, for example, give rise to distortion in an FM system. Travelling wave tubes are found to be a source of AM–PM conversion.

6.9 Group delay

It has been mentioned earlier (Section 6.7.1) that phase information may be necessary as well as gain. An important concept is that of **group delay**, τ, related to the phase characteristic by $\tau = -\mathrm{d}\phi/\mathrm{d}\omega$, ϕ being regarded as a phase advance. Consider an amplitude modulated carrier being handled by the amplifier. The input waveform can be written in the form

$$\begin{aligned} v_1 &= A\cos\omega t(1 + m\cos pt) \\ &= A(\cos\omega t + \tfrac{1}{2}m\cos(\omega + p)t + \tfrac{1}{2}m\cos(\omega - p)t) \end{aligned}$$

The output will be composed of these three constituent waveforms each shifted by the appropriate phase. Thus we find for the output waveform

$$\begin{aligned} v_2 = A[&\cos(\omega t + \phi) + \tfrac{1}{2}m\cos\{(\omega + p)t + \phi(\omega + p)\} \\ &+ \tfrac{1}{2}m\cos\{(\omega - p)t + \phi(\omega - p)\}] \end{aligned}$$

For small enough values of modulation frequency p we may approximate by taking

$$\phi(\omega \pm p) = \phi(\omega) \pm p\phi'(\omega) \quad \phi'(\omega) = \mathrm{d}\phi/\mathrm{d}\omega$$

Assembling the terms we find

$$v_2 = A\cos(\omega t + \phi)[1 + m\cos\{p(t + \phi')\}]$$

Thus the envelope has suffered a time delay equal to the group delay:

$$\tau = -\mathrm{d}\phi/\mathrm{d}\omega$$

It is necessary to equalize the group delay over the width of a communication channel in order to avoid distortion.

Measurement of group delay is accomplished by realizing the process just discussed. The amplifier is used with an amplitude modulated carrier: input and output samples are taken, each is demodulated and the phase difference between the two measured. If this difference is denoted by $\Delta\phi$, the group delay is approximately equal to $\Delta\phi/p$. Choice of p may present problems, since the measured phase shift is proportional to p. Too small a value leads to errors in the phase measurement, too large may cause error if the phase characteristic is too curved.

Summary

This chapter has introduced the constructional features of microwave bipolar and field-effect transistors together with methods of design of amplifier modules.

The characterization of amplifiers for gain, bandwidth and noise properties has been developed, and the effects of non-linearity considered.

Amplifier specification relates to: gain; frequency response; bandwidth; group delay; noise figure; and dynamic range.

EXERCISES

6.1 Determine the outputs to all ports in the circuit of Figure 6.8 when the amplifiers are non-identical but unilateral and all ports are terminated in matched loads. Hence show that when the amplifiers are identical, the input reflection coefficient is zero, and the forward gain is that of one amplifier.

6.2 Table 6.2 shows measured values for the input to an amplifier required to maintain constant output as the frequency is altered. Determine

(1) the bandwidth to 3 dB,

(2) the noise bandwidth.

6.3 The amplifier for which the gain-frequency response is given in question 2 is equipped with a true power measuring device on the output, and a CW signal generator on the input is tuned for maximum response. With the generator switched off the output is 1.36 mW; a signal of −94 dBm increases the output to 5 mW. Estimate the gain, noise figure and noise temperature of the amplifier. It may be assumed that the signal generator reading is of available power.

6.4 A single-stage amplifier has a gain of 9 dB and noise figure of 1.5 dB. Input and output are matched to 50 Ω. Determine the noise figure of two such stages in cascade. Determine also the noise measure of the amplifier.

6.5 The single stage of question 4 is used as a preamplifier to a high gain amplifier with noise figure 8 dB. Determine the overall noise figure. What improvement would be obtained by using a preamplifier comprising two single stages in cascade?

Table 6.2

Frequency (MHz off centre)	−6	−5	−4	−3	−2	−1	0	1
Input (pW)	3340	925	244	103	62.8	57.8	57.8	61.0
Frequency	2	3	4	5	6	7	8	9
Input	62.7	64.5	64.5	68.2	110	306	1480	5220

6.6 A satellite antenna is connected to an amplifier through a feeder which introduces attenuation. The feeder attenuation is 0.2 dB, the amplifier has a noise temperature of 80 K and bandwidth 50 MHz. Calculate the effective noise temperature of this combination. Calculate also the signal power from the antenna required to give a signal-to-noise ratio at the amplifier output of 20 dB when the antenna noise temperature is 55 K. It may be assumed that the feeder is at 290 K.

6.7 A particular amplifier has a substantially uniform response in the range 100–200 MHz. Equal inputs at frequencies of 120 and 125 MHz of magnitude −50 dBm produce in the output −10 dBm of each component at 120 and 125 MHz, and −90 dBm at 115 and 130 MHz. Estimate the gain of the amplifier and the coordinates of the third order intermodulation product intercept. Assuming a minimum detectable signal equal to noise power, estimate the spurious-free dynamic range if the noise figure of the amplifier is 3 dB.

6.8 With the amplifier described in question 7, the strengths of the two components of input signal are made unequal. Show that the strengths of the third order intermodulation products may be calculated from the following formulae:

$$\text{frequency} \quad 2\omega_1 - \omega_2 \qquad P_i - 3p_i + 2p_1 + p_2$$
$$2\omega_2 - \omega_1 \qquad P_i - 3p_i + p_1 + 2p_2$$

in which p_1 and p_2 are the strengths of the input components at frequencies ω_1 and ω_2 respectively, and p_i and P_i are the coordinates of the intercept point, all in dBm. Hence determine the components in the output when the input is −45 dBm at 120 MHz and −50 dBm at 125 MHz.

6.9 If only the terms A_1 and A_3 in equation (6.6) are significant, determine the position of the 1 dB compression point in relation to the intercept point.

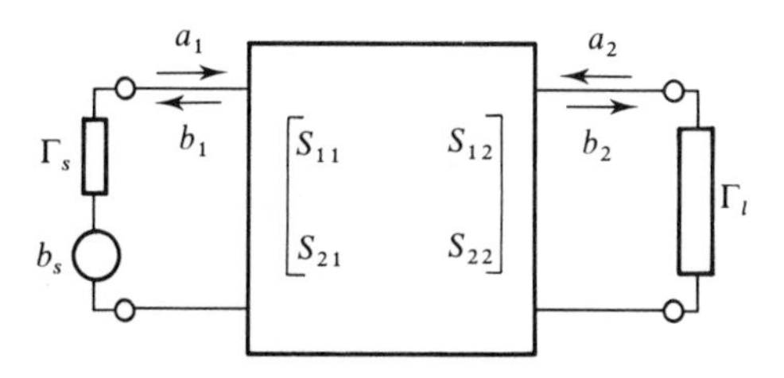

Figure 6.16 Notation for question 6.10.

The design of transistor amplifiers concerns the design of suitable circuits at input and output by which 50 Ω source and sink are transformed to present the required reflection coefficients to the transistor. The techniques use scattering matrix representation for the transistor, and presentation on the Smith Chart. The following problems illustrate the application of the ideas of Section 2.11 in this context, forming a self-learning exercise.

6.10 A transistor amplifier is represented in Figure 6.16. The transistor is described by the 2-port scattering matrix

$$\begin{bmatrix} S_{11} & S_{12} \\ S_{21} & S_{22} \end{bmatrix}$$

The output, port two, is connected to a load Z_l. In the case when S_{12} may be assumed negligibly small (i.e. the transistor is unilateral), show that the power delivered to the load when a wave of complex amplitude a_1 is incident on port one may be expressed in the form

$$P_0 = \tfrac{1}{2}|S_{21}|^2 \frac{1 - |\Gamma_l|^2}{1 - |S_{22}\Gamma_l|^2} |a_1|^2$$

6.11 Show that in the unilateral case considered in question 6.10 the actual power gain, G, and the transducer power gain, G_T, may be expressed in the forms

$$G = \frac{|S_{21}|^2(1 - |\Gamma_l|^2)}{(1 - |S_{11}|^2)|1 - S_{22}\Gamma_l|^2}$$

$$G_T = \frac{1 - |\Gamma_s|^2}{|1 - S_{11}\Gamma_s|^2} |S_{21}|^2 \frac{1 - |\Gamma_l|^2}{|1 - S_{22}\Gamma_l|^2}$$

6.12 Show that the maximum unilateral gain is given by the expression

$$G_{um} = |S_{21}|^2[(1 - |S_{11}|^2)(1 - |S_{22}|^2)]^{-1}$$

occurring when both source and load are conjugate matched.

6.13 The transistor to be used in a single stage amplifier has scattering parameters with the values shown in Table 6.3. The component S_{12} may be assumed to be negligibly small.

Table 6.3

f(GHz)	1		2		3		4		5	
	$\|S\|$	$\angle S°$	$\|S\|$	$\angle S°$	$\|S\|$	$\angle S°$	$\|S\|$	$\angle S°$	$\|S\|$	$\angle S°$
S_{11}	0.95	−20	0.90	−42	0.81	−65	0.71	−86	0.63	−108
S_{21}	3.55	161	3.38	143	3.23	126	2.86	110	2.57	94
S_{22}	0.82	−5	0.79	−8	0.75	−15	0.75	−17	0.73	−21

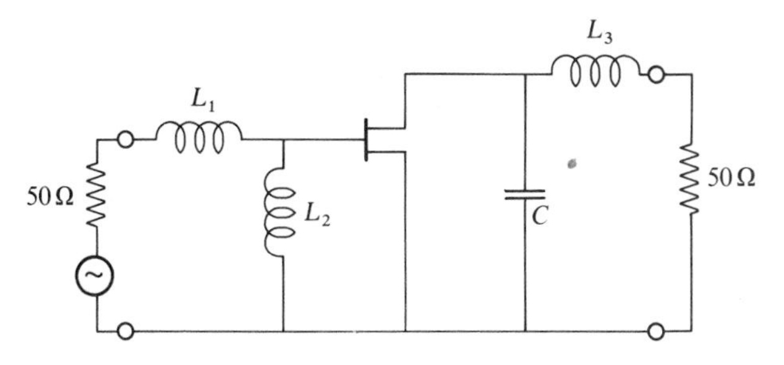

Figure 6.17 Circuit for question 6.13.

Determine the maximum unilateral gain for the above transistor at a frequency of 2 GHz. Show that component values may be chosen so that this gain would be achieved with the circuit of Figure 6.17, in which source and load impedances are $50\,\Omega$. (The matching networks can be chosen in many ways: that shown uses lumped components.)

6.14 The expression for transducer gain in question 6.11 contains factors which may be associated with input and output matching, both of the form

$$G = \frac{1 - |\Gamma|^2}{|1 - \Gamma_0 \Gamma|^2} \qquad 0 < G < (1 - |\Gamma_0|^2)^{-1}$$

in which Γ may be either Γ_s or Γ_l and Γ_0 either S_{11} or S_{22}. Show that on a Smith Chart (Argand diagram in the Γ-plane) the locus for a constant value of G is a circle, centre $G\Gamma_0^*/(1 + G|\Gamma_0|^2)$, radius $(1 + G|\Gamma_0|^2 - G)^{\frac{1}{2}}/(1 + G|\Gamma_0|^2)$. Prepare separate charts at frequencies of 1, 2, 3, 4 GHz for input circuit gains of 0, 2, 4, 6 dB and output gains of 0, 2, 3 dB. Investigate the gain of the amplifier designed in question 13 at frequencies other than the design value. (A programmable calculator can be used as an alternative to these graphical techniques.)

6.15 Achievement of optimum noise figure at 2 GHz requires the source to have a reflection coefficient of $0.6 \angle 40°$. Determine the maximum gain then attainable, and design a two-element input circuit fed from $50\,\Omega$.

6.16 The formula of Section 6.7.5 relates noise figure to source impedance. Show that a locus of constant F on a Smith Chart is a circle, centre $\Gamma_n/(1 + \gamma)$, radius $[\gamma^2 + \gamma(1 - |\Gamma_n|^2)]^{\frac{1}{2}}/(1 + \gamma)$ in which

$$\gamma = (F - F_0)|1 + \Gamma_n|^2/4r_n$$

Γ_n is the reflection coefficient corresponding to the admittance $-(G_0 + jB_0)$ and $r_n = R_n/Z_0$.

This set of curves can be used in conjunction with the input gain chart to find a suitable compromise between gain and noise figure.

6.17 For the transistor detailed earlier, at a frequency of 2 GHz, $F_0 = 1.5$ dB, $\Gamma_n = 0.60 \angle 40°$, $r_n = 0.54$. Prepare an input chart with noise figure contours with $F = 2.0$, 3.0 dB. Suggest a suitable compromise between gain and noise figure.

References

Circuit theory

Cattermole, K. W. (1964). *Transistor Circuits.* London: Temple Press Books.

Microwave amplifier design

Ha, T. T. (1981). *Solid-state Microwave Amplifier Design.* New York: Wiley.

Vendelin, G. D. (1982). *Design of Amplifiers and Oscillators by the S-parameter Method.* New York: Wiley.

Transistor fundamentals

Carroll, J. E. (1974). *Physical Models for Semiconductor Devices.* London: Arnold.

Lindmayer, C. Y. (1965). *Fundamentals of Semiconductor Devices.* New York: Van Nostrand Reinhold.

Microwave transistors

Howes, M. J. and Morgan, D. V. (1976). *Microwave Devices.* New York: Wiley.

Howes, M. J. and Morgan, D. V. (1980). *Microwave Solid-state Devices and Applications.* Stevenage: Peter Peregrinus.

Pengelly, R. S. (1982). *Microwave Field-effect Transistors—Theory, Design and Applications.* New York: Research Studies Press (Wiley).

Watson, H. A., Ed. (1969). *Microwave Semiconductor Devices and their Applications.* New York: McGraw-Hill.

White, M. H. and Thurston, M. O. (1970). 'Characterisation of microwave transistors'. *Solid-state Electronics*, **13**, 524–42.

CHAPTER 7

MICROWAVE RECEIVERS

OBJECTIVES

This chapter examines further the superheterodyne receiver introduced in Chapter 1. With the exception of the mixer stage, the constituent parts have been studied in previous chapters. This omission is made good, and the specification and calculation of overall performance investigated.

7.1 Introduction

A schematic diagram of a typical superheterodyne type microwave receiver is shown in Figure 7.1. Because receivers are usually required to detect weak signals, amplification is necessary. Sometimes initial amplification is used at the microwave frequency (RF amplification), but in general amplification is more easily provided by a high gain amplifier at a frequency much lower than the carrier. In the diagram this amplification is represented by the 'intermediate frequency amplifier'. The modulation on the incoming radio frequency carrier is translated to a carrier at a lower frequency (the intermediate frequency, IF) in a mixer stage: the new frequency is the difference between the radio frequency and a local oscillator frequency. Such a stage is sometimes referred to as a 'down-converter'. Because the intermediate frequency is fixed, the frequency response can be accurately determined by appropriate filters. The IF amplifier is followed by a detector, or demodulator, to translate the information on the signal into base-band form for signal processing as required by the particular application. Prior to the mixer, the amplifier at the signal frequency may be used both to reduce reception of spurious signals and to improve the reception of weak signals.

It must be stressed that this description is of a 'simple' receiver, and in so far as emphasis has been placed on single frequency reception, is virtually a high gain amplifier which may well be linear down to the demodulator. In practice, however, much more may be demanded of a receiver. For example, it may be required to respond to a weak signal in the presence of a strong one at a neighbouring frequency; it may be required to 'search' for signals by using a swept-frequency local oscillator; it may be required to have a highly non-linear response to signal amplitude, or a very linear response. In later sections consideration will be given to the basic building blocks illustrated in Figure 7.1; particular system requirements will be considered in the relevant chapters.

7.2 Receiver parameters

It is apparent that a receiver is very similar to a high gain amplifier, and can be characterized in the same way, so the discussion in Section 6.7 applies largely as it stands. The important parameters are input impedance, gain, bandwidth and noise figure.

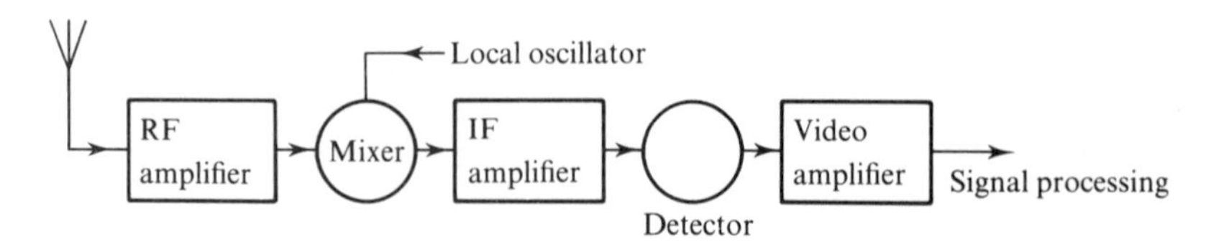

Figure 7.1 Block diagram of superheterodyne receiver.

In some circumstances the gain has to be specified closely, but usually in a receiver a gain control is provided, normally in the IF amplifier, which may be set or may be automatically controlled by a feedback loop from the output. Such a loop could, for example, be used to maintain a constant average value of output signal. It might equally be arranged to act very rapidly in order to handle large signals which would otherwise overload the final stages of the system. Input impedance has the same significance as with a unilateral amplifier. The overall gain-frequency response of a receiver is of the form shown earlier, in Figure 6.11. The bandwidth specified for a receiver is usually the difference between frequencies of the half-power points, the **3 dB bandwidth**. It is determined primarily in the IF amplifier, either by tuned circuits or by a filter. The bandwidth specified for a receiver is chosen so that any modulated carrier is not appreciably distorted. A bandwidth is also associated with the post-detector video circuits. This is chosen to handle the demodulated signal and will not in general be the same as the IF bandwidth. Each of the amplifier blocks in Figure 7.1 needs its own specification. This fact has implications for measurement: gain and bandwidth refer to a linear amplifier handling a range of sinusoidal signals, and thus the non-linear demodulator has to be circumvented. This can be done by, for example, an auxiliary pre-detector output, or using a narrow band modulation which can be monitored at the output.

A most important receiver parameter is the noise figure. This is defined as in Section 6.7 referring to the linear pre-detector part of the receiver. The RF amplifier, the mixer and the IF amplifier may all contribute to the overall noise figure.

7.3 Mixers

The frequency changing stage, or mixer, in the block diagram of Figure 7.1 has traditionally assumed a more important place in microwave technology than at lower frequencies: the reason for this has been the difficulty of producing amplifiers at carrier frequency which improve the signal-to-noise ratio rather than degrade it. (The importance of this was emphasized in Section 6.7.2.) Although it is now possible to have amplifiers with noise figures that are better than that of a simple mixer, the mixer still affects the overall performance, and for some requirements the amplifier may be undesirable on grounds of complexity or vulnerability to damage. Thus the design of good mixers is important. Some of the more important considerations will now be outlined.

7.3.1 The mixing process

Since the mixing process concerns a change of frequency between input and output signals it must involve a non-linear device. In general terms, a circuit

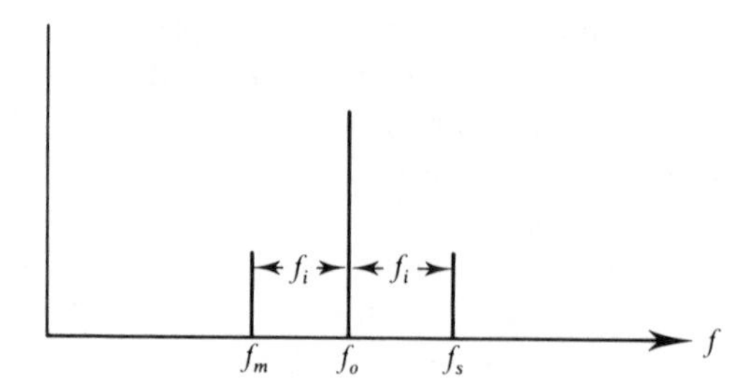

Figure 7.2 Showing disposition of frequencies in a mixer: f_s is the signal frequency; f_o is the local oscillator frequency; f_i is the intermediate frequency; and $f_m = 2f_o - f_s$ is the image frequency.

containing a non-linear element is driven by the signal, of frequency f_s, in the presence of a local oscillator drive, frequency f_o. The resulting current will contain, in varying proportions, components of all frequencies of the form

$$mf_s \pm nf_o$$

in which m and n are integers.

It is usually convenient to make the local oscillator lower in frequency than the signal, in which case the intermediate frequency f_i, is given by

$$f_i = f_s - f_o$$

Another frequency which will give the same intermediate frequency is the **image frequency**

$$f_m = 2f_o - f_s$$

The relationship between these various frequencies is indicated in Figure 7.2. Although mixing is a non-linear process, the output at intermediate frequency is linearly proportional to the signal input provided that the latter is much smaller in amplitude than the local oscillator. The action of mixing can then be taken as a perturbation on the state obtaining when only the local oscillator is applied.

A basic mixer circuit is shown in Figure 7.3. The sources of signal and local oscillator together with IF load are in series with a non-linear element. A simple approximation is to regard the strong local oscillator as switching the diode on and off and thus connecting and disconnecting the signal source to the IF load at the local oscillator frequency. Thus the voltage applied to the IF load may be written as

$$V_s \cos \omega_s t f(\omega_o t)$$

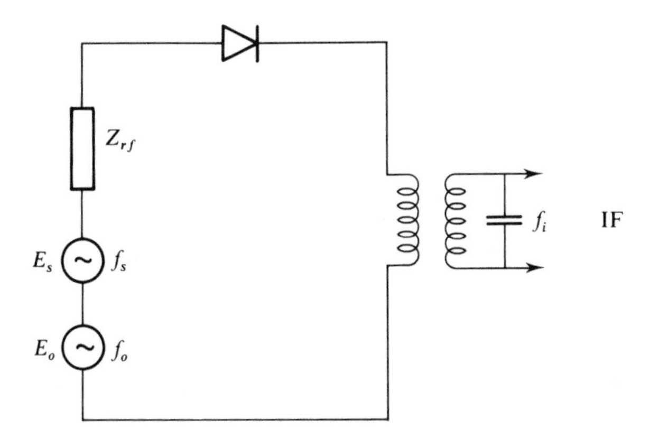

Figure 7.3 Simple mixer circuit.

where

$$f(\omega_o t) = 1 \qquad 2m\pi < \omega_o t < (2m+1)\pi$$
$$= 0 \qquad (2m-1)\pi < \omega_o t < 2m\pi$$

This may be written as

$$V_s \cos \omega_s t \left(\frac{1}{2} + \frac{2}{\pi} (\sin \omega_o t + \tfrac{1}{3} \sin 3\omega_o t + \ldots) \right)$$

The IF output will be the component at the difference frequency, which is seen to be

$$v_i = \frac{V_s}{\pi} \sin (\omega_o - \omega_s) t$$

This demonstrates linearity and, although many factors have been left out of account, it hints at the important fact that performance is largely independent of local oscillator strength, providing this is large enough.

Since under this condition the mixer is linear, its performance may be specified by a **conversion gain** (less than unity). This is customarily defined as the ratio of available power out to available power in, and thus depends on source impedance. Conversion gain may be compared with the available power gain as defined in Section 6.7. The linear theory of mixing is considered further in Section 7.8.

7.3.2 Mixer rectifiers

The basic circuit of Figure 7.3 shows a diode as the non-linear element. Although any non-linear element can be used in a mixer, at microwave

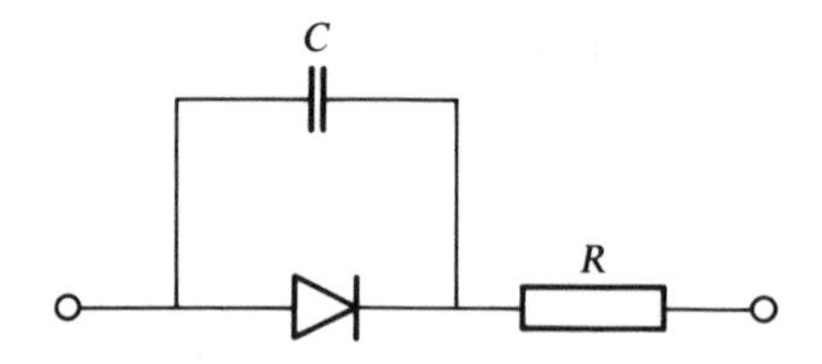

Figure 7.4 Equivalent circuit of mixer diode.

frequencies semiconductor diodes have been found to be the most satisfactory. These were initially 'crystal rectifiers', consisting of a silicon crystal with a tungsten 'cat's whisker' to form the contact. More recently Schottky barrier diodes have given better performance. Figure 7.4 shows the equivalent circuit of such a diode, incorporating the unavoidable parasitics: the capacity in parallel with the junction and the resistance of the bulk material in series. It can be seen that these two components act to reduce current flow through the non-linear element, and thus to reduce the efficiency of the device as a mixer.

A parameter which affects the parasitics and hence performance is the bias under which the diode operates. A direct current will flow under the action of the local oscillator, and biasing can be controlled both by local oscillator strength and by series resistance. Provision has to be made in a mixer circuit for DC to flow, and it is useful to monitor this current.

7.3.3 Mixer circuits

A realization of the circuit of Figure 7.3 is illustrated in Figure 7.5 in microstrip. This circuit suffers from the disadvantage that the signal and local oscillator circuits are not isolated, leading to loss of signal and radiation of local oscillator power. A filter or an isolator (Section 5.10.1) can be added to reduce signal loss. The circuit also responds equally to signal and image frequencies unless a high-Q filter is added, making the circuit highly frequency sensitive. Improvement can be obtained by using a balanced mixer circuit, an example of which is shown in Figure 7.6 using the waveguide 'magic-T' discussed in Section 5.7.1. In this circuit signal and local oscillator are isolated, and the IF outputs are balanced with respect to earth. An additional advantage is that noise on the local oscillator appears in phase on the two outputs and is thus removed when they are combined. These advantages can be more conveniently realized by use of the double-balanced diode ring shown in Figure 7.7, in which signal, local oscillator and IF ports are mutually isolated. The symmetry also causes the suppression of a number of unwanted intermodulation products, thus avoiding a waste of signal power and giving improved conversion efficiency. Such a ring would be difficult to realize at microwave frequencies using separately packaged diodes, but is quite realizable using Schottky barrier diodes in an integral form. The trans-

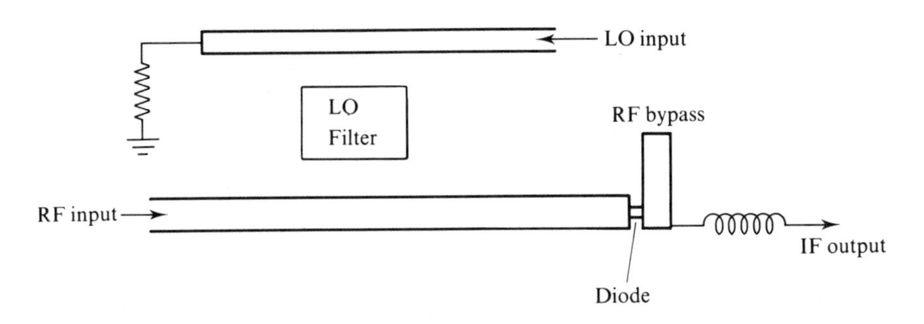

Figure 7.5 Microstrip realization of mixer.

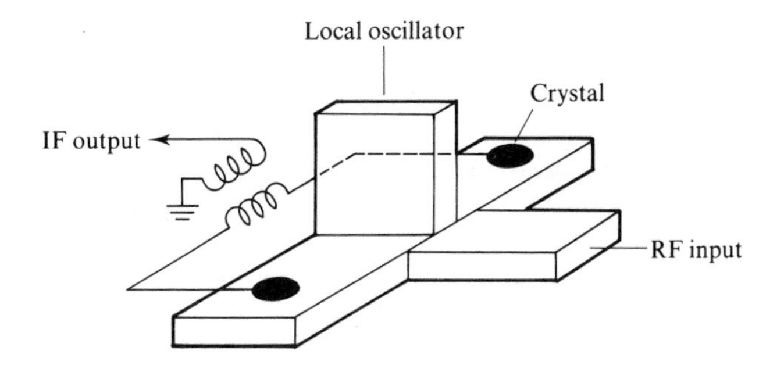

Figure 7.6 Balanced mixer using waveguide magic-T.

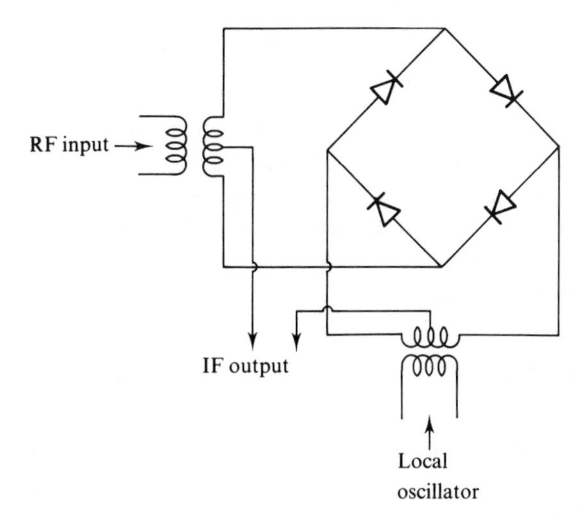

Figure 7.7 Double-balanced ring mixer.

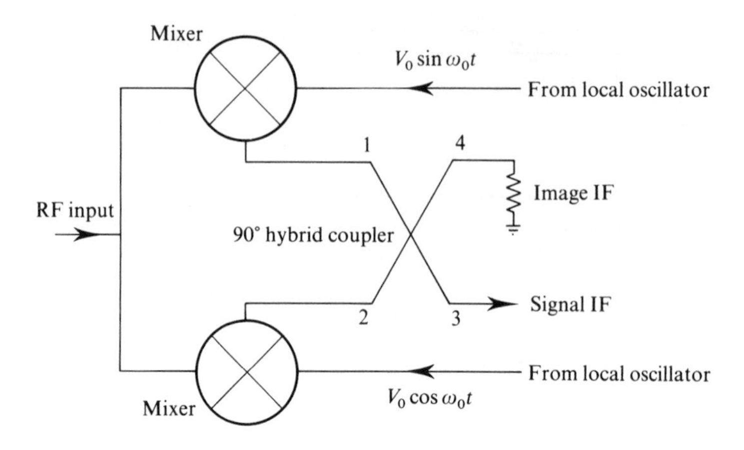

Figure 7.8 Image suppression mixer.

formers can be realized using balanced to unbalanced transitions (**baluns**). A simple theory of operation is to regard B as at the potential of P when $\cos \omega_0 t > 0$, and D at the potential of P when $\cos \omega_0 t < 0$. Thus $v_{PQ} = \pm v_s$ according to whether $\cos \omega_0 t \gtrless 0$, from which it may be shown that the IF output is

$$\frac{2V_s}{\pi} \cos (\omega_s - \omega_0)t \tag{7.1}$$

The above circuits treat signal and image frequencies alike, and further improvement can be obtained if the mixer circuit is designed to reject the image frequency. In this way conversion gain can be increased and the reception of spurious signals will be reduced. Another advantage gained by image rejection is the elimination of the noise otherwise introduced into the IF output by the image frequency channel. An image rejection circuit using two diode rings and a hybrid is shown in Figure 7.8. In this diagram the hybrid has been numbered in conformity with equation (5.5) referring to Figure 5.22(a). An advance of phase angle ϕ in the local oscillator input of Figure 7.7 will change equation (7.1) to

$$\frac{2V_s}{\pi} \cos [(\omega_s - \omega_0)t - \phi] \tag{7.2}$$

If the input is taken to be an image frequency signal

$$v_m = V_m \cos (2\omega_0 - \omega_s)t$$

the output will be

$$\frac{2V_m}{\pi}\cos\left[(\omega_s - \omega_0)t + \phi\right] \tag{7.3}$$

Thus a phase change on the local oscillator gives equal and opposite changes on the IF outputs produced by signal and image frequency inputs. The separation of the two channels then follows from the action of the hybrid.

7.3.4 The GaAs FET as mixer

The circuits described in the previous sections have used diodes exclusively as the non-linear elements, and at the present time this is the normal practice. However, field-effect transistors may be used as non-linear elements and from their performance as amplifiers it might be expected that they would also make good mixers. Although good conversion gain has been achieved, the noise figures so far obtained have been inferior to those for a good diode mixer.

7.4 IF amplifiers

IF amplifiers will not be discussed in great detail. The intermediate frequency should be much lower than the microwave signal frequency. The choice of intermediate frequency depends on several factors, and is not in any way critical: amplification must be easy to obtain; the required bandwidth should be a small fraction of the midfrequency. High IF means wide separation between signal, image and local oscillator frequencies, helping circuit design in the mixer. Since the choice of IF is not critical, standard frequencies are accepted offering well proven design techniques, such as the 70 MHz region. It is sometimes desirable to have two frequency changes, the second IF perhaps permitting easier filter design.

Although in Figure 7.1 the mixer and IF amplifier are shown separated, the first stage of IF amplification is often built as an integral part of the mixer. The circuit design of the IF input for good noise figure is not then complicated by the need to transform to the impedance of a connecting link. Such a unit has enough gain to render subsequent noise sources unimportant.

7.4.1 Logarithmic amplifiers

In some applications such as radar, a receiver may be required to handle signals at the limit of detectability at the same time as very large signals, so that a large dynamic range is required. One way of achieving this is shown

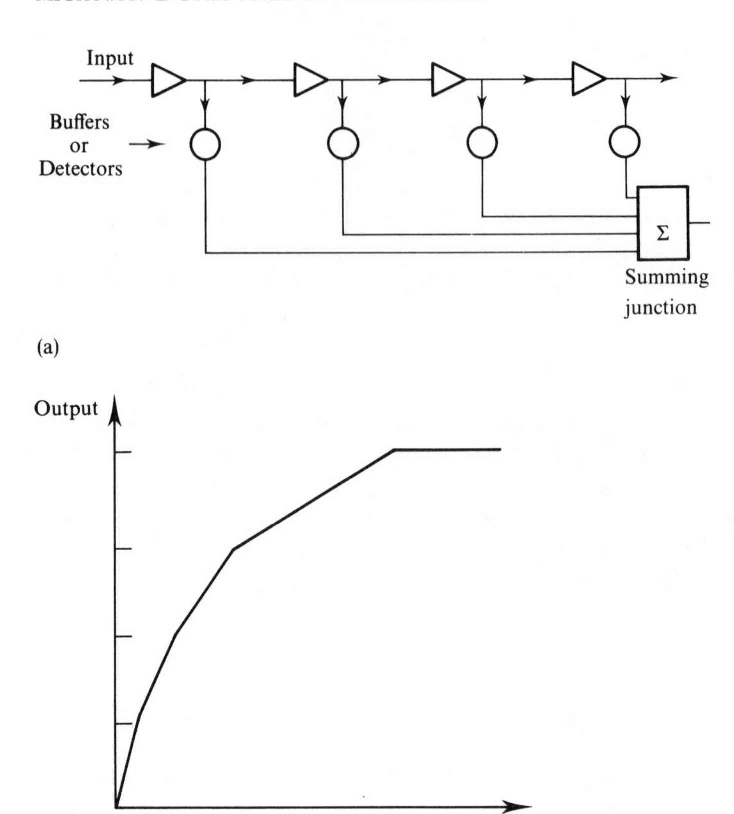

Figure 7.9 Logarithmic amplifier: (a) schematic circuit; (b) ideal input–output characteristic.

schematically in Figure 7.9(a). Each stage is designed to limit, so that a large signal will cause, for example, the last two stages to limit while the first is operating with linear gain. The outputs after each stage are summed, giving a resultant characteristic of the general form shown in Figure 7.9(b). The summation may be at IF giving an IF output, or each stage can be fitted with a detector and summation performed at the base band. There is obviously potential for unwanted feedback loops, necessitating the incorporation of buffer elements to avoid instability.

7.5 RF amplifiers

In the block diagram of Figure 7.1 an amplifier is shown preceding the mixer. It has been emphasized that the purpose of frequency changing is to achieve high gain and defined bandwidth in as simple a manner as possible, so that the RF amplifier has somewhat different functions. It serves two main purposes:

(1) to confine response to the requisite signal frequency, reducing sensitivity to the image frequency and other spurious frequencies;

(2) to improve the noise figure.

7.5.1 Filter action

It has been pointed out that a simple mixer will respond to the image frequency as well as the signal frequency. The many harmonic signals present in the mixer also make it possible for a number of other frequencies to give an IF output. The image frequency is spaced from the signal frequency by twice the intermediate frequency, so that a fairly broad RF pass band suffices to reject it. This band can be maintained even if a variable centre frequency is required. It must be observed that for some applications the widest possible RF bandwidth may be needed, in which case the RF amplifier would not have the filter action.

7.5.2 Noise figure

The noise figure of the receiver in a system can be traded off against other parameters, notably transmitter power: a reduction of noise figure by 3 dB reduces the required transmitter power by a factor of two, with perhaps a major saving in design effort and cost. Whether or not such a reduction is possible depends on the available amplifiers. It is only in recent years that the development of microwave transistors, particularly the GaAs FET, has produced high gain, low noise devices. Other amplifiers that have found use are travelling wave tubes, parametric amplifiers and masers. The principle of the travelling wave tube was outlined in Section 4.2.3, and parametric amplifiers and masers are briefly considered in Sections 7.6 and 7.7.

SOLID-STATE AMPLIFIERS

The advantages to be gained by using solid-state devices are as great at microwave frequencies as they are in other regions of the frequency spectrum. Performance alone has led to the continuing use of the other devices mentioned, such as travelling wave tubes and parametric amplifiers. Low noise travelling wave tubes have already been replaced by transistor amplifiers and without doubt parametric amplifiers with all their attendant operational problems will follow suit. GaAs field-effect transistors provide noise figures in the region of 1.5 dB at 3 GHz, rising to 4–5 dB at 10 GHz.

Specially constructed transistors are already commercially available with noise figures better than 1 dB at 3 GHz. Lower figures can be achieved with cooled transistors, and new concepts of transistor design (such as the high-electron-mobility transistor) will in due course replace the parametric amplifier completely. Details of equipment described later in this book in connection with operational systems should be viewed with this replacement prospect in mind.

TRAVELLING WAVE TUBES

The mechanism of travelling wave amplification makes possible broad band amplifiers which are electrically robust against overload. Since at one time they provided almost the only convenient method of microwave amplification, development effort resulted in mechanically robust, high gain, low noise tubes which took their part in otherwise solid-state systems. Noise figures at S-band are of the order of 4 dB for narrow-band tubes, increasing to 11 dB for wide-band tubes. They are commercially available up to 18 GHz.

PARAMETRIC AMPLIFIERS

The principle of parametric amplification is described briefly in Section 7.6. Parametric amplifiers are essentially negative resistance amplifiers (cf. Section 4.9). This negative resistance at the signal frequency is obtained by the action of a 'pump', at a higher frequency than the signal, on a variable reactance. Low noise operation is possible, and noise figures as low as 1.5 dB can be obtained at S-band, with 20 dB gain. Performance is somewhat worse at X-band. Bandwidths of the order of 150 MHz are attainable. By cooling to cryogenic temperatures further improvement can be made.

Parametric amplifiers are difficult to adjust for optimum performance, and have the disadvantage of requiring substantial pump power at a frequency much greater than the signal frequency: for X-band, a pump frequency of 35 GHz is used.

MASERS

The principle of the maser (Microwave Amplification by Stimulated Emission of Radiation) is described briefly in Section 7.7. It is a highly specialized device used only in very particular applications. Its merit is that extremely low noise temperatures (less than 10 K) are attainable, with gain better than 20 dB over a narrow bandwidth. To obtain such a noise temperature operation at liquid helium temperature is necessary, with all the attendant complications this entails. As a result, such amplifiers are only of value when the antenna temperature is also very low, as with deep space telemetry.

The significance of these values for noise figure and gain have to be seen against the values for a mixer plus IF. The noise figure for a simple mixer

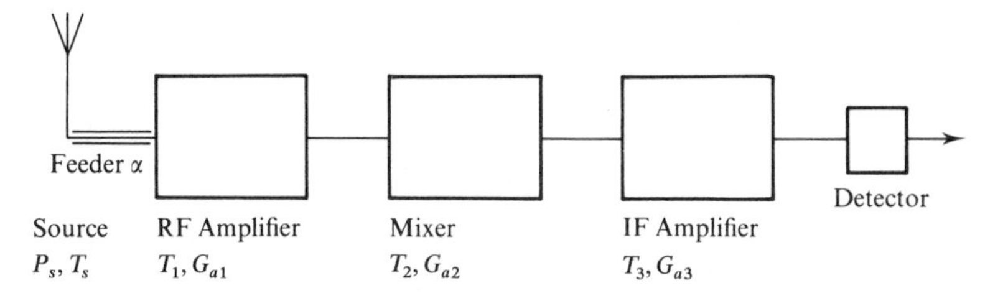

Figure 7.10 Noise sources in a receiver.

followed by an IF with 1.5 dB noise figure is about 8 dB; by using the more complicated image rejection mixer outlined in Section 7.3.3 this can be reduced to between 4 and 5 dB.

7.5.3 Overall performance

Figure 7.10 shows the signal and noise sources relevant to system performance. As was pointed out in Section 6.7.2, the significant parameter is signal-to-noise ratio. In Figure 7.10 the antenna temperature T_s will in general be different from T_0, room temperature: for example, a microwave antenna looking into a quiet region of the sky will have $T_s \cong 4\,\text{K}$; a radar antenna looking at the earth's surface will have $T_s \cong T_0$. The performance of the system has to be judged by the effect on the ratio P_s/kT_sB_n.

The overall noise figure was derived in Section 6.7.4 as equation (6.5):

$$F = F_1 + \frac{F_2 - 1}{G_{a1}} + \frac{F_3 - 1}{G_{a1}G_{a2}} + \cdots$$

In terms of noise temperature, Section 6.7.3, this may be written

$$T = T_1 + T_2G_{a1}^{-1} + T_3(G_{a1}G_{a2})^{-1} + \ldots \tag{7.4}$$

which is to be compared directly with T_s. In Figure 7.10 the second element is a lossy feeder of power gain $\alpha < 1$, at temperature T_0. Assuming it operates from a source equal to its characteristic impedance, the noise figure may readily be shown to be α^{-1}, so that

$$T_1 = (\alpha^{-1} - 1)T_0$$

Thus for a feeder loss of 0.1 dB, $T_1 \cong 7\,\text{K}$. If the final block represents a mixer and IF amplifier with a noise figure of 6 dB, T_3 is found to be 865 K. Assuming the feeder is low loss G_{a1} will not significantly affect the second and third terms in equation (7.4), so that the last term has a value $865/G_{a2}$. Thus an amplifier gain of 20 dB will give for this last term about 9 K addition to

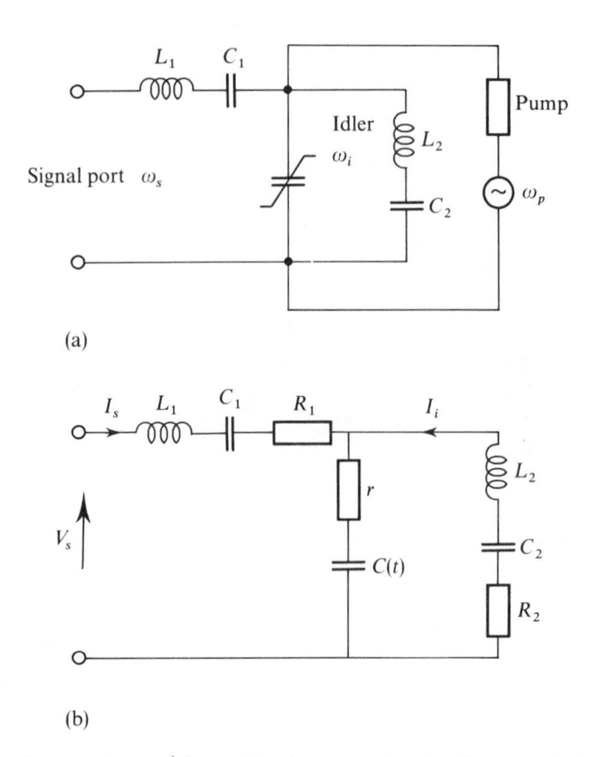

Figure 7.11 Parametric amplifiers: (a) schematic circuit; (b) equivalent circuit using time-varying capacitor.

overall noise temperature, comparable with contribution of the feeder. An amplifier with noise figure 1.5 dB has noise temperature 120 K. We may conclude that with an antenna resistance at temperature T_0, improvement gained by bettering the figure of 1.5 dB is marginal. For an antenna temperature of 4 K a maser amplifier would give a great improvement. Other configurations should be assessed in the same way.

7.6 Parametric amplification

A schematic diagram for a parametric amplifier is given in Figure 7.11(a). C represents a non-linear reactance, in practice a p–n junction diode (varactor), which is driven by a pump at angular frequency ω_p. Also in parallel with the varactor are the signal circuit, frequency ω_s, and the idler circuit usually of frequency $\omega_p - \omega_s$. If the pump is much greater in amplitude than the signal, the circuit of Figure 7.11(a) may be reduced to that of Figure 7.11(b) incorporating a time-varying capacitance, varying at the pump frequency. It will be shown in the following simplified analysis that at the signal frequency the terminals AB may be represented by a negative resistance. It is assumed that the tuned circuits restrict possible currents to those at the signal frequency in the one mesh and idler frequency in the other.

The time-varying capacitance may be defined by

$$C^{-1}(t) = \sum_{-\infty}^{\infty} \gamma_n \exp(jn\omega_p t) \tag{7.5}$$

in which $\gamma_{-n} = \gamma_n^*$ and γ_0 is real. The current through the diode has two terms only, and can be expressed in the form

$$i(t) = \tfrac{1}{2}[I_s \exp(j\omega_s t) + I_s^* \exp(-j\omega_s t) + I_i \exp(j\omega_i t) + I_i^* \exp(-j\omega_i t)] \tag{7.6}$$

where $\omega_i = \omega_p - \omega_s$.

The voltage across the component $C(t)$ is given by

$$v_c(t) = C^{-1}(t) \int i \, dt \tag{7.7}$$

Carrying out the integration, forming the product of equation (7.7) and selecting the terms in signal and idler frequencies only we find

$$v_c(t) = \text{Re}[\{\gamma_0(I_s/j\omega_s) + \gamma_1(I_i/j\omega_i)^*\} \exp(j\omega_s t) + \{\gamma_0(I_i/j\omega_i) + \gamma_1(I_s/j\omega_s)^*\} \exp(j\omega_i t)]$$

Thus, if we write

$$v_s = \tfrac{1}{2}[V_s \exp(j\omega_s t) + V_s^* \exp(-j\omega_s t)]$$

then for the signal mesh we have

$$V_s = [Z_1 + r + (\gamma_0/j\omega_s)]I_s - (\gamma_1/j\omega_i)I_i^* \tag{7.8}$$

where Z_1 represents the impedance of L_1, C_1 at the signal frequency. For the idler mesh we have

$$0 = -(\gamma_1/j\omega_s)I_s^* + [(\gamma_0/j\omega_i) + r + Z_2]I_i \tag{7.9}$$

in which Z_2 is the impedance in the idler arm.

Assuming that the components in Z_1 and Z_2 are adjusted to tune out all reactances, I_s can be found by taking the complex conjugate of equation (7.9) and subsequent elimination of I_i^*. We find

$$\frac{V_s}{I_s} = R_1 + r - \frac{|\gamma_1|^2}{\omega_s \omega_i (R_2 + r)} \tag{7.10}$$

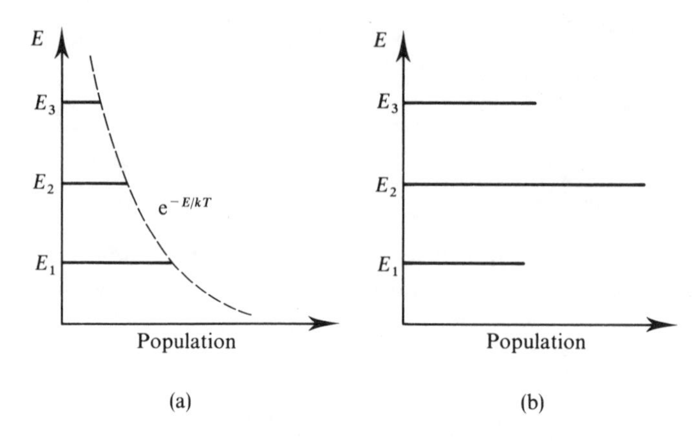

Figure 7.12 Population states in a maser amplifier.

Thus, providing the idler circuit is of sufficiently low loss, the signal sees a negative resistance.

If R_1 and R_2 are made small, the added noise comes from the diode resistance r. Noise at the signal frequency appears in series with the signal input, and there is also a component at the idler frequency transformed back into the signal circuit. Analysis shows that the latter component is proportional to ω_s/ω_i. Thus, good noise performance can be achieved by making the idler frequency much higher than the signal frequency, and also by cooling the amplifier.

7.7 Masers

The amplification process depends upon the behaviour of excited states in atoms. A solid-state maser consists typically of a sparse solution of suitable paramagnetic atoms in a host lattice, such as chromium ions in a lattice of aluminium oxide (ruby). Electrons in each atom can only take certain defined energy levels, and in equilibrium the populations in the appropriate levels will be as indicated in Figure 7.12(a). Excitation of the system will result in population distributions in which the higher levels contain more electrons than they should, and re-establishment of equilibrium takes place by transitions between levels accompanied by emission of quanta of radiation. Thus the transition dropping from state three to state two is accompanied by emission of a quantum of frequency $(E_3 - E_2)/h$, in which h is Planck's constant. These downward transitions can occur spontaneously or they can be triggered off by the presence of a quantum of the same energy from another source. The former process is spontaneous emission, the latter stimulated emission. In the 3-level maser the system is 'pumped' with a frequency corresponding to the transition between levels one and three, $(E_3 - E_1)/h$. Strong pumping can raise many atoms to state three, with a consequent reduction in the population of state one, and also an increase in state two because of rapid transitions from state three to state two. The precise situation will depend on material properties such as state lifetimes. In the conditions shown in Figure 7.12(b) a population inversion has been achieved between levels one and two: in equilibrium level two would inevitably be less populated than level one. An external quantum of frequency $(E_2 - E_1)/h$ will trigger the transition and cause all the excess states to be emptied, with the emission of many quanta at the same frequency. Because of the quantum mechanical nature of stimulated emission, these quanta are all cophased, forming a coherent signal. Thus the initial signal is amplified. The energy of a quantum of frequency 10 GHz, hf, is 6.6×10^{-24} J. At room temperature $kT = 4 \times 10^{-21}$ J, so that it can be appreciated that maser action requires the crystal to be cooled to liquid helium temperature (4 K). The energy levels used are paramagnetic levels and can be varied by means of an external magnetic field. The maser behaves as a negative resistance device and is used with a circulator.

7.8 The linear theory of mixing

The purpose of this section is to indicate the way in which a detailed theory of the mixing process may be developed. The diode shown in the circuit of Figure 7.3 may be considered as a non-linear resistance, in association perhaps with other linear components, for which the characteristic may be written

$$i = f(v) \tag{7.10}$$

A small change in v is associated with a small change in i in accordance with

$$\delta i = \frac{\mathrm{d}f}{\mathrm{d}v}\,\delta v$$

In the presence of the large local oscillator signal the operating point changes with time, so that the ratio $\delta i/\delta v$ is also a function of time. We may write

$$\delta i = \delta v \sum_{n=-\infty}^{\infty} c_n \exp(\mathrm{j}n\omega_0 t) \tag{7.11}$$

in which $c_{-n} = c_n^*$ since $f(v)$ is real. It will be presumed that the change δv in the voltage across the diode in Figure 7.3 contains only components at the signal, image and intermediate frequencies of small amplitude. Thus we may write

$$\delta v = \tfrac{1}{2}[V_s \exp(j\omega_s t) + V_m \exp(j\omega_m t) + V_i \exp(j\omega_i t)] + \tfrac{1}{2}[V_s \exp(j\omega_s t) + V_m \exp(j\omega_m t) + V_i \exp(j\omega_i t)]^* \quad (7.12)$$

Substituting in equation (7.11) and collecting terms we find that the change in current may be expressed in the form

$$\delta i = \tfrac{1}{2}[I_s \exp(j\omega_s t) + I_m \exp(j\omega_m t) + I_i \exp(j\omega_i t)] + \tfrac{1}{2}[I_s \exp(j\omega_s t) + I_m \exp(j\omega_m t) + I_i \exp(j\omega_i t)]^* + \text{components at other frequencies}$$

where

$$I_s = V_s c_0 + V_m^* c_2 + V_i c_1$$
$$I_m = V_s^* c_2 + V_m c_0 + V_i^* c_1$$
$$I_i = V_s c_1^* + V_m^* c_1 + V_i c_0$$

The second of these equations also implies that

$$I_m^* = V_s c_2^* + V_m^* c_0 + V_i c_1^*$$

Hence we may define an admittance matrix relating signal, image and IF ports:

$$\begin{pmatrix} I_s \\ I_m^* \\ I_i \end{pmatrix} = \begin{bmatrix} c_0 & c_2 & c_1 \\ c_2^* & c_0 & c_1^* \\ c_1^* & c_1 & c_0 \end{bmatrix} \begin{pmatrix} V_s \\ V_m^* \\ V_i \end{pmatrix} \quad (7.13)$$

If the voltage waveform driving the diode at the local oscillator frequency is symmetrical about some time origin, the coefficients c_n in equation (7.11) can

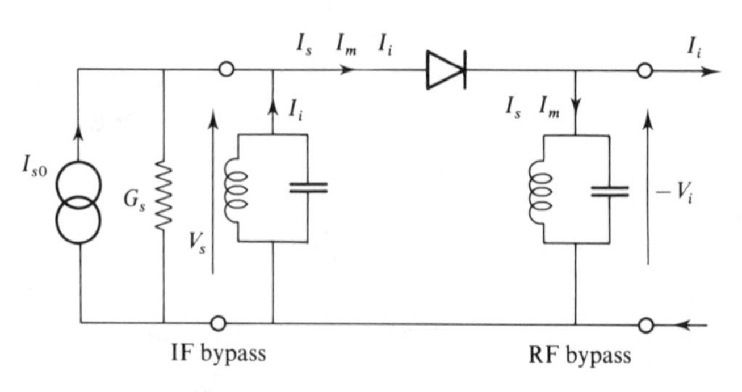

Figure 7.13 Current components in a mixer circuit.

be made real, when the matrix in equation (7.13) is real and symmetrical. In a practical circuit of the form of Figure 7.3, the RF impedance is negligibly small at the intermediate frequency, and the IF impedance negligibly small at the signal and image frequencies. This enables us to draw the circuit as shown in Figure 7.13. Frequently, the RF circuits are low-Q, so that the image and signal frequency currents both see the same conductance, G_s. We can then write

$$\begin{aligned} I_s &= I_{s0} - G_s V_s \\ I_m^* &= -G_s V_m^* \end{aligned} \tag{7.14}$$

Elimination of the signal and image components between equations (7.13) and (7.14) yields, assuming the matrix components are real,

$$I_i = \frac{c_1 I_{s0}}{c_0 + c_2 + G_s} + V_i\left(c_0 - \frac{2c_1^2}{c_0 + c_2 + G_s}\right) \tag{7.15}$$

This provides expressions for the short-circuit current and output conductance at the IF port, and shows how they depend on the RF source impedance. Other conditions on the image frequency component may be similarly treated.

Summary

This chapter has outlined the general form of a microwave superheterodyne receiver and the characteristics of the various parts considered. The principles of parametric and maser amplification have been briefly described.

Receiver specification relates to:

- RF amplifier,
- mixer,
- IF amplifier,
- demodulator,
- local oscillator.

EXERCISES

7.1 Design, in the form of a block diagram, the input section of a microwave heterodyne receiver system (up to the mixer output) suitable for satellite communications. Rectangular waveguide components should be used where appropriate. Discuss briefly the properties of the microwave components used in the design and explain their functions. Pay particular attention to the way in which the signal-to-noise power ratio is maximized. Detailed noise figure calculations are not required. (It may be assumed that the system operates at 6 GHz and requires a bandwidth of 50 MHz.)

7.2 Using the result of equation (7.15), obtain an expression for the conversion loss of the mixer (available signal power/available IF power). Find the minimum value and the source conductance for which this occurs. Derive also the results equivalent to equation (7.15) when the impedance to the image frequency is

(a) zero, $V_m = 0$,

(b) infinite, $I_m = 0$.

Note: In all cases equation (7.15) can be written in the form

$$I_i = \frac{AI_{s0}}{G_s + a} + BV_i\left(1 - \frac{b}{G_s + a}\right)$$

giving

$$L = \frac{B}{A^2 G_s}(G_s + a)(G_s + a - b).$$

The minimum of L occurs for $G_s = (a(a - b))^{\frac{1}{2}}$ when

$$L_m = \frac{B}{A^2}[2(a(a - b))^{\frac{1}{2}} + 2a - b]$$

7.3 The idealized characteristic of a mixer diode is

$$i = \alpha v^2 \quad v > 0, \quad i \equiv 0 \quad v < 0.$$

Determine the coefficients c_n of equation (7.11) when a local oscillator voltage $E \cos \omega_0 t$ is present across the diode. In a particular situation $\alpha = 10^{-2}\,\text{AV}^{-2}$, and $E = 0.7\,\text{V}$. Determine the optimum values of conversion loss in the cases considered in question 7.2. Determine also the power absorbed in the diode from the local oscillator.

7.4 The individual amplifiers in the logarithmic amplifier circuit shown in Figure 7.9(a) have a voltage gain of A, and saturate at an input V_s volts. The rectified output current increases linearly from zero to a value I_s at saturation. Currents from all stages are summed by flowing through a common low impedance load. Show that the total summed current from a cascade of N stages takes on the values

$$I = I_s\left(N - n + \frac{1}{A^n}\frac{A^n - 1}{A - 1}\right)$$

when $V_i = V_s A^{-n}, n = 0, 1, \ldots, N$.

Determine the total characteristic for the case $N = 6$, $V_s = 2$, $A = 4$ and $I_s = 1\,\text{mA}$.

7.5 Show that on the basis of equations (7.2) and (7.3), the outputs of the hybrid coupler in Figure 7.8 will be $2^{\frac{1}{2}}V_m \cos \omega_i t$ and $2^{\frac{1}{2}}V_s \sin \omega_i t$, corresponding to signal and image inputs $V_s \cos \omega_s t$ and $V_m \cos \omega_m t$ respectively.

7.6 A negative resistance amplifier uses a circulator, as shown in Figure 4.22. Show that the gain is given by

$$\left(\frac{1 + |g|}{1 - |g|}\right)^2$$

where g is the normalized negative conductance presented to port two of the circulator.

7.7 The antenna of a communication system has a noise temperature of 200 K. The received signal is connected to a mixer by a waveguide having attenuation 1.2 dB. The mixer has a noise figure of 4.5 dB and conversion loss of 3.5 dB. The mixer output feeds an IF amplifier of noise figure 1.5 dB and limiting bandwidth 20 MHz. Calculate the magnitude of signal which must be supplied by the antenna to achieve an output signal-to-noise ratio of 5 dB. What would this figure become if either

(a) an amplifier of noise figure 1.5 dB, gain 8 dB, or

(b) a travelling wave tube, noise figure 4 dB, gain 22 dB

were available as signal amplifiers? Comment on the results of this comparison.

References

Howes, M. J. and Morgan, D. V. (1980). *Microwave Solid State Devices and Applications.* Stevenage: Peter Peregrinus.

Howes, M. J. and Morgan, D. V. (1978). *Variable Impedance Devices.* New York: Wiley.

Howson, D. P. and Smith, R. B. (1970). *Parametric Amplifiers.* New York: McGraw-Hill.

Niederleithner, J. (1979). 'Noise figure microwave mixer' *Marconi Review* **42**, 153–78.

Pound, R. V. (1948). *Microwave Mixers.* New York: McGraw-Hill.

Siegman, A. E. (1964). *Microwave Solid State Masers.* New York: McGraw-Hill.

Torrey, H. C. and Whitmer, C. A. (1948). *Crystal Rectifiers.* New York: McGraw-Hill.

Van Voorhis, S. N. (1948). *Microwave Receivers.* New York: McGraw-Hill.

CHAPTER 8
RADAR

OBJECTIVES

This chapter is concerned with radar systems. It describes the basic principles of radar, and gives configurations appropriate to various different types. The factors which affect radar performance are examined in relation to the limitations imposed by the subsystems involved. Attention is given to the statistical theory of detection and the considerations of the chapter are illustrated by detailed description of an operational airport radar.

8.1 Introduction

The physical phenomenon underlying radar is that electromagnetic waves are, to some extent, scattered by any object on which they fall. This scattered radiation can be detected by sensitive receiving equipment, and its characteristics provide certain information about the object. In particular, the use of a directional antenna will provide bearing and elevation of the scattering object, and time delay will provide range. It is evident that the magnitude of the received signal will depend not only on the nature of the scatterer but also on the strength of the electromagnetic wave field illuminating the object. Thus early experiments designed to prove the possibility of radar as a viable defence system used what we now regard as relatively low frequencies simply because high power transmitting tubes and sensitive detectors were available at those frequencies.

The separation of the reflections from different targets requires resolution in both range and angle. Good angular resolution requires a well defined, narrow beam of radiation, more easily achieved at microwave frequencies, since the antenna size for a given resolution will be proportional to the wavelength. Since the development of suitable sources of microwave power, the majority of radar systems operate at microwave frequencies.

8.2 A basic radar system: the radar equation

A simple form of pulsed radar system is shown schematically in Figure 8.1(a). A transmitting antenna irradiates a region of space; the scattered radiation is picked up by a receiving antenna (shown here as separate, but in practice often the same antenna switched to a different mode). A simple form for the radiated wave is indicated in Figure 8.1(b): a sequence of wave trains of duration τ, separated by period T. The interval τ is termed the **pulse length** and the frequency $1/T$ the **pulse repetition frequency**. The waveform appearing at the receiver output is indicated in Figure 8.1(c): because of the finite receiver bandwidth the pulses appear rounded rather than rectangular and random noise, principally generated in the receiver, is also present. The time delay, T_r, between emission and reception allows the range to be determined through the equation

$$T_r = 2r/c$$

in which c is the velocity of light in free space.

The estimation of the received signal power proceeds as follows: the electromagnetic wave incident on the target will be quasi-plane and of power density p_i given by

$$p_i = \frac{P_t}{4\pi r^2} G_t \qquad (8.1)$$

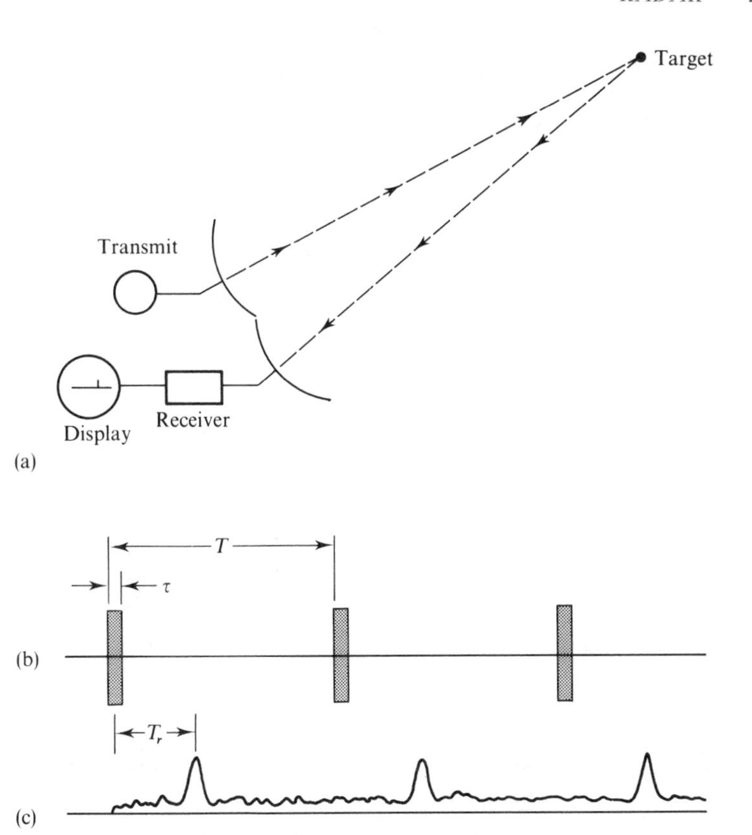

Figure 8.1 Schematic diagram of a radar system: (a) disposition of transmitter and receiver; (b) characteristics of emitted radio frequency pulse train; (c) typical form of received signal, showing delay, pulse distortion and noise background.

in which P_t is the power radiated during the pulse, G_t the gain of the transmitting antenna and r the range of the target.

The signal received back at the antenna will have a power flux of density p_s which may be related to the incident power flux by an equation of the form

$$p_s = \frac{p_i \sigma}{4\pi r^2} \tag{8.2}$$

This equation defines the **scattering cross-section**, σ, for the specific orientation of the target with respect to the radar beam. The value of σ will depend on both the target aspect and the frequency. (Scattering cross-section as defined here should be distinguished from that used in optical theory, which refers to total scattered power rather than back-scattering.) Introducing the effective area of the antenna, the received power is given by

$$P_r = \frac{P_t}{(4\pi r^2)^2} \sigma G_t A_r$$

If the same antenna is used for both transmission and reception, as is usually the case, the use of equation (3.8), Section 3.6, leads to the expression

$$P_r = \frac{P_t}{(4\pi)^3 r^4} \sigma G^2 \lambda^2 \tag{8.3}$$

or

$$P_r = \frac{P_t}{4\pi r^4} \frac{\sigma A^2}{\lambda^2} \tag{8.4}$$

Equations (8.3) and (8.4) are alternative forms of the **radar equation**. If the various parameters are known and a value, S_m, can be assigned to the minimum detectable signal, an expression for the maximum range of the radar may be obtained:

$$r_m = (P_t \sigma A^2/(4\pi S_m \lambda^2))^{\frac{1}{4}} \tag{8.5}$$

This equation applies to situations where the transmission path does not suffer from ground reflections, as in ground-to-air radar. Such effects, and those associated with propagation in the earth's atmosphere, are not considered in this chapter but will appear in connection with terrestrial communication links in Chapter 9. In assigning values to P_t, P_r and S_m it will in general be necessary to make allowance for losses, such as those in antenna feeds, for example. The various factors involved in the radar equation will now be briefly considered.

8.2.1 Antenna gain

The general factors involved in determining antenna gain and effective area have been considered in Chapter 3. In considering the frequency dependent equations (8.3) and (8.4) it should be remembered that the gain G is also a function of frequency. At microwave frequencies the effective receiving area A in equation (8.4) is typically about 0.6 times the physical area of the aperture, the dimensions of the latter being restricted by factors such as mechanical construction, mobility, and so on. The numerical values given to gain or area will refer to the maximum values obtained by choosing the optimum direction. The radiation pattern of the antenna will be dictated by the uses for which the radar is designed. This is considered further below.

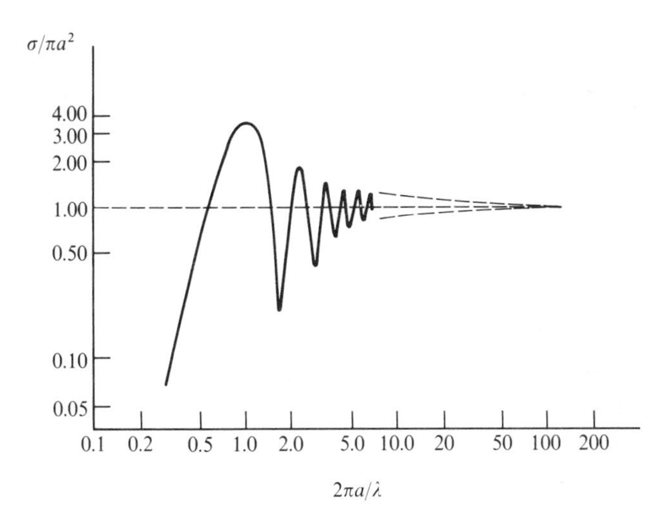

Figure 8.2 The radar cross-section of a conducting sphere as a function of frequency. Sphere radius a, wavelength λ, radar cross-section σ.

8.2.2 Scattering cross-section

It is only for very simple objects that a theoretical value can be obtained with any ease. Figure 8.2 shows the theoretical variation of scattering cross-section with frequency for a perfectly conducting sphere of radius a. Because of the symmetry, the cross-section is independent of aspect. For wavelengths greater than twice the circumference, the cross-section is proportional to λ^{-4}; when the wavelength is much less than the circumference the 'optical' cross-section, $\sigma = \pi a^2$, is found. In between, the power reflection varies between 5.7 dB above and 4 dB below that corresponding to the 'optical' cross-section. For an asymmetrical object large compared to the wavelength, the cross-section will be a complex function of aspect exhibiting very rapid fluctuations, down to virtually zero. In practice, the aspect of a moving target is continually changing, so that conditions for a small return will not last long. Change in radar frequency can also be used to avoid such low returns. The radar cross-section of complicated objects can be estimated by regarding the scattered wave as emanating from a number of simpler shapes, and reasonably accurate values can be obtained. It will be found, however, that although a typical value may be taken, as the target aspect changes σ may vary over a range of the order of ± 10 dB.

The scattering cross-section depends on the material from which the target is made as well as its shape: it will be much less for a dielectric sphere than for a conducting one, although reflectivity of a metal structure may be reduced by coating it with suitable microwave absorbent material.

8.2.3 Transmitter power

The radar equation shows clearly that in order to get long range detection high transmitter power is an advantage, in the same way that large antenna area or low minimum detectable signal are advantageous. The various devices considered in Chapter 4 are available, and the choice will be made as part of the optimization of the entire system. In some systems the transmitter tube might be a pulsed magnetron, in others a travelling wave tube or klystron amplifier used with pulse compression (Section 8.6). In a phased array radar (Section 8.7), the power is distributed over many elements, each of which would incorporate a solid-state amplifier fed from a common source. At one end of the scale, a big klystron might run at a peak power of 5 MW; at the other end the peak power generated by a single solid-state amplifier might be of the order of tens of watts, with many amplifiers contributing to the total power.

8.2.4 Minimum detectable signal

The receiver output will contain noise, as depicted in Figure 8.1(c), and as discussed in earlier Sections 6.7.2 and 7.5.2. Depending on the method of detection employed, a noise pulse can be mistaken for a signal pulse, or a noise pulse can reduce a signal pulse below detectability, in the one case giving a false alarm, in the other a missed target. Since noise power is proportional to bandwidth, it is desirable to reduce the bandwidth as far as is consistent with signal handling. For a radar emitting a rectangular pulse, an approximate expression for the IF bandwidth for optimum signal-to-noise ratio is

$$B \cong 1.2/\tau$$

in which τ is the pulse duration. In practice, the output will not be rectangular, and the average duration of a noise pulse (time between zero crossings) is also about B^{-1}, so that at low level noise and signal pulses can be confused.

The simplest detection process is to use an operator watching an '*A*-scan'. This displays receiver output (after demodulation) on a cathode ray oscilloscope with time base duration less than the repetition period, as indicated in Figure 8.3. A target echo appears as a pulse at a position determined by the range. All that can really be specified about the detection process is probability of detection for a given signal-to-noise ratio, and with an operator this has to be determined experimentally. Automatic detection methods usually work by comparing the signal level with a predetermined threshold level: pulses above the threshold are classed as targets. If the threshold is set too low, noise pulses will cause frequent false alarms; too high and targets will be missed. This is illustrated in Figure 8.3(c). A constant false

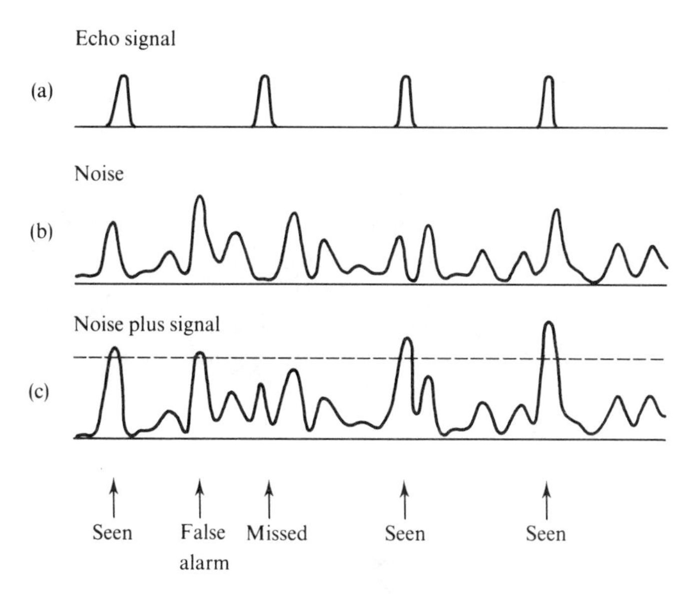

Figure 8.3 Illustrating the effect of noise on a threshold detector: (a) echoes from different targets in absence of noise; (b) the noise signal; (c) combination of signal and noise showing the threshold and indicating false alarm and missed targets.

alarm rate can be achieved by controlling the noise output to constant level. The statistics of the process will be considered in Section 8.8.

A factor that can affect the receiver performance is the necessity of protecting the receiver from the effect of the transmitted pulse. Special duplexer circuits (Section 8.5) are needed to ensure that the power injected into the receiver at transmission is not large enough to damage the receiver components. This particularly applies to solid-state components, which may suffer progressive degradation or catastrophic failure.

From these considerations it is clear that the radar equation should be regarded only as providing useful design guidelines, and various other factors must be allowed for in more exact calculations of radar performance.

8.3 Characteristics of the radar signal

In the discussion above of a simple pulse radar only one characteristic of the return signal is used, its delay time. Other characteristics also provide information about the target.

8.3.1 Pulse shape

The duration of the pulse emitted by the transmitting antenna has one very straightforward bearing on the information that may be derived: a pulse of duration τ has a spatial length of $c\tau$, c being the velocity of light. Thus a 0.1 μs pulse has a length of 30 m. If such a pulse falls onto two targets, the return signals will be confused unless the targets are separated by more than one-half of a pulse length. High resolution requires short pulse length (in the context of the simple pulse radar considered in this section).

In a less obvious way the shape of the returning pulse provides information about the target: for example, multiple targets too close for separation will nevertheless cause the return pulse to be lengthened compared to the transmitted pulse. In this way it might be deduced, for example, that the return indicated several aircraft in formation. In a more detailed sense a particular reflector will have a characteristic transient response, which might be of use to discriminate between classes of target.

8.3.2 Signal amplitude

The radar equation indicates that for a given target scattering cross-section, the return power decreases as the fourth power of range. On this basis, the amplitude could be used to estimate range. From what has been said of the parameters in the radar equation, it is clear that amplitude measurements could not be used to derive range with any certainty. The variation of return power with range does however lead to a perhaps unwanted emphasis on small targets at close range. This may be compensated for by automatically varying the receiver gain during the period between pulses, from a lower value immediately after transmission to a higher level at the time of returns from distant targets. This facility is known as **swept-gain** or **sensitivity-time-control (STC)**.

8.3.3 Frequency

The frequency of the return signal will only be the same as that of transmission if the relative position of target and radar is fixed. If the distance between target and radar is changing, the return will be shifted in frequency by an amount proportional to the line-of-sight velocity. This is known as the Doppler effect. A derivation suitable at non-relativistic speeds is as follows: the phase of the return signal at the radar may be written

$$\phi = \omega(t - 2r/c)$$

Hence the instantaneous frequency is given by

$$f_i = \frac{1}{2\pi}\frac{d\phi}{dt} = f - \frac{2f}{c}\frac{dr}{dt}$$

Thus the return from a target receding with radial velocity v_r suffers a reduction of frequency given by

$$f_d = 2fv_r/c$$

For example, an aircraft receding at 200 m s^{-1} will produce a Doppler shift of 4 kHz on a carrier of frequency 3 GHz. This change of frequency can be detected, allowing distinction to be made between fixed and moving targets.

8.3.4 Signal spectrum

The description of a signal by its transient waveform is, for some purposes, replaced by a definition in terms of the way its energy is distributed in frequency. Although the mathematics of spectral analysis is not utilized in this book, the underlying concept is of great importance in the design of systems. In particular, the choice of bandwidth in any part of a system is governed by the spectral content of the expected signal. The formal definition of spectral density is through the use of Fourier Transforms. A signal $s(t)$, supposed to be of finite energy content, can be related to a spectral function $S(f)$ by the Fourier Transform pair

$$s(t) = \int_{-\infty}^{\infty} S(f) \exp(2\pi \mathrm{j} f t)\, \mathrm{d}f$$

$$S(f) = \int_{-\infty}^{\infty} s(t) \exp(-2\pi \mathrm{j} f t)\, \mathrm{d}t$$

In characterizing a signal it is often convenient to think of the instantaneous voltage $s(t)$ appearing across a resistor of 1 Ω value, so that the instantaneous power is $(s(t))^2$ and the total energy is given by

$$E = \int_{-\infty}^{\infty} (s(t))^2\, \mathrm{d}t$$

By a well known theorem in Fourier analysis, Parseval's theorem, this may be written

$$E = \int_{-\infty}^{\infty} |S(f)|^2\, \mathrm{d}f$$

allowing the interpretation of $w(f) = |S(f)|^2$ as the power spectral density. Examples are shown in Figure 8.4.

The choice of system bandwidth results from a need to pass most of the signal energy, i.e. to have a pass-band including frequencies for which $w(f)$ is significant, but at the same time as narrow as possible in order to reject unwanted noise signals, which are widely distributed in frequency. A rough

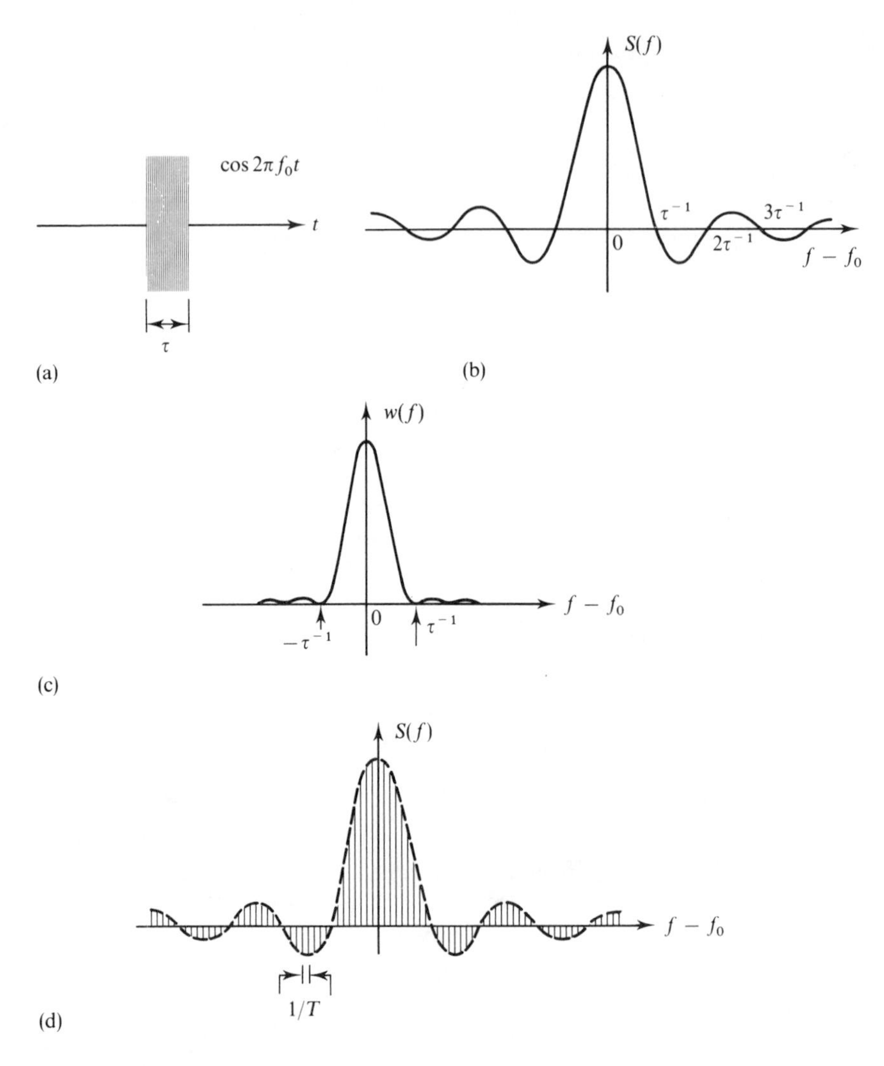

Figure 8.4 Spectrum of a carrier modulated with rectangular pulses. (a) Single pulse waveform. (b) Spectral density, $S(f)$. (c) Power spectral density, $w(f)$. (d) Line spectrum for repetitive modulation, period T.

and ready rule makes the bandwidth necessary to handle a signal of duration τ of the order of τ^{-1}. It must be appreciated, however, that power spectral density concerns only the modulus of the complex function $S(f)$: it also has a phase part. Prediction of the fine details of pulse response require complete knowledge of $S(f)$ and of the exact shape of the pass-band, together with the phase response.

The examples shown in Figure 8.4(a), (b) and (c) concern a single pulse, and it is seen to have a continuous spectrum. In a radar system pulses are emitted at regular intervals, so that the spectrum can be expressed through a

Fourier Series exhibiting discrete spectral lines at all harmonics of the repetition frequency. In such a case when the interval between pulses is much greater than the length of a single pulse, analysis shows that the spectrum consists of lines whose amplitudes follow the envelope of the spectral density for a single pulse, as indicated in Figure 8.4(d).

8.4 Types of radar

The differences in characteristics of the emitted and returned signal discussed in Section 8.3 can be used in a number of ways to detect the presence of targets, and to gain some information about them. It is to be noticed that the simple pulse radar described in Section 8.2 measures range but not velocity, whereas a CW radar using the Doppler effect, Section 8.3.3, would measure velocity but not range: in order to measure range some time marker on the signal is necessary. This means the signal must be modulated and hence will have, of necessity, a finite bandwidth. In fact, the greater the bandwidth occupied by the signal, the more accurate range measurement becomes. Thus various types of radar systems are possible, some of which will now be briefly described.

8.4.1 CW Doppler radar

A block diagram of a CW Doppler radar is shown in Figure 8.5. A transmitter feeds an antenna which illuminates the target volume: a moving target will return a scattered wave shifted by the Doppler frequency. This,

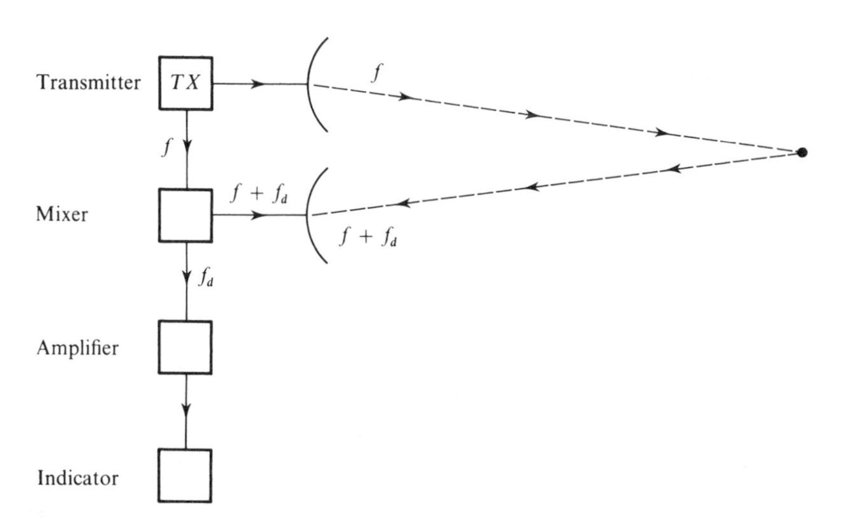

Figure 8.5 Simple CW Doppler radar, using zero frequency IF. Carrier frequency f, Doppler shift f_d.

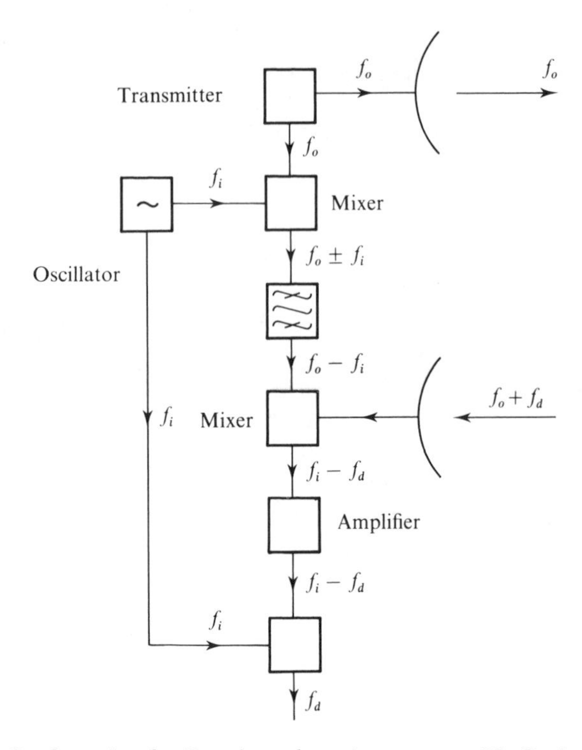

Figure 8.6 Configuration for Doppler radar using non-zero IF. Carrier frequency f_0, intermediate frequency f_i.

together with some of the transmitted waveform, goes to a mixer, which provides an output of the difference frequency f_d. This output is amplified and actuates an indicator. A very simple system can use one antenna for both transmission and reception, and in this form has practical use for speed checking and for intruder alarms: the required range is small so that a low power of the order of 10 mW might suffice at a frequency of the order of 20 GHz.

A high power system for ranging over longer distances can be made, but certain problems have to be noted. Firstly, some spillover from the transmitting antenna to the receiving antenna is inevitable, and this has two effects:

(1) the spillover power may be enough to damage the receiver;

(2) the transmitter will have some noise modulation on it, which can give signals in the range of Doppler shifts to be expected, reducing sensitivity.

The system of beating transmitter frequency with the return is referred to as a **homodyne** system, or zero IF system. With this system, the signal-to-

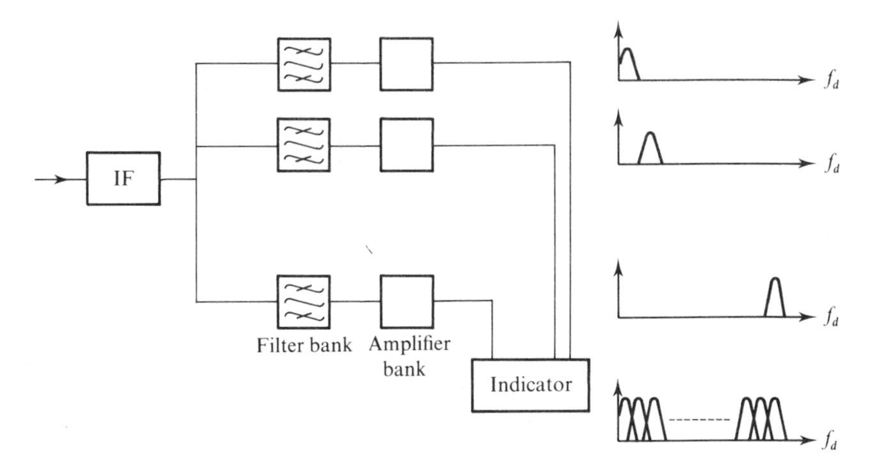

Figure 8.7 Showing the use of filters in a CW radar to separate returns with different Doppler shifts.

noise ratio deteriorates as the Doppler frequency decreases, since the noise produced in the mixer and Doppler amplifier has a frequency spectrum for which the power spectral density is proportional to f^{-1}. This is **flicker noise**, and gives way to white thermal noise at frequencies above about 1 kHz. Improvement can be obtained at the expense of complexity by using a **heterodyne** system, as indicated in Figure 8.6. The improvement in signal-to-noise ratio may be as much as 30 dB.

Processing of the output signal at the Doppler frequency depends on the requirements. The expected range of frequency shift determines the maximum bandwidth required in the Doppler amplifier. If a specified velocity is being sought, a much narrower filter can be used. This width cannot be infinitely small, if for no other reason than that the target will only be illuminated for a finite time, giving a rectangular pulse at the Doppler frequency. A bank of filters to improve the signal-to-noise ratio can be used, as indicated in Figure 8.7. In the diagram, these are shown as IF filters, and the output from each filter will denote both the particular velocity range and also whether the target is approaching or receding. Filters may be used after demodulation, with similar results, except that the sign of v_r can no longer be deduced. It may however be determined by a circuit of the form of Figure 8.8. Analysis of the circuit will show that when $v_r > 0$ Channel B lags by 90° on Channel A, and when $v_r < 0$, Channel B leads by 90°. The circuit may be compared with that of Figure 7.8.

An alternative to the use of analogue filters is to digitize the output, either at IF or at video and to carry out frequency analysis in a special purpose computer by means of the Fast Fourier Transform algorithm. The viability of this approach depends on the time available and speed of processing.

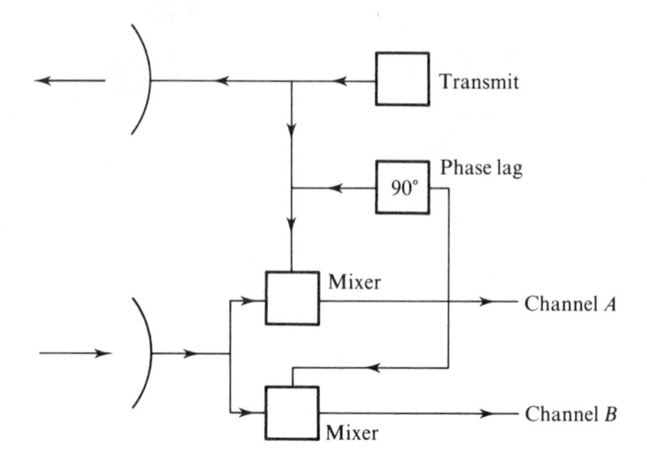

Figure 8.8 Circuit configuration permitting determination of line-of-sight velocity in systems with post-detection filtering. For receding targets channel B lags channel A by 90°; for approaching targets channel B leads channel A.

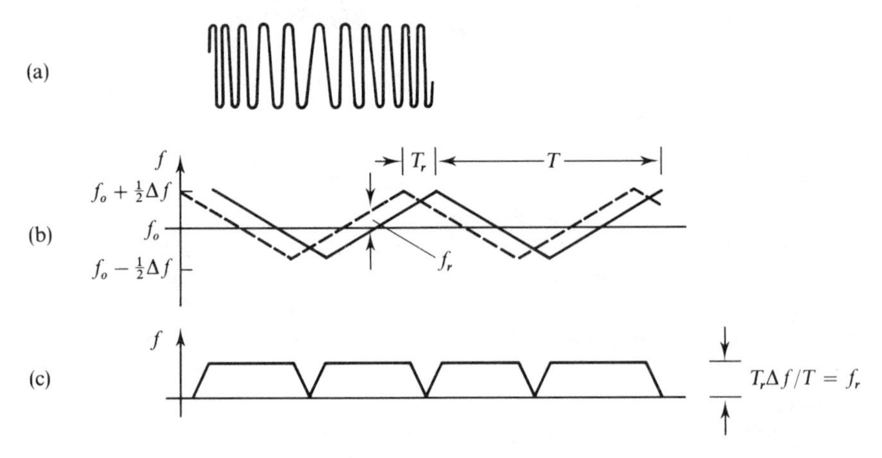

Figure 8.9 Waveforms in an FM-CW radar: (a) frequency modulated waveform; (b) instantaneous frequency versus time (pecked line transmitted wave, full line received wave); (c) showing frequency difference between transmission and echo from a stationary target.

8.4.2 FM–CW radar

It was mentioned earlier that in order to measure range, some modulation had to be imposed on the carrier. One possibility for a CW transmission is to modulate the beam in frequency. In Figure 8.9(a) a waveform with a 'triangular' FM is shown. The return signal from range r will have suffered a delay $T_r = 2r/c$, so that the frequency of the received signal varies as indicated by the dotted line in Figure 8.9(b), and the beat frequency as in Figure 8.9(c).

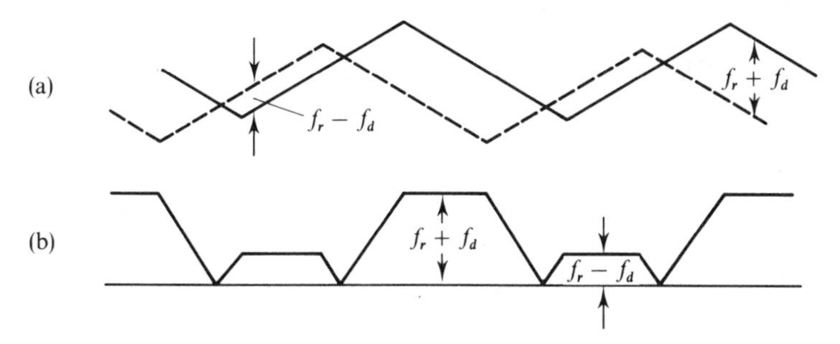

Figure 8.10 The signal in an FM-CW radar for a moving target: (a) transmitted, pecked line, received, full line; (b) frequency difference.

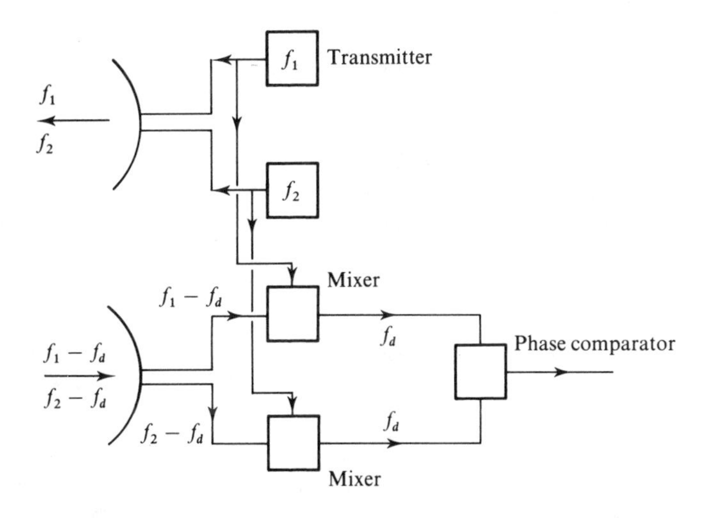

Figure 8.11 Configuration for a two frequency CW radar.

Measurement of beat frequency thus provides a measure of range. If the target is moving, a Doppler shift is imposed, and the new situation is shown in Figure 8.10. Measurement of the average beat frequency over the two halves of the cycle still provide range. The difference gives target radial velocity. The diagram shows the situation where $f_r > f_d$. If the opposite is the case, the roles of average and difference are reversed. Such schemes are used in radio altimeters carried in aircraft.

8.4.3 Two frequency CW radar

Range can also be determined if returns from two different frequency CW signals are available. Consider the situation illustrated in Figure 8.11, in which the target is moving, giving a Doppler shift f_d which, providing f_1 and

f_2 are nearly equal, will be virtually the same for both frequencies. At frequency f_1 we have

$$\textit{Transmit} \quad \sin \omega_1 t$$
$$\textit{Receive} \quad \sin [(\omega_1 - \omega_d)(t - T_r)]$$

Mixing these two signals yields

$$\sin [\omega_d t - (\omega_1 + \omega_d) T_r]$$

For the emission at f_2 a similar process yields

$$\sin [\omega_d t - (\omega_2 + \omega_d) T_r]$$

Comparison of these two signals yields the phase difference

$$\phi = (\omega_1 - \omega_2) T_r = 4\pi(f_1 - f_2) r/c$$

The maximum unambiguous range of ϕ is 2π, and thus the maximum unambiguous range is given by

$$r_m = c/2(f_1 - f_2)$$

For a frequency difference of 1 MHz

$$r_m = 150\,\text{m}$$

As described, a moving target has been assumed. The process has been used as the basis of distance measurement to a stationary target by arranging a transponder at the far end which re-emits with a frequency shift.

In some ways the use of two neighbouring frequencies may be likened to the action of a Vernier scale: two scales of slightly different pitches are juxtaposed.

8.4.4 Pulse radar

The basic ideas of a radar emitting pulses of single frequency have already been described in Section 8.2, and these ideas will be extended in this and subsequent sections.

CHOICE OF FREQUENCY, PULSE LENGTH AND REPETITION FREQUENCY

There is considerable interplay between these parameters. The pulse length is perhaps the simplest, because it directly determines resolution between multiple targets, as discussed in Section 8.3.1, although in Section 8.6 it will be shown how intra-pulse modulation enables long pulses to be used. Directional information about a target must come from the properties of the antenna: maximum illumination produces maximum return signal. The directional properties for a given physical size of aperture are enhanced as wavelength decreases, although the necessary tolerance on reflector surface accuracy becomes tighter at the same time. Thus a given beam width and hence angular discrimination is easier to achieve at higher frequencies. Another factor influencing choice of frequency is transmitter power required. In general, the power from available sources is greater at lower frequencies, and all power sources will have a maximum level of peak power. The mean power depends on repetition frequency:

$$P_{av} = P_t \tau f_p$$

where P_t is the peak power, τ the pulse duration and f_p the pulse repetition frequency.

The mean power is generally limited by problems of cooling, so that a fixed installation may be run at higher power than a mobile one. Whilst increase of mean power may be had by better cooling, maximum peak power is generally inherent to the class of source and cannot simply be increased. The pulse length too cannot be extended beyond certain values, determined by effects in the transmitting device. Thus the long pulses used in pulse compression systems (Section 8.6) are limited to about 100 μs. In electron beam devices, unwanted electronic modes may develop or temperature rise become excessive: in solid-state devices the rise in junction temperature during the pulse has to be contained.

The repetition frequency also determines the maximum unambiguous range: the maximum delay that can be accommodated without ambiguity is f_p^{-1}, so that the maximum unambiguous range is given by

$$r \leqslant c/2f_p$$

Thus a repetition frequency of 1 kHz limits the range to 150 km.

RADIATION PATTERN

As mentioned earlier, directional information depends on the antenna radiation pattern, and the required directional properties depend on the purpose of the radar. A surface-to-surface radar designed for detecting targets on land or on the sea need only have discrimination in azimuth, and a basic design of the type shown in Figure 3.22(b) might be appropriate. An airport radar, on the other hand, has to pick up aircraft at any height within its operational range. It also needs discrimination in azimuth but must illuminate the whole volume within its scope; further, a method of height finding may be

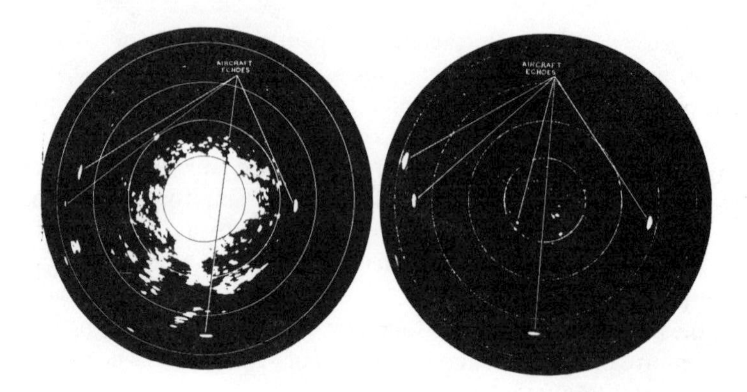

Figure 8.12 The plan position indicator (PPI) form of display: (a) a straightforward radar map, showing aircraft echoes and ground clutter; (b) same radar signal processed to show only moving targets, eliminating ground clutter.

an operational requirement. This can be done with the aid of a separate height finding radar with its radiation pattern narrow in elevation (the 'nodding' height finder), or as in current practice, by use of a multiple beam system with the receive beams stacked in elevation (Section 8.7.3). For a specific target, a pencil beam would give both elevation and azimuth, but the problem of acquiring the target is much greater than if all heights are covered simultaneously.

For any directional beam to cover a volume of space, it is necessary to vary the direction of the beam. In many systems this is done by rotation of the entire antenna, and in some cases by variation of elevation as well. This factor influences antenna design: the large antenna of a surveillance radar weighs a great deal and has to be rotated at a few rpm. In other cases when only limited coverage is required it may be that the antenna feed can be moved with respect to the reflector, giving change in direction of maximum gain while producing only minimal change in beam shape.

The overall radiation pattern is influenced also by presence of the Earth. At high frequencies the Earth behaves as a good conductor, with the result that the antenna pattern is considerably modified at low angles of elevation: it is difficult to propagate the beam at such low angles, with the result of poor radar coverage at low heights. (The effects of ground reflections on propagation will be considered further in Section 9.4.4.)

DISPLAYS

A number of visual displays have found application in radar systems. The simplest is the A-scan described in Section 8.2.4: it is useful where echoes from a fixed direction are to be observed, or when slow manual change of observation direction is adequate.

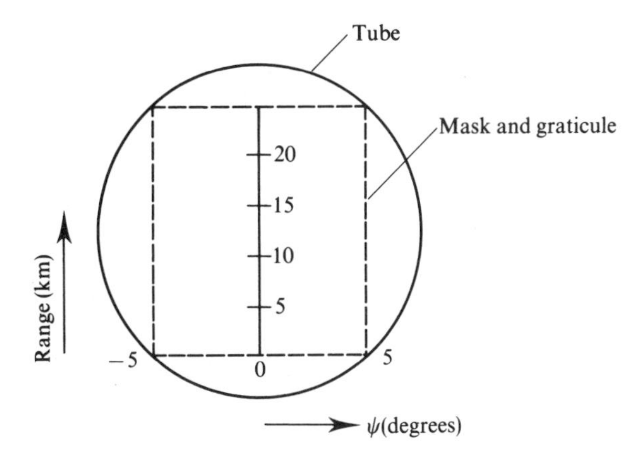

Figure 8.13 The *B*-scope display, range versus limited range of azimuth.

For 360° observation, as in an airport radar, a **plan-position indicator (PPI)** is used. In this display, the radar output is used to modulate the brightness of a cathode ray tube beam. The deflection system applies a normal time base sweep from centre to periphery in a direction which rotates in synchronism with the radar antenna. The sweep time corresponds to maximum range. The brightness is adjusted so that the tube is dark except where a returning pulse brightens the display. As the threshold is reduced, the noise output causes spots which are random in time and position, usually referred to as 'snow'. A target will only cause a bright spot over one small region of the PPI, whilst the antenna is pointing in its direction. Once the antenna has moved off, the indication will die. To provide continuity a long persistence phosphor is used so that visible evidence of a target remains until the next revolution of the antenna and consequent refreshment of the PPI. This display thus presents a map or plan of the surrounding region, with targets displayed according to direction and range. In a simple PPI, certain echoes will be fixed. These may come from local obstacles, or topographical features such as hills: these echoes are referred to as **clutter**. An example is shown in Figure 8.12.

In some cases a limited cover of azimuth is required, and a brightness modulated display called a B-scan is used. In this display one coordinate is related to antenna scan and gives azimuth, the other gives range. This is indicated in Figure 8.13.

This discussion has related to the display of analogue data. In an Automatic Detection and Tracking (ADT) system, the space represented on the PPI is divided up into resolution cells based on quantizing in azimuth and range. The return from each cell is examined by comparison with a threshold to determine whether or not a target is present. Such information can be portrayed on a synthetic display, to which can be added, using computer

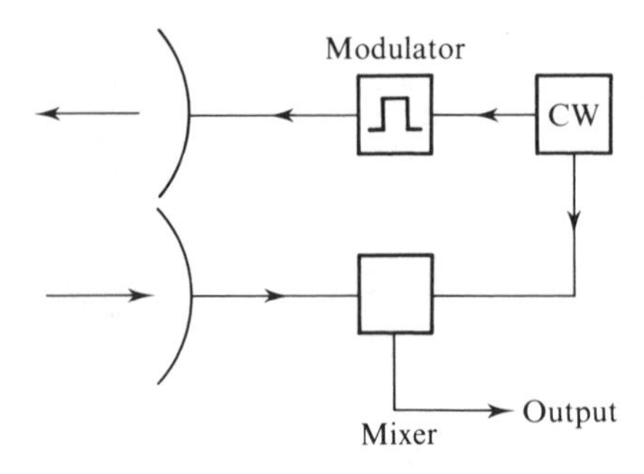

Figure 8.14 Circuit configuration for deriving Doppler shift with a pulse radar.

technology, data such as labels, height or speed as determined from previous returns. Such an arrangement has the added benefit that the useful information from the radar can be transmitted over circuits having much smaller bandwidth than the radar video channel.

8.4.5 Pulse Doppler radars

One of the problems arising in particular with ground based radar is the clutter return, which renders a moving target difficult to see in certain areas. The fact that targets of interest are moving allows use of the Doppler shift to separate them from clutter. It is often useful to know directly that a target is moving without having to plot its position on a PPI display. The phrase **Moving Target Indicator (MTI)** is also used to describe this type of radar, although the two terms have slightly different connotations. A configuration which can detect the Doppler shift in a pulsed emission is shown in Figure 8.14. The output from a continuous running, stable source is modulated to form the emitted pulse (usually after amplification) and also to act as the local oscillator in the mixer.

Consider a target which is receding with radial velocity v_r, so that the radial distance is given by

$$r = r_0 + v_r t$$

This target is illuminated by a pulse $s(t) \sin \omega t$, $s(t)$ representing the short pulse modulation. The return signal will be of the form

$$s(t - 2r/c) \sin \omega(t - 2v_r t/c - 2r_0/c) = s(t - 2r/c) \sin ((\omega - \omega_d)t - \phi_0) \quad (8.6)$$

The action of the mixer is to multiply this expression by the local oscillator input $\sin \omega t$ and to select the low frequency component of the resulting

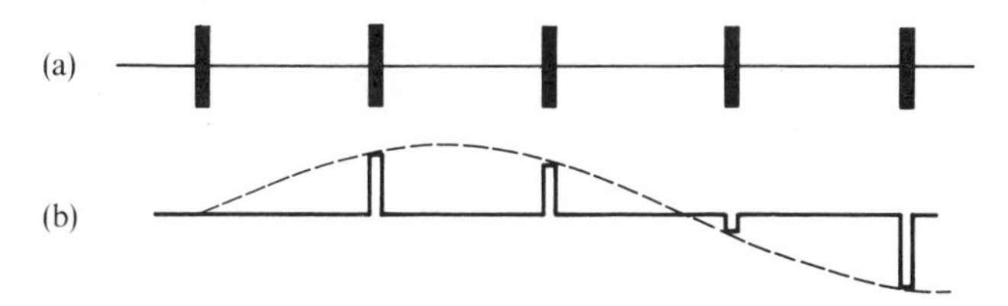

Figure 8.15 Form of signals in the circuit of Figure 8.14: (a) received RF signal; (b) output from mixer when Doppler shift is smaller than the pulse repetition frequency.

waveform, which can be shown to be

$$s(t - 2r/c)\cos(\omega_d t + \phi_0)$$

The pulse length is normally much less than the Doppler period, so that this output is effectively a sampled signal of the form

$$\cos(\omega_d(nT - T_r) + \phi_0) \tag{8.7}$$

in which $T_r = 2r/c$, T is the repetition period f_p^{-1}, and n an integer. This state of affairs is indicated in Figure 8.15: the presence of the moving target is indicated by variation of amplitude between successive pulses. If the Doppler shift is much higher, as might occur with reflections from satellites, for example, several cycles of Doppler frequency would appear within each pulse. From expression (8.7) it is clear that if $\omega_d T$ is an integer multiple of 2π, the output will not show the pulse-to-pulse variation and thus not indicate a moving target. In terms of velocity, these are termed **blind speeds** and are given by

$$v_r = nf_p\lambda/2 \qquad n = 1, 2, \ldots$$

in which λ is the wavelength of the transmission. Since the maximum unambiguous range is given by $c/2f_p$ (Section 8.4.4), there is a conflict between this range, r_m, and the lowest blind speed which is expressed by the equation

$$r_m v_r = c\lambda/4 = c^2/4f$$

This equation is displayed graphically in Figure 8.16 for various frequencies.

The value of the lowest blind speed may be considerably increased by using the technique of staggered pulse repetition frequency. Consider two radars operating with different pulse repetition frequencies, which will have different blind speeds. If the two frequencies are f_{p1}, f_{p2} a common blind speed will only occur when integers n_1, n_2 exist for which $n_1 f_{p1} = n_2 f_{p2}$. Its value will be given by

$$v_r = n_1 f_{p1}\lambda/2 = n_2 f_{p2}\lambda/2$$

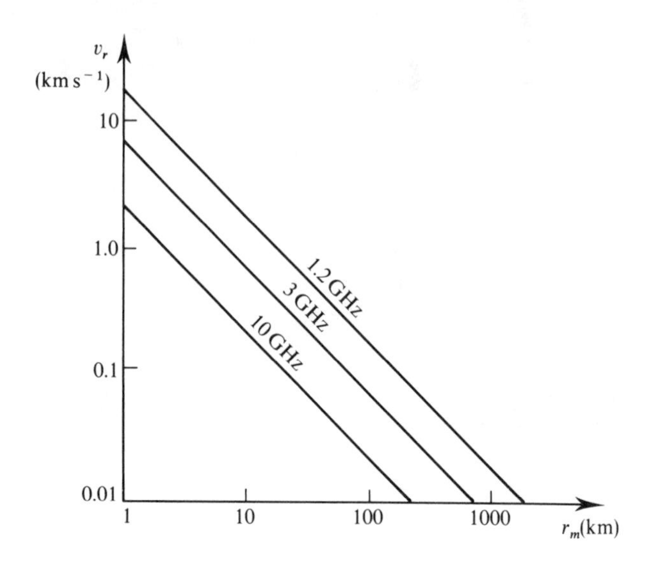

Figure 8.16 The relationship between lowest blind speed and maximum unambiguous range in a pulse Doppler radar for various carrier frequencies.

and can be considerably higher than the lowest value for either radar individually. The same effect can be obtained with a single radar by timing successive pulses according to the different repetition frequencies. Systems using more than two different frequencies are used.

There is also the possibility of **blind phase**. This can occur when $\omega_d T = \pi$ or $v_r = \lambda f_p/4$, half the value of the lowest blind speed. When this condition is satisfied, the expression of equation (8.7) becomes $(-1)^n \cos(\omega_d T_r - \phi_0)$, which may vanish for all n if the phase $\omega_d T_r - \phi_0$ happens to be an odd multiple of $\pi/2$. This situation can be remedied by forming in addition to the in-phase channel (I) represented by equation (8.7) a quadrature channel (Q) generated by mixing the return signal of equation (8.6) with $\cos \omega t$ rather than with $\sin \omega t$. This gives an output $\sin(\omega_d(nT - T_r) + \phi_0)$, which when $\omega_d T = \pi$ becomes $(-1)^n \sin(\omega_d T_r - \phi_0)$. The I and Q channels for the blind phase condition are shown in Figure 8.17, from which it can be seen that if both channels are combined an output is always obtained.

It is usually necessary to accept ambiguity in either range or blind speeds. An MTI radar normally accepts blind speeds; a pulse Doppler radar is one which uses a high enough pulse repetition frequency to avoid velocity ambiguity, but accepts range ambiguity.

The comparison of successive pulses has to be done by delaying a pulse for at least one repetition period, which may be several milliseconds. In analogue systems delays of this order of magnitude are achieved with ultrasonic delay lines, but in modern radars this comparison is usually

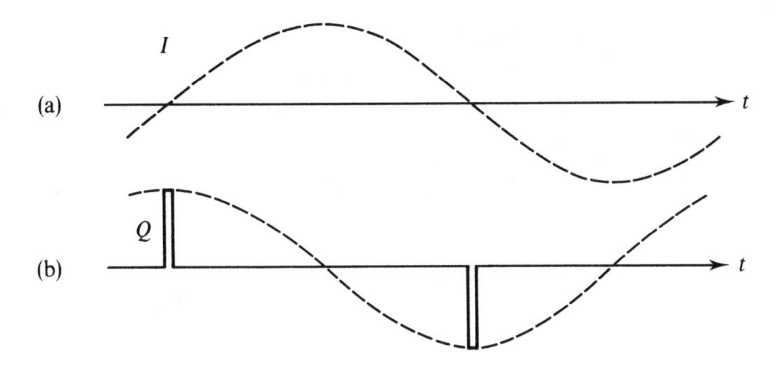

Figure 8.17 The blind phase condition can occur when $f_d = \frac{1}{2} f_p$: (a) output from the in-phase channel, I; (b) output from the quadrature channel, Q.

performed by digital signal processing: the mixer outputs are digitized and stored for the requisite period. The flexibility of digital methods makes much more sophisticated methods of processing possible, making schemes such as staggered pulse repetition frequency easier to implement.

Comparison of successive pulses rejects clutter returns with zero frequency shift, but in practice clutter does have a finite frequency spectrum arising, for example, from finite illumination time or movement of leaves and trees. The use of more complicated filters will reduce this return also.

RANGE GATING

It was mentioned in Section 8.4.4 when discussing displays that in an automatic system, range intervals were defined to form resolution cells. A way of doing this is indicated in Figure 8.18: the returns from different range intervals are separated by passing the return signal to a bank of gates in parallel, with each gate being opened for the interval of time corresponding to a particular range cell. The length of a cell is usually that of one pulse.

FREQUENCY AGILITY

Frequency agility is the ability of some radar systems to change the frequency of transmission rapidly, perhaps on a pulse-to-pulse basis. As has been mentioned earlier (Section 8.2.2), the scattering cross-section of a target is dependent on frequency: therefore a change in frequency will reduce the chance of a target being missed because the scattering cross-section corresponding to a particular aspect and frequency is low. A further advantage for military radars is that frequent change of frequency makes narrow-band jamming more difficult.

It is convenient to mention here the use of frequency synthesizer techniques (see Section 11.2.1) in this respect. These enable all the various

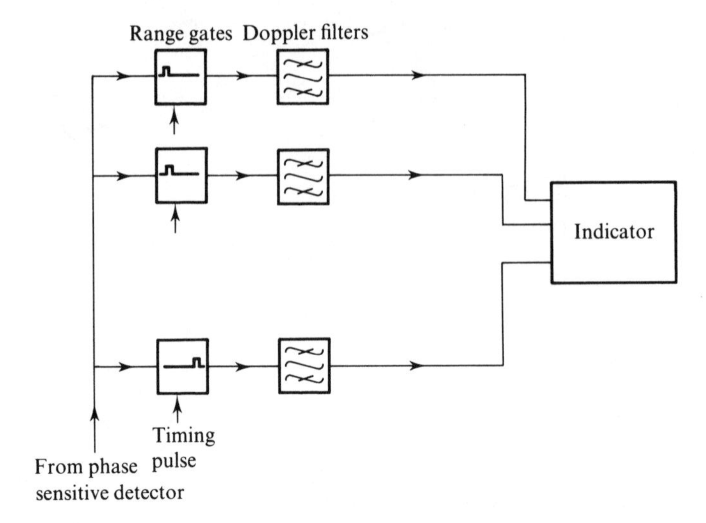

Figure 8.18 The use of range gates in a moving target indicator radar. The filters are the same in each channel, encompassing the maximum Doppler shift expected.

frequencies used in the radar (e.g. RF, local oscillator, pulse repetition) to be firmly related to a single, stable oscillator. Furthermore, the selection of the various frequencies can be under computer control, providing flexibility and switching in the order of microseconds.

8.4.6 Tracking radar

A particular target may be tracked using position coordinates determined, for example, by a surveillance radar; the use of data from successive scans allows some prediction of its course to be made. However, it is for some purposes necessary to track continuously. Automatic tracking can be obtained if a signal indicating the sign and magnitude of the pointing error can be derived. One method by which this can be done is the **monopulse** technique, illustrated in Figure 8.19. The antenna is equipped with two feeds which give radiation patterns whose peaks lie in slightly different directions, the patterns overlapping at about the half-power points as shown in Figure 8.19(a). Combining the two received signals in, say, a 'magic-T' hybrid junction, the equivalent of sum and difference patterns can be obtained, as shown in Figure 8.19 (c) and (d). The 'sum' and 'difference' signals are converted to IF, amplified and combined in a phase-sensitive detector to give an output whose magnitude and sign are related to the pointing error. This can then be used to control the antenna via a servomechanism until the angle error is reduced to zero. Thus the position of the target is obtained as a function of time, and the plot can be used to predict its course. This principle can be extended to the problem of tracking in both planes.

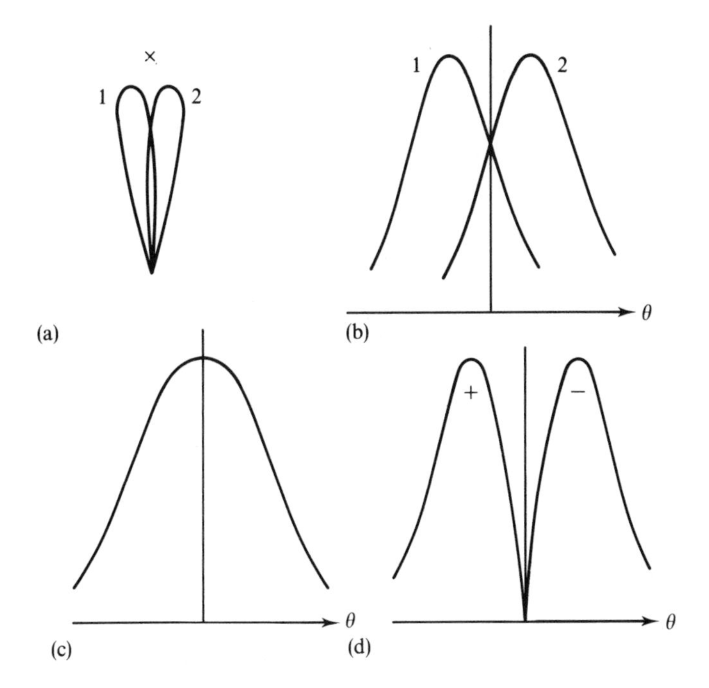

Figure 8.19 Beam shapes in a monopulse tracking radar: (a) beams in relation to target, polar form; (b) Cartesian plot; (c) sum pattern; (d) difference pattern.

Other methods of automatic tracking are available. For example, a single pencil beam with its axis conically scanned about the presumed target direction would provide information for tracking.

8.5 Duplexing

The problem of leakage of transmitter power into the receiver has been briefly mentioned earlier (Sections 8.2.4 and 8.4.1). It is most severe in a high power radar using the same antenna for both transmission and reception: leakage of transmitter power into the first stages of the receiver will damage semi-conductor components unless kept to a very low level. The protection circuits traditionally used involve gas filled cells, known as **TR cells** (transmit–receive). On the application of high power, the gas ionizes, producing a virtual short circuit. Two duplexer circuits are shown in Figure 8.20 (a) and (b). In the former a branch-line system is shown: breakdown of the TR and ATR (anti-TR) cells results in a through connection from transmitter to antenna; when the cells represent open circuits a combination of quarter-wave transformers opens the path to the receiver and isolates the transmitter, preventing loss of returned signal power.

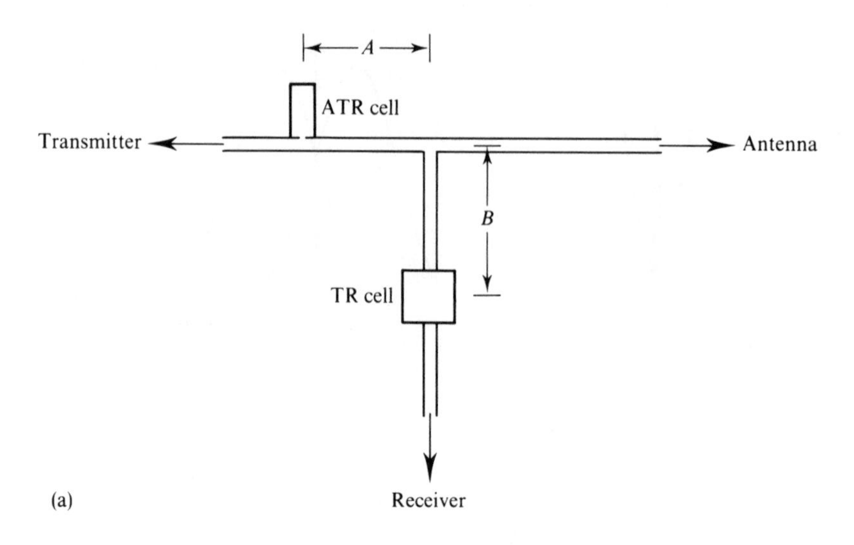

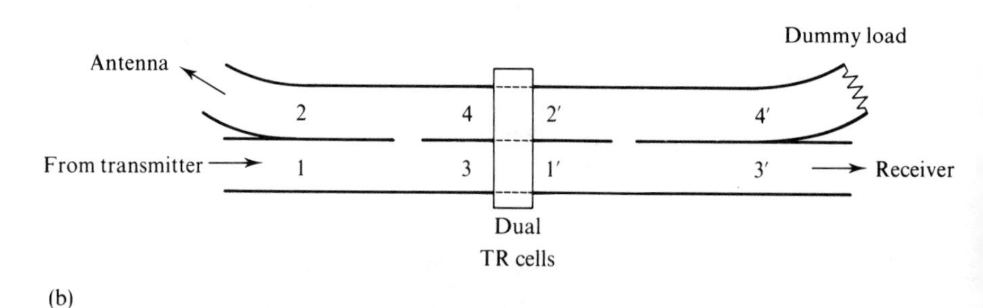

Antenna

Transmitter

Receiver

(c)

Matched load

Figure 8.20 Duplexing circuits. (a) Branch line form using series junction: A is nominally a multiple of $\lambda/4$, B a nominal multiple of $\lambda/2$. In transmission the receiver arm presents a short-circuit at the junction; in reception the transmitter arm presents a short. (b) Balanced duplexer. The waveguides are in form of 90° hybrid couplers, numbered corresponding to equation (5.5). (c) A 4-port circulator used as a duplexer. The fourth arm increases isolation between receiver and transmitter.

In Figure 8.20(b) a balanced duplexer using two 3 dB hybrid waveguide couplers is shown. (The ports are numbered corresponding to the scattering matrix of equation (5.5).) When the dual TR cells provide short circuits, transmitter power is directed to the antenna; any leakage through the TR cells is directed to the dummy load. When the TR cells are transparent, power from the antenna is directed to the receiver. Such balanced duplexers can operate at higher power and over wider bandwidth than the branch-line type.

At lower powers a circulator may be used as a duplexer, as shown in Figure 8.20(c). If the effect of reflections from the receiver on the transmitter are unimportant, a 3-port circulator can be used.

8.5.1 TR cells

The construction of typical TR and ATR cells is shown in Figure 8.21. In the TR cell the gaps and associated irises are resonant structures designed to pass the relevant frequency range; the primer electrode is supplied with DC, and maintains a low level discharge to assist in rapid ionization when a transmitter pulse arrives. The discharge starts off at the primed gap and then also at the other gap, and at high power to the neighbourhood of the window, creating a virtual short across the guide. The ATR cell has no resonant structure. It is designed so that at low power a field maximum appears across the aperture in the waveguide wall, presenting a low admittance to the incoming signal and consequent reflection. On ionization, the aperture becomes conducting and forms part of the waveguide wall, allowing transmitter power to pass. The cells used in the balanced duplexer of Figure 8.20(a) are similar to the TR cells except that they have gaps but no resonant structures and incorporate the (radioactive) gas tritium in the filling to promote early ionization in place of a priming electrode.

All duplexing arrangements allow some leakage of RF power, which can damage the receiver. As ionization builds up, a spike of energy leaks through; after ionization a low level breakthrough exists for the duration of the transmitter pulse. Further protection for the receiver can be given by a 'protector' tube between duplexer and receiver. This configuration also gives protection against pulses picked up from neighbouring high power radars.

The build up of ionization and its effect on leakage have been mentioned, but recovery time is also important: it determines the minimum range at which the radar can function. Small losses are incurred with all duplexing arrangements, usually of the order of 1 dB.

8.5.2 Solid-state limiters

The action of a TR cell is essentially that of a rapid action switch. Although gas cells are used for the highest powers, *p–i–n* diodes can be used up to some 300 kW peak, 5 kW mean. The diode is switched from high impedance to low

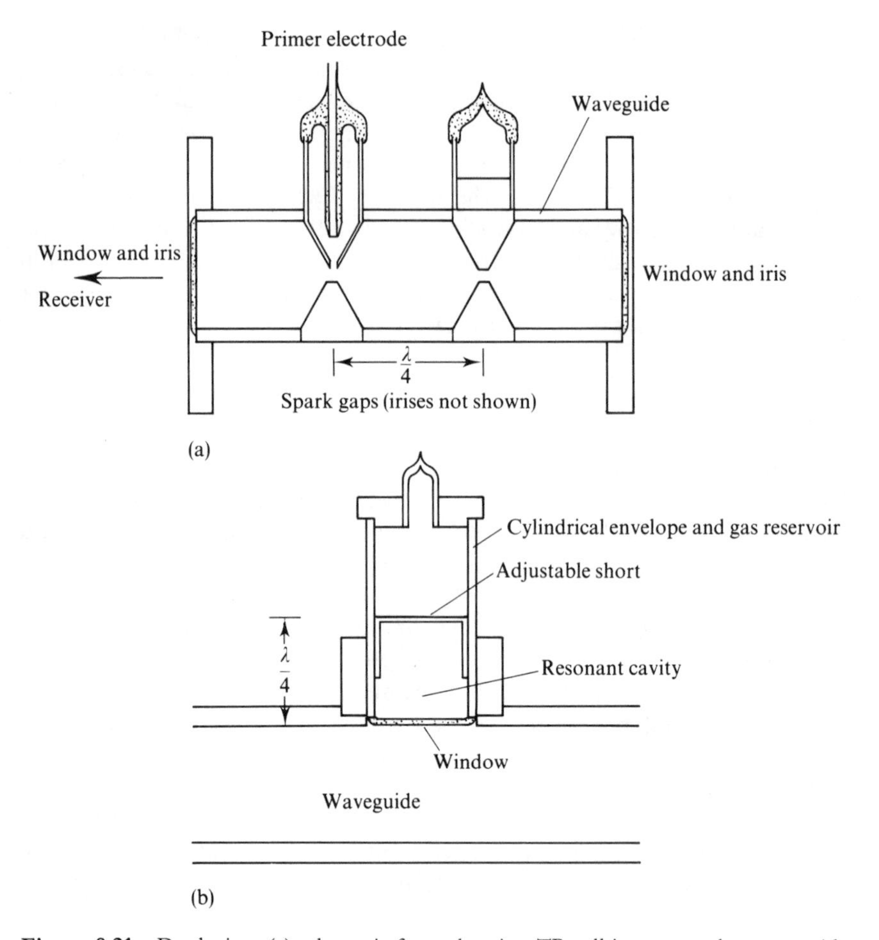

Figure 8.21 Duplexing: (a) schematic form showing TR cell in rectangular waveguide, narrow dimension in plane of diagram; (b) ATR cells for rectangular waveguide.

by the passage of a direct current: this is provided by a pulse in synchronism with the transmitter pulse, starting slightly earlier to reduce the spike energy. Because this switch is not self-activating, it does not protect against stray pulses. A self-acting limiter can be provided by a varactor diode: by using the diode in a self-biasing circuit, the rectified current produced when RF power is applied alters its capacitance and brings about a reflection. Varactor diodes can handle about 10 W peak power.

8.6 Pulse compression

It has been mentioned earlier that peak transmitter power is limited by the type of power source whereas mean power is rather more dependent on heat

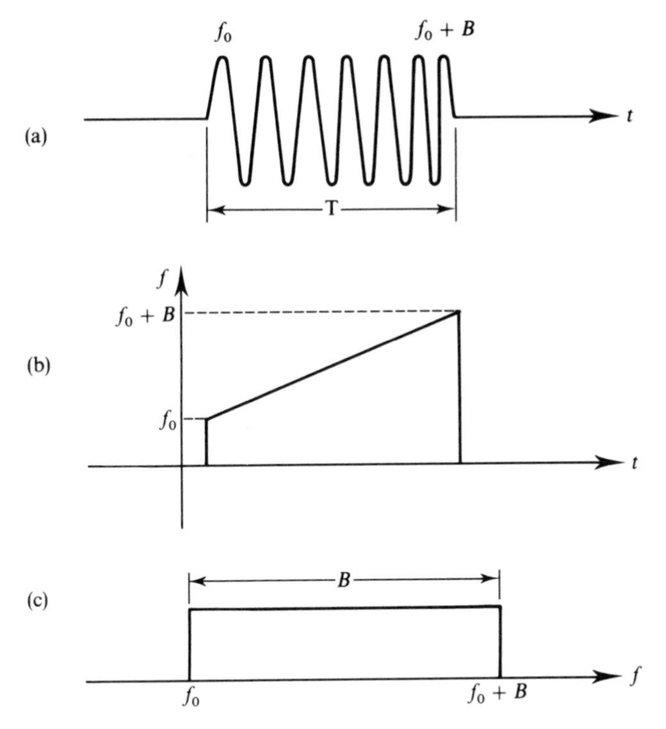

Figure 8.22 Linear modulation on pulse: (a) waveform; (b) variation of frequency during pulse; (c) approximate form of spectral density when $BT \gg 1$.

extraction. Pulse compression is a technique which allows the transmitter tube to be used at peak power for a longer pulse whilst retaining short pulse range accuracy. The simplest scheme is **Linear Frequency Modulation On Pulse (LFMOP)**, for which the pulse waveform is shown in Figure 8.22 (a) and (b). This is loosely referred to as a 'chirp' waveform. In a simplistic way this waveform may be regarded as a sequence of pulses with that of frequency f being transmitted at a time $(f - f_0)T/B$ after the beginning of the wave train, so that delay occurs in transmitting the high frequencies. Provided BT is much greater than unity the waveform would be expected to have significant spectral components only in the range $f_0 < f < f_0 + B$. On reception the waveform is passed through a pulse compression filter which compensates for this delay on the higher frequencies, so that all the components of the waveform arrive in phase at the output. The form of transfer function the filter must have may be seen as follows. If the phase of the transfer function is denoted by $\phi(\omega)$, the group delay is given by $-\mathrm{d}\phi/\mathrm{d}\omega$, and the required compensation is achieved when

$$\frac{\mathrm{d}\phi}{\mathrm{d}\omega} = (f - f_0)\frac{T}{B}$$

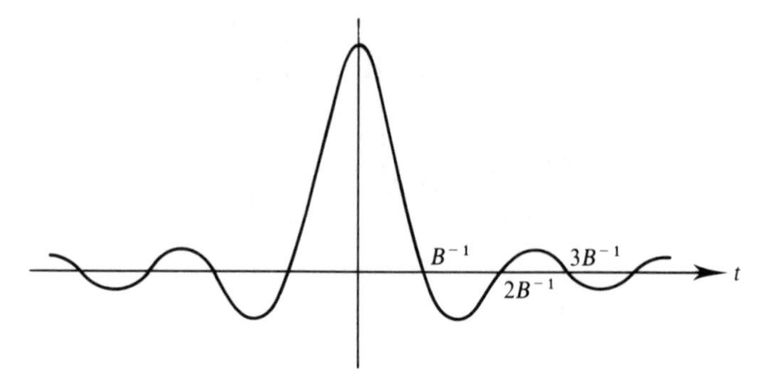

Figure 8.23 Idealized form of output given by chirp waveform passing through a matched filter.

This gives

$$\phi = \pi(f - f_0)^2 \frac{T}{B}$$

The bandwidth of the signal occupies only the range $f_0 < f < f_0 + B$, so that the filter must have this phase characteristic only over a restricted frequency range, a characteristic which can be approximately realized. Analysis shows that the output from such a filter is typically of the form

$$(TB)^{\frac{1}{2}} \frac{\sin \pi Bt}{\pi Bt}$$

This is illustrated in Figure 8.23. The compressed pulse is of the order of B^{-1} in duration, with an effective power gain of BT. This latter product is termed the **pulse compression ratio**.

Although the theory just outlined provides a ready interpretation of the process of pulse compression, it is in fact more satisfactory to regard the 'chirp' pulse as having been generated by the passage of a short pulse through an appropriate filter. Pulse compression is then achieved when the lengthened pulse is passed through a 'matched' filter. This approach will be considered briefly in Section 8.6.1, and in greater detail in Section A.10.

The problems are

(1) generating the chirp waveform,

(2) designing the pulse compression filter,

(3) the subsidiary maxima in the pulse output (Figure 8.23) which indicate false targets.

The chirp waveform is usually formed at low power level at a suitable frequency and subsequently shifted in frequency and amplified for transmission. A voltage controlled oscillator might be used, or the waveform might be constructed by a frequency synthesizer using digital techniques. The waveform may also be produced as the response of a suitable filter to a short pulse, as mentioned above. Filters are now frequently made using surface-acoustic-wave (SAW) devices. Acoustic waves can be generated in suitable materials by the piezo-electric effect, and a very wide range of characteristics can be produced by shaping and disposition of the launching and receiving electrodes on the surface.

The unwanted peaks (usually termed **range side-lobes**, by analogy with antennas) may be reduced by suitable design of filter characteristic, at the expense of somewhat broadening the compressed pulse and slightly reducing the output power relative to noise. The maximum pulse length that can be used is limited by the transmitter characteristics, and does not normally exceed 100 μs. Pulse compression ratios of 1000:1 are attainable.

In a general sense, the chirp waveform distinguishes between the subintervals in a long pulse, so it will provide accurate range information. Thus any coding which breaks down the pulse length into assignable intervals can be used, with the appropriate filter.

8.6.1 Matched filters

The general theory of pulse compression is related to that of matched filters used to optimize signal-to-noise ratio. We may regard any signal $s(t)$ as the impulse response of some filter of transfer characteristic $S(f)$, related by the Fourier Transform pair of Section 8.3.4. The matched filter for this signal has a transfer characteristic $S^*(f)$. The impulse response of the matched filter is $s(-t)$, and since $s(t)$ is itself an impulse response, it follows that the matched filter is strictly unrealizable. However, if a delay T is permitted, approximate realization is possible. It can then be shown that the output is maximum at $t = T$, and at this instant the signal-to-noise ratio is equal to

$$2 \times \frac{\textit{signal energy}}{\textit{noise energy in 1 Hz}}$$

It can also be shown that the instantaneous output from the matched filter can be expressed as the convolution integral

$$v(t) = \int_{-\infty}^{\infty} s(u)s(T - t + u)\,\mathrm{d}u$$

The application of this to practical systems depends on either analogue realization of the matched filter, or on digital evaluation of the convolution integral. The theory is considered further in Section A.10.

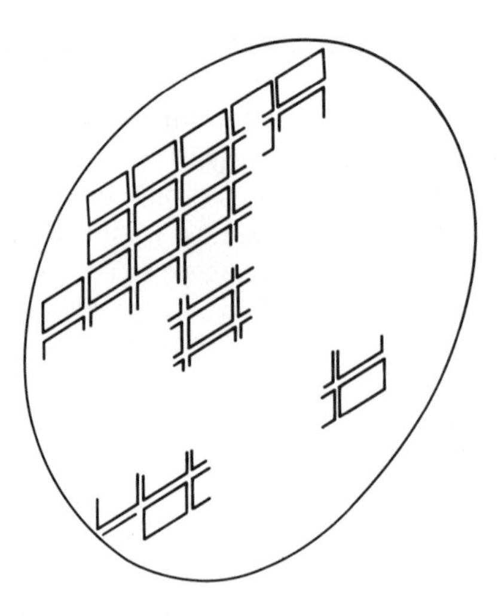

Figure 8.24 Phased array antenna.

8.7 Phased-array radars

The discussion in earlier sections has concentrated largely on systems using reflector antennas. These are in some ways inflexible: the radiation pattern is predetermined with respect to the reflector, and beam scanning has to be achieved by mechanical movement which can only take place at a strictly limited rate. In principle, the reflector can be replaced by an assembly of individual sources separately controlled in amplitude and phase. (Linear arrays were considered in Section 3.10.4.) However, to replace a reflector antenna might require of the order of 10^4 elements. Such arrays would be expensive not only because of the number of elements but because of the complexity of the signal processing requirements. The advantages are in the flexibility provided by electronic control of beam direction or of the radiation pattern: for example, the same array can be programmed to give different radiation patterns in transmission and reception. An advantage of the large number of elements is that failures will produce a gradual deterioration rather than complete failure. Part of such an array is represented in Figure 8.24. The size and spacing of the individual radiators are each of the order of one-half wavelength: they might be open ended waveguides, small waveguide horns or, in a coaxial line system, dipoles. In order to set up a particular radiation pattern, the amplitude and phase of the excitation to each element has first to be calculated and then established. To ensure correct phasing, all the individual excitations must be referred to the same source, either by passive power splitting or by using a collection of amplifiers fed from a common source. A configuration for a drive module to implement such a scheme is shown in Figure 8.25. In this, a common, low level signal is

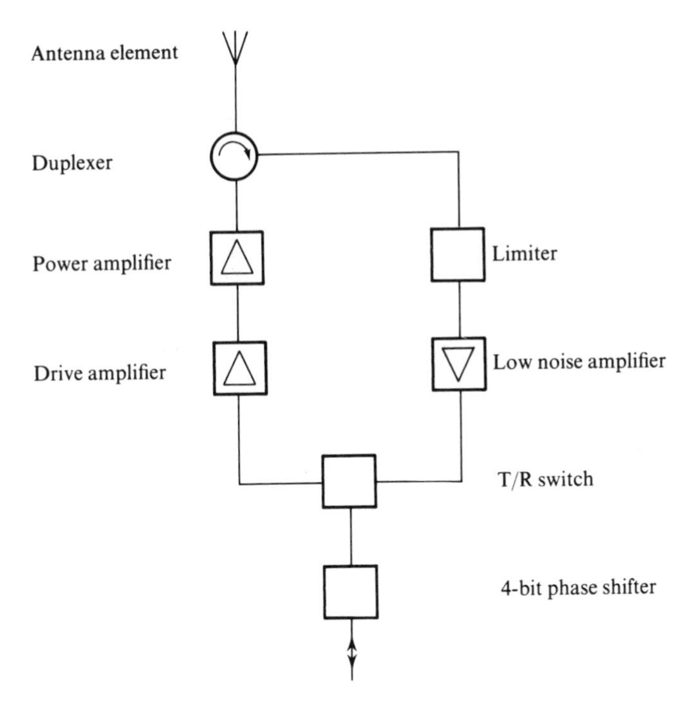

Figure 8.25 Possible form of circuit element for a phased array radar.

amplified for transmission, providing coherence between elements. Each element contains a duplexer and a switch, with a limiter to protect the low noise amplifier. The RF amplifier would be solid-state, using Si or GaAs transistors as dictated by frequency. Peak power could be between 5 W and some tens of watts. The whole element would be solid-state and compact enough to be formed directly into the array. The calculation of the required element amplitudes and phases and the control signals implementing them would require a dedicated special purpose computer using parallel computation, as would the signal processing on reception.

Although a planar array can be flexible in radiation pattern, a single planar array cannot provide a pencil beam to cover the complete hemisphere in front of it: the beam will degrade as its axis approaches the plane of the array. A cone of semi-angle 45° represents a practical limit. Such arrays can be mechanically scanned to provide wide angle coverage, or several arrays can be used.

It is evident from these considerations that an independent element phased array contains much hardware in the form of phase shifters and switches as well as in power distribution. Any general move to the phased array and exploitation of the potential advantages depends on the development of compact, mass-produced elements.

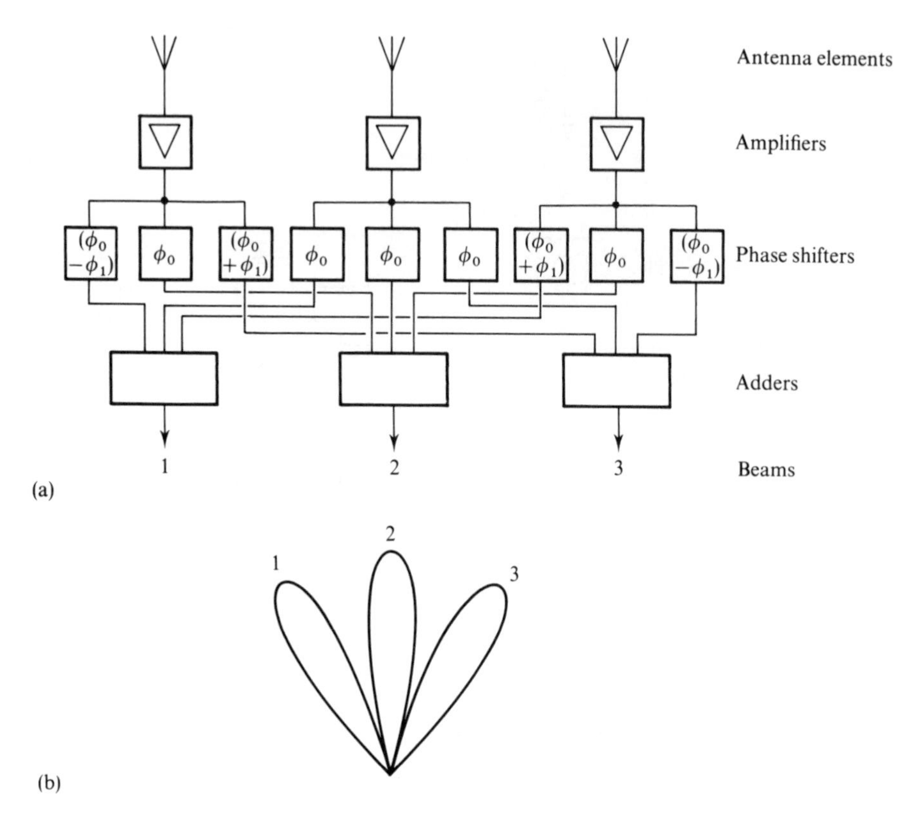

Figure 8.26 The use of a number of elements to produce a multiple beam antenna: (a) circuit configuration; (b) radiation pattern.

8.7.1 Multi-beam reception

The idea of producing more than one beam was referred to in the previous section. A simple example is shown in Figure 8.26: by altering the phases in which the outputs from the three separate antennas are combined, three distinct beams are formed in reception. More complicated beam forming networks allow a larger number to be produced.

8.7.2 Frequency scanning antennas

One form of phased array is derived from arrays such as that depicted in Figure 3.21. The geometrical spacing between elements is fixed but the electrical spacing depends on frequency and on the dispersive properties of the waveguide, so that beam swinging may be accomplished by changing frequency. A single linear array of this type would produce a fan beam with its narrow dimension in a plane containing the waveguide axis. A stack of such arrays could produce directivity in the perpendicular direction and thus a pencil beam.

8.7.3 Synthetic aperture radar (SAR)

The principle of the phased array discussed above concerns the combination of simultaneous coherent pulses from a large number of elements, giving the effect of a single large antenna. There is, in principle, no reason why the transmission and reception for each element should not occur sequentially provided pulse-to-pulse coherence is maintained and return pulses stored, keeping phase information. This provides the basis of the synthetic aperture radar. Consider a radar carried in straight level flight: pulses will be emitted at equally spaced intervals from the radar antenna, providing target returns from the ground. Storage and recombination of the return pulses will simulate an antenna with a much longer dimension in the direction of travel than that of the radar antenna itself. With this technique resolution in azimuth of the same order as that in range can be obtained. The heart of SAR is in the signal processing and maintenance of coherence in transmission, as well as allowing for deviations from straight level flight.

8.8 Detection statistics

It has been pointed out that the detection of a target has necessarily associated with it a probability, because of the presence of noise. A theory cannot be worked out unless the method of demodulation and detection process is specified. For automatic detection, a threshold method is usually used, whereby signals exceeding a specified level are registered as targets. As pointed out in Section 8.2.4, noise can produce false alarms and the false alarm rate, or time between alarms, is an important consideration in setting the threshold. It will be assumed in the following presentation that the demodulator is an 'envelope detector', and that the IF noise is Gaussian of bandwidth B.

8.8.1 False alarm rate

The probability of Gaussian noise having magnitude between v and $v + \mathrm{d}v$ is given by

$$p_1(v)\,\mathrm{d}v = (2\pi\sigma)^{-\frac{1}{2}} \exp(-v^2/2\sigma)\,\mathrm{d}v$$

It may be verified using standard integrals that σ is the mean square voltage, defined as

$$\overline{v^2} = \int_{-\infty}^{\infty} v^2 p(v)\,\mathrm{d}v$$

Passage of this noise signal through an envelope detector may be shown to give a probability of the video noise voltage being between V and $V + \mathrm{d}V$ of

$$p_2(V)\,\mathrm{d}V = V\sigma^{-1}\exp(-V^2/2\sigma)\,\mathrm{d}V$$

The probability of V exceeding a threshold V_T and thus creating a false alarm is given by

$$P_a = \int_{V_T}^{\infty} p_2(V)\,\mathrm{d}V = \exp(-V_T^2/2\sigma)$$

During the interval between false alarms, T_a, only a single noise pulse exceeds threshold. The duration of this pulse is of the order of B^{-1}, so that an alternative way of expressing P_a is

$$P_a \cong B^{-1}/T_a$$

Hence

$$T_a \cong B^{-1}\exp(V_T^2/2\sigma)$$

or

$$V_T^2 = 2\sigma\ln(BT_a)$$

If a value for T_a of 1000 s is acceptable, with a bandwidth of 1 MHz, the ratio $V_T^2/2\sigma$ is found to be 20.7, or, since the mean square video noise may be shown to be 2σ, the ratio of threshold to rms video is 4.5. The probability of false alarm for this threshold is 10^{-9}.

8.8.2 Probability of detection

In the presence of a sine wave carrier of amplitude A at the centre of the IF bands, the probability of the video voltage lying between V and $\mathrm{d}V$ may be shown to be

$$p_3(V)\,\mathrm{d}V = V\sigma^{-1}I_0(AV/\sigma)\exp((-V^2 + A^2)/2\sigma)\,\mathrm{d}V$$

in which I_0 is the modified Bessel function of the first kind. The probability of

V exceeding V_T is then given by

$$P_s = \int_{V_T}^{\infty} p_3(V)\,\mathrm{d}V \tag{8.8}$$

The significance of equation (8.8) is illustrated in Figure 8.27 for several values of A^2/σ. Each figure shows the function $p_2(V)$, together with $p_3(V)$ for a

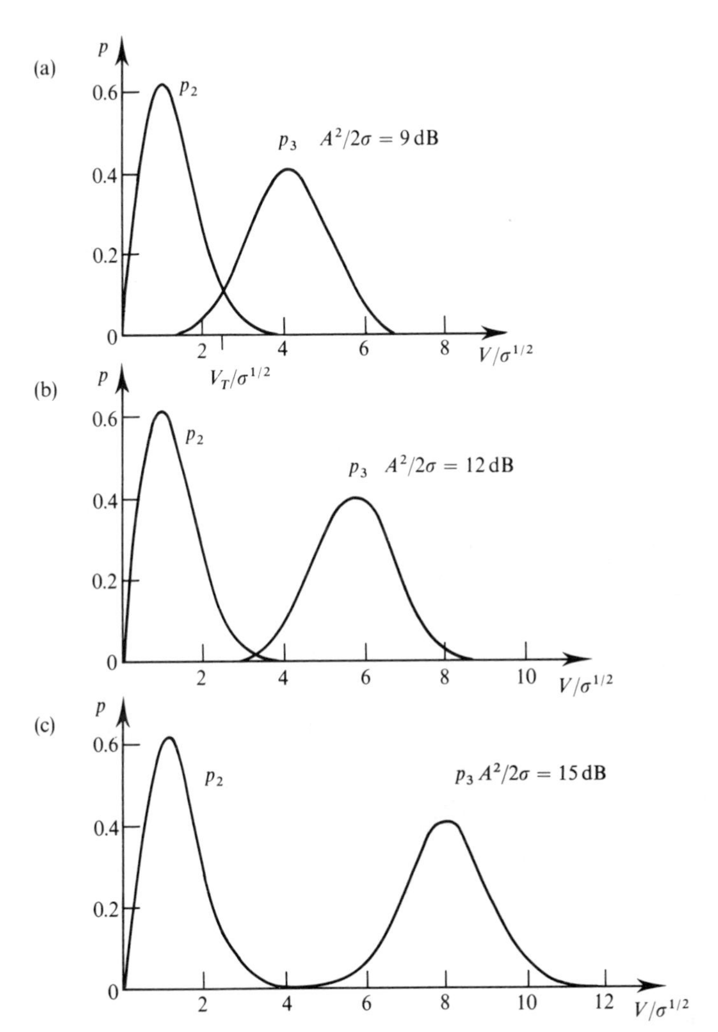

Figure 8.27 Probability distribution of the video signal output from an envelope detector. In all graphs $p_2(v)$ is the probability density for pure noise, and $p_3(v)$ the density in the presence of a sinusoidal carrier of peak amplitude A. The signal-to-noise ratio $A^2/2\sigma$ has the values (a) 9 dB (b) 12 dB and (c) 15 dB. The threshold indicated in (a) corresponds to $V_T/\sigma^{\frac{1}{2}} = 2.5$.

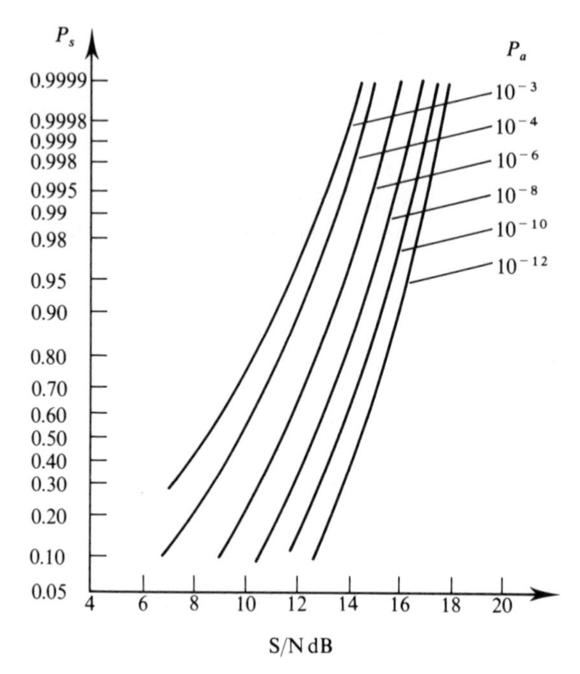

Figure 8.28 Graphs showing variation of probability of detection as a function of signal-to-noise ratio for various false alarm probabilities.

specific value of the IF signal-to-noise ratio, $A^2/2\sigma$. The area under any curve is unity. The effect of the threshold V_T is indicated in Figure 8.27(a) for $V_T/\sigma^{\frac{1}{2}} = 2.5$. The probability of detection is given by the area under $p_3(V)$ to the right of the line $V_T/\sigma^{\frac{1}{2}} = 2.5$, the probability of false alarm similarly by the area under $p_2(V)$ to the right of this line.

From this illustration it will be seen that the threshold affects both probability of false alarm and probability of detection. For the example given above, probability of false alarm 10^{-9}, the ratio $V_T/\sigma^{\frac{1}{2}}$ is 6.4: the effect on probability of detection can be realized with the aid of the three graphs. For any given false alarm probability there is a defined relationship between signal-to-noise ratio and probability of detection, given by equation (8.8). This integral cannot be simply evaluated, but results are presented graphically in Figure 8.28.

8.8.3 Integration

If n pulses from the same constant target are added coherently, the signal amplitude increases by a factor equal to n. Random noise, which adds on a mean power basis, will increase in amplitude by $n^{\frac{1}{2}}$. Thus the signal-to-noise power ratio will increase by a factor of n. (In practice target fluctuations will have to be considered.) This process is known as **integration**. A number of pulses are available in the return from a target even in a scanning radar, because of the finite beam width, so that integration can be used to improve performance. This integration process can be carried out in the IF or after demodulation.

For integration, some form of storage is needed, which may be analogue or digital. A human operator viewing a long-persistence cathode ray tube can approach ideal post-demodulation improvement under good conditions.

8.9 Practical systems

Pulsed radar systems are used in such variety as to render classification difficult. The principles remain the same, but power, frequency and pulse length, to say nothing of reflector size, represent compromises between purpose and type of installation. Considerations such as whether the installation is mobile or fixed, is land based or airborne, is permissible in size and weight, the state of the art in component design, and available budget, all enter into the final system design. Very large, low frequency systems have been used for planetary investigations; very high frequency, small systems are carried in missiles. A large class of radar is associated with air traffic control, and a brief consideration will be given to these. Long range surveillance radars tend to operate in D-band, using pulse compression techniques giving compressed pulses of the order of a few microseconds. Such radars would use a travelling wave tube or a klystron as a transmitter at power levels up to a few megawatts. Large D-band antennas have been made with extreme dimensions as much as 20 m horizontally and 6 m vertically, giving azimuthal resolution better than 1°. Such an antenna might rotate at 5 rpm.

Airport approach radars are more likely to operate at E/F-band, using a shorter pulse, and to have shorter range and better range resolution. Such a radar, the Marconi Radar S512, will be described in some detail in Section 8.10.

As the frequency increases reflectors can become smaller, making J-band attractive for airborne equipment or for small boat radars.

8.10 An air traffic control radar

Radar is widely used for air traffic control and this section describes the main microwave features of a typical airport surveillance radar, based on a simplified description of the S512 system manufactured by Marconi Radar Systems. An outline specification appears in Table 8.1.

Plate 2 The Marconi S512 radar. (a) The antenna, comprising the precision reflector and dual feed-horn. The waveguide behind each horn contains a linear/circular polarizer. (b) Picture typical of the medium power, 3 GHz, amplifier used for feeding the travelling wave tube. An input transistor (on the left) feeds four output transistors, the outputs of which are combined to give an output of 250 W peak power. The Wilkinson couplers (Section 5.6.1, Figure 5.21(a)) used for splitting and combination are visible. (*Photographs courtesy of Marconi Radar Ltd.*)

(a)

(b)

Table 8.1 Outline specification of radar for airport surveillance.

Frequency		3 GHz
Antenna	Gain	34 dB
	Horizontal beamwidth	1.5°
	Vertical pattern	modified cosec^2
	Rotation speed	15 rpm
Transmitter	Peak power output	160 kW
	Mean power output	2 kW
	Pulse width (transmitted)	12 μs
	Pulse width (compressed)	0.6 μs
	Pulse repetition frequency	650–1000 Hz
Receiver	Noise figure	2 dB

8.10.1 Antenna

The antenna is made from metallic coated moulded carbon fibre composite, to reduce weight and to improve the finish of the reflecting surface. The overall dimensions of the reflector are approximately 5 m horizontally and 3 m vertically. The rotating mass is about 500 kg. The 'modified cosec^2' pattern referred to in the specification is often used in air traffic radars in order to make the echo strength independent of the separation of target and radar. This is possible because aircraft are normally limited in flight altitude, and it allows the overall efficiency of the antenna to be optimized. Over a restricted range of the elevation angle $\theta_m > \theta > \theta_0$ the pattern approximates to the function

$$G(\theta) = G(\theta_0)\left(\frac{\operatorname{cosec}\theta}{\operatorname{cosec}\theta_0}\right)^2$$

To obtain such a pattern suitable reflector shaping must be used, taking into account the need for low side lobes in both vertical and horizontal radiation patterns. The polarization of the radiated beam can be either circular or linear vertical. Circular polarization can be used to reduce echoes from rain clouds: the reflection from spherical raindrops is of the opposite sense of polarization, and can be filtered out by the antenna. This component will enhance rain echoes, and suitable signal processing will provide a weather map to aid the air traffic controller. It is of interest to note that the gain of a uniformly illuminated aperture of the same area as the actual reflecting surface, about 12 m^2, would at 3 GHz be 42 dB; design for shaping to the required pattern and low side lobes incurs a loss in peak gain of some 8 dB.

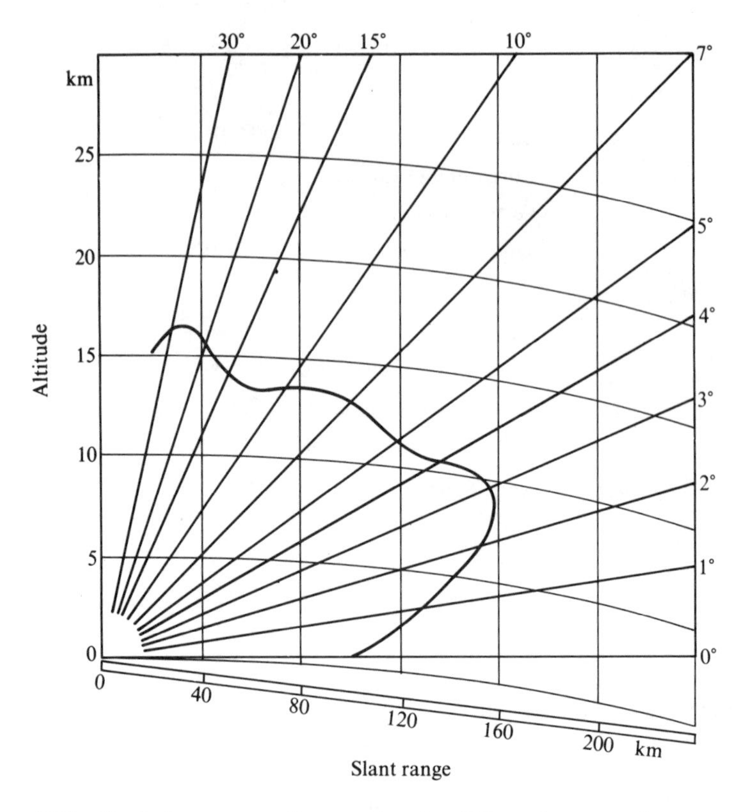

Figure 8.29 Vertical coverage diagram for the S512 radar, assuming a target of scattering cross-section 2 m^2.

The shape of the vertical pattern is indicated by the coverage diagram of Figure 8.29.

8.10.2 Transmitter

This radar uses pulse compression (Section 8.6), and the final power stage in the transmitter is a travelling wave tube (Section 4.3). At such high peak powers the tube has to be switched on for a time only slightly longer than the duration of the RF input. This involves applying a negative pulse of 100 kV, duration about 14 μs, at several amperes current to the cathode of the tube. A basic circuit for producing such a pulse is shown in Figure 8.30. It uses a pulse-forming network (in essence an artificial delay line) and electronic switch. In the design under discussion field-effect transistors are used as the switch in a circuit producing pulses of 600 V magnitude. To obtain the required current at 100 kV, the outputs from 40 identical units, each with its own pulse-forming network and FET switch, are combined in a multi-input

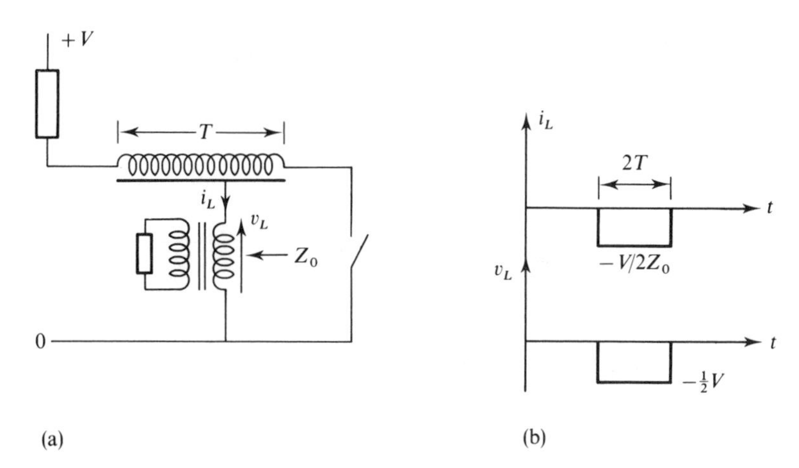

Figure 8.30 Pulse modulator circuit. (a) Single section using a pulse forming network of delay T, characteristic impedance Z_0. (b) Ideal waveforms.

step-up transformer. The multiplicity of units has the advantage that failure of one unit does not disable the whole system.

The gain of the TWT used is about 27 dB, requiring an input of some 300 W at 3 GHz. This is produced by combining the outputs of eight Class-C amplifier stages using 60 W bipolar transistors. The overall transmitter configuration is shown schematically in Figure 8.31. The original 'chirp' signal is derived by exciting a surface-acoustic-wave expanding filter with a short (a few cycles) pulse of a 45 MHz waveform. The lengthened pulse has a frequency excursion of 2 MHz. This pulse is frequency shifted first to 155 MHz, by mixing with 110 MHz derived from a crystal controlled oscillator, and then to carrier frequency by mixing with a local oscillator derived by frequency multiplication of a 100 MHz crystal controlled source. The RF output is amplified and applied to the TWT as described earlier. The output from the TWT is taken to the antenna via a duplexer using a 4-port circulator, together with a TR cell (Section 8.5). The local oscillators at 3.2 GHz and 110 MHz are also used in the receiver. A similar radar without pulse compression, the S511, uses a magnetron delivering a pulse of duration 0.85 μs and peak power 650 kW. The simplicity of this system has to be traded off against improved stability and frequency agility possible in the S512 coherent system.

8.10.3 Receiver

The receiver has a low noise front end using FETs. The gain is about 20 dB and the noise figure 2 dB. It is necessary with such a solid-state amplifier to keep breakthrough from the transmitter down to a few milliwatts of RF power. This is accomplished with the aid of a gas cell, reducing the

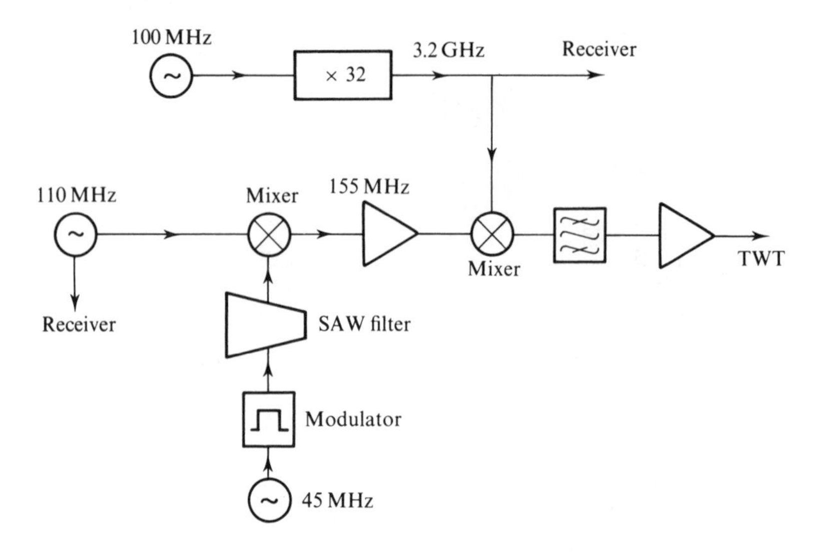

Figure 8.31 Overall block diagram for the transmitter of the S512 radar.

breakthrough to less than 200 mW, and is backed up by a solid-state limiter such as a varactor diode. A block diagram of the receiving system is shown in Figure 8.32. The receiver is a double superheterodyne system with a first intermediate frequency of 155 MHz, followed by a second frequency change to 45 MHz. The signal then goes to a pulse compression SAW device, which compresses the chirp waveform to 0.6 μs with side lobes better than 40 dB below peak. After compression a bipolar video output is formed by comparison with the original 45 MHz. This signal is then converted to digital form for feeding to the signal processor. The high first IF assists in image rejection and in reducing breakthrough by the local oscillator to the antenna. The first mixer is of the image rejection type described in Section 7.3.3.

8.10.4 Range calculations

The coverage diagram for the radar shown in Figure 8.29 indicates detection of a target of 2 m^2 cross-section out to about 160 km. The signal-to-noise ratio at this range can be estimated by use of the radar equation introduced in Section 8.2. The return power is given by equation (8.3)

$$P_r = \frac{P_t}{(4\pi)^3 r^4} \sigma G^2 \lambda^2$$

or

$$10 \log P_r = 10 \log P_t + 20 \log G + 10 \log(\sigma\lambda^2/(4\pi)^3 r^4)$$

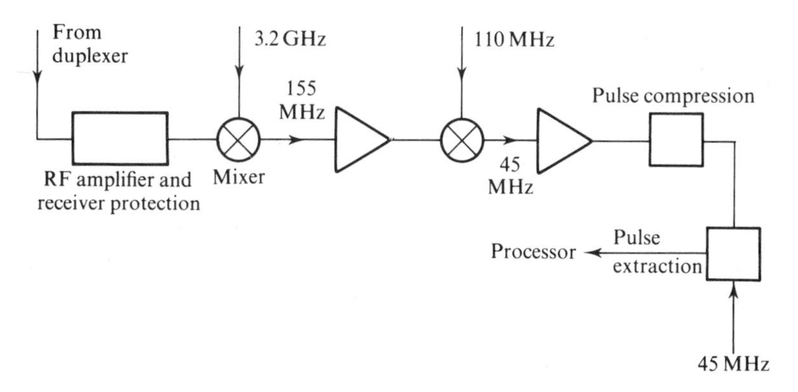

Figure 8.32 Overall block diagram for the receiver of the S512 radar.

Assuming the scattering section of 2 m^2 and a wavelength of 0.1 m, at a range of 160 km the last term is found to be -258 dB. The maximum antenna gain is given in Table 8.1 as 34 dB, and the effective peak power may be estimated as $160 \times 12/0.6$ kW $= 3.2$ MW, or 65 dB W. Inserting these values we find for P_r a value of -125 dB W. Allowing a bandwidth of 2 MHz and a noise figure of 2 dB, $NkTB$ has the value -139 dB. Thus the estimate of S/N is $+14$ dB. From Figure 8.28 this corresponds to about 95% probability of detection at a false alarm probability of 1 in 10^6. As emphasized in Section 8.2, a more accurate calculation must allow for such losses as have not been included in the figure given for antenna gain, and for any improvement because of signal integration. However, the order of magnitude is what might be expected.

NOTE ON THE COVERAGE DIAGRAM

The vertical coverage diagram of Figure 8.29 is basically a Cartesian representation at the point on the earth's surface occupied by the radar, but with an enlargement of scale in the vertical direction (in the case of the figure, a factor of 8 has been used). Thus, low angles of elevation are expanded. Curvature of the earth's surface is important, and an allowance is also made for the effect of the refractive index of the atmosphere by using an effective radius of 4/3 times the actual. (This device will be discussed in Section 9.4.1.) Lines of constant range are correctly ellipses elongated in the vertical direction, but except for the smaller ranges they appear parallel to the vertical axis.

Summary

This chapter has presented the concepts of radar and given the derivation of the radar equation. The various parameters in that equation have been also discussed. Various types of radar have been described, illustrating uses of the various characteristics of the radar return signal. The principles of pulse compression and of phased arrays have been presented. Finally the application of statistical methods to radar detection has been illustrated.

EXERCISES

8.1 A particular radar operates at a wavelength of 10 cm. It emits pulses of 1 μs duration at a peak power of 0.8 MW. A return power of 1.2×10^{-13} W can just be detected. The gain of its antenna is 34 dB. Estimate the range at which a target of scattering cross-section 10 m^2 can be detected. Show that a pulse repetition frequency of 500 Hz will encompass this range. What will be the mean power?

8.2 The antenna of the radar of question 8.1 can 'see' down to an elevation of 3° above the horizon. At what altitude will a target at 200 km range be lost if the radar is at ground level on a smooth Earth, radius 6380 km?

8.3 It is reported that a certain radar can detect a copper plated ping-pong ball at a range of 50 miles! The diameter of such a ball is 3.5 cm, and the wavelength was 23 cm. The radar cross-section of a thrush at this wavelength is about 20 cm^2. At what range might a flock of 20 thrushes be seen on the radar?

8.4 The scattering cross-section, in square metres, of a raindrop of diameter d m is given approximately by the formula $288\, d^6/\lambda^4$. To calculate the radar return from a continuous distribution, it may be assumed that the echo from range r comes from a cell occupying radial depth $c\tau/2$ of the beam to half-power points, τ being the pulse length.

A radar operates at 3 cm with a pulse length of 0.5 μs and peak power 50 kW. The antenna produces an axially symmetrical beam 0.6° between half-power points and has gain 47 dB. Estimate the power of a return echo from a rain cloud containing drops 1 mm in diameter and with 10 drops per cubic metre at a range of 30 km, assuming the cell is entirely filled with cloud.

8.5 A radar used for monitoring aircraft movements often has a radiation pattern approximating to the form

$$G(\theta) = G(\theta_0)\frac{\operatorname{cosec}^2\theta}{\operatorname{cosec}^2\theta_0^2} \qquad \theta_m > \theta > \theta_0$$

in which θ is the angle of elevation. Outside this range of values, $G(\theta)$ has the characteristics of a normal pattern. Discuss the effect this has on aircraft echoes seen on the radar.

8.6 Obtain expressions for the power spectral density of single pulses of carrier, frequency f_0, having the following envelopes:

(a) rectangular, duration τ;

(b) $\cos^2(\pi t/\tau)$, $|t| < \tau/2$;

(c) $(t/\tau)\exp(-t/\tau)$.

Compare the results for the three waveforms.

8.7 The antenna of a ground observation radar has a (power) radiation pattern $F(\phi)$, in which ϕ is the azimuthal angle. The antenna rotates once in T seconds. Show that the return echo from a stationary target has the form

$$s(t) \sum_{-\infty}^{\infty} c_n \exp(j2\pi nt/T)$$

in which $s(t)$ is the periodic pulse train emitted by the radar, and relate the coefficients c_n to the function $F(\phi)$.

For a particular antenna $F(\phi)$ may be approximated by the expression

$$\begin{aligned} F(\phi) &= \exp(-\phi^2/\phi_0^2) \qquad && |\phi| < 3\phi_0 \\ &= 0 && \pi > |\phi| > 3\phi_0 \end{aligned}$$

Show that

$$c_n = \tfrac{1}{2}\phi_0\pi^{\frac{1}{2}}\exp(-n^2\phi_0^2/4)$$

Determine the value of n at which c_n has fallen to 10% of its maximum value when the antenna has a half-power beamwidth of 1° and rotates once in 15 s. Find also the corresponding frequency n/T.

What relevance has this calculation to the case of a moving target?

8.8 Calculate the Doppler shift at a wavelength of 3 cm for reflections from

(a) a car moving at 40 mph,

(b) an aircraft at 500 mph.

A satellite in a circular orbit passes through the zenith at the observation point. Show that the maximum Doppler shift occurs as the satellite reaches the horizon, when it is given by

$$f_d = 4\pi a/T\lambda$$

in which a is the Earth radius, λ the carrier wavelength and T the period of revolution of the satellite. Assume that T is much less than 24 h, so that earth rotation can be ignored. Evaluate for the case $T = 2$ h, $\lambda = 25$ cm. Find also the rate of change of Doppler frequency as the satellite goes through the zenith.

8.9 The receiver of the radar of question 8.1 has a bandwidth of 1.2 MHz and a noise figure of 4 dB. By reference to Figure 8.28, estimate the return power that would be necessary to ensure a probability of detection of 0.5 with a false alarm rate period of 500 s. Compare this with the value given in question 8.1.

8.10 Discuss the design of a small boat radar satisfying the following requirements:

maximum range	20 km
range discrimination	15 m
angular resolution	1°

Assume a typical target to have a cross-section of 1 m^2.

References

The general subject area of radar is comprehensively covered in a number of textbooks, so that specific references have not been made. A selection of books is given below, together with some references in more specialized areas.

Berkowitz, R. S. (1965). *Modern Radar*. New York: Wiley.

Brookner, E. (1977). *Radar Technology*. Norwood, Mass.: Artech House.

Clarke, J., Ed. (1985). *Advance in Radar Techniques*. Stevenage: Peter Peregrinus.

Harger, R. O. (1970). *Synthetic Aperture Radar Systems*. London: Academic Press.

Lynn, P. A. (1987). *Radar Systems*. London: Macmillan.

Matthews, H., Ed. (1977). *Surface Wave Filters*. New York: Wiley.

Skolnik, M. I. (1980). *Introduction to Radar Systems*. New York: McGraw-Hill.

Smullin, L. D. and Montgomery, C. G. (1948). *Microwave Duplexers*. New York: McGraw-Hill.

Swords, S. S. (1986). *History of the Beginning of Radar*. Stevenage: Peter Peregrinus.

CHAPTER 9

TERRESTRIAL COMMUNICATIONS

OBJECTIVES

This chapter considers the design and performance of communication channels operating by propagation within the Earth's atmosphere. The factors associated with two types of system are studied: firstly, microwave point-to-point systems, which are restricted to line-of-sight; secondly, attention is paid to troposcatter systems used for beyond-the-horizon services. The chapter is illustrated by detailed descriptions of three operational systems.

9.1 Introduction

This chapter concerns the use of microwave radio transmission to establish communication links. The microwave spectrum is only a part of the total spectrum used for communication purposes, and the chief features which make microwaves suitable for certain applications are

(1) the directivity that can be obtained with antennas of modest dimensions,
(2) the large bandwidths and consequent high information rates possible with the high frequency carrier.

The various authorities have allocated bands over most of the microwave spectrum. Operation at frequencies between 2 and 20 GHz is current, and design for higher frequencies in progress.

The wide bandwidths available make microwave links suitable components of trunk communication routes. At frequencies below 10 GHz, frequency division multiplex (FDM) is used to assemble channels. The carrier is frequency modulated. Higher frequencies have proved more suitable to digital modulation techniques, so that these links use time division multiplex (TDM).

The choice of type of modulation depends on many factors: basically the aim is to have transmission free from distortion in the minimum bandwidth. Whatever method is used the end product for the microwave engineer is a specification of minimum signal-to-noise ratio for adequate system performance. Overall, the receiver design must depend on the modulation technique used, but most of the processing proper to modulation and demodulation will be carried out at lower frequencies followed by up-converters or down-converters as appropriate. These aspects are not dealt with in this book, and standard works on the theory and practice of communications should be consulted (e.g. Stremler, 1982; Feher, 1981).

9.2 System gain

In Section 3.7.2 a power budget, equation (3.14), was established from which the received power in a free space link could be calculated. For a terrestrial link this equation must be modified by the inclusion of loss terms arising from several causes, such as atmospheric attenuation, fading because of variability of the atmosphere, and the presence of the Earth's surface or obstacles. The power budget equation then becomes, assuming decibel measures,

$$P_r = P_t + G_t + G_r - L_s - L_p \qquad (9.1)$$

In this equation P_r(dB W) is the received power; P_t(dB W) the transmitted power; L_s(dB) the free space loss, $20 \log(4\pi r/\lambda)$; G_t and G_r the transmitting and receiving antenna power gains; and L_p(dB) the additional propagation loss. For adequate performance, the received power must exceed a certain minimum level determined by the noise level and the acceptable error rate or equivalent criterion. Denoting this minimum power by C_m(dBW), the **system gain**, G_s, is defined by the equation

$$G_s = P_t - C_m = L_s + L_p - G_t - G_r \tag{9.2}$$

Of these terms the extra path loss L_p is the most uncertain and is moreover variable, so that in order to satisfy equation (9.2) an estimate of the maximum loss must be made. To satisfy equation (9.2) at all times, allowing for the most extreme conditions, would be very difficult, so that the requirement is relaxed to apply for all except a certain fraction of time. This fraction will depend on the use to which the link is put: it might be of the order of 0.01 per cent or as little as 0.0001 per cent. To meet such a requirement it is necessary to know the various factors included in the extra-loss term and to assess the associated probabilities. The chief factors involved are:

(1) attenuation in the medium;
(2) variable propagation characteristics;
(3) presence of the Earth's surface and of other obstacles.

The last of these is invariant and dealt with in path planning, the first two are liable to variations.

9.3 Attenuation

Attenuation of a beam of radiation traversing a gaseous medium is characterized by an attenuation constant, equivalent to that discussed in Section 2.5 in connection with transmission lines. In principle, the attenuation along a transmission path between transmitter and receiver can be calculated from the formula

$$\int_T^R \alpha(s)\,\mathrm{d}s \tag{9.3}$$

in which $\alpha(s)$ contains contributions from gaseous absorption and the effects of rain or snow (precipitation). The former is usually negligible below 15 GHz: values for oxygen and water vapour are shown in Figure 9.1. Whilst oxygen content is fairly well defined, that of water vapour is subject to wide variations. Attenuation by precipitation is usually more important, because of both the frequent severity and the variability. Some average values for the attenuation caused by rain are shown in Figure 9.1(b). The figures of 0.01 per

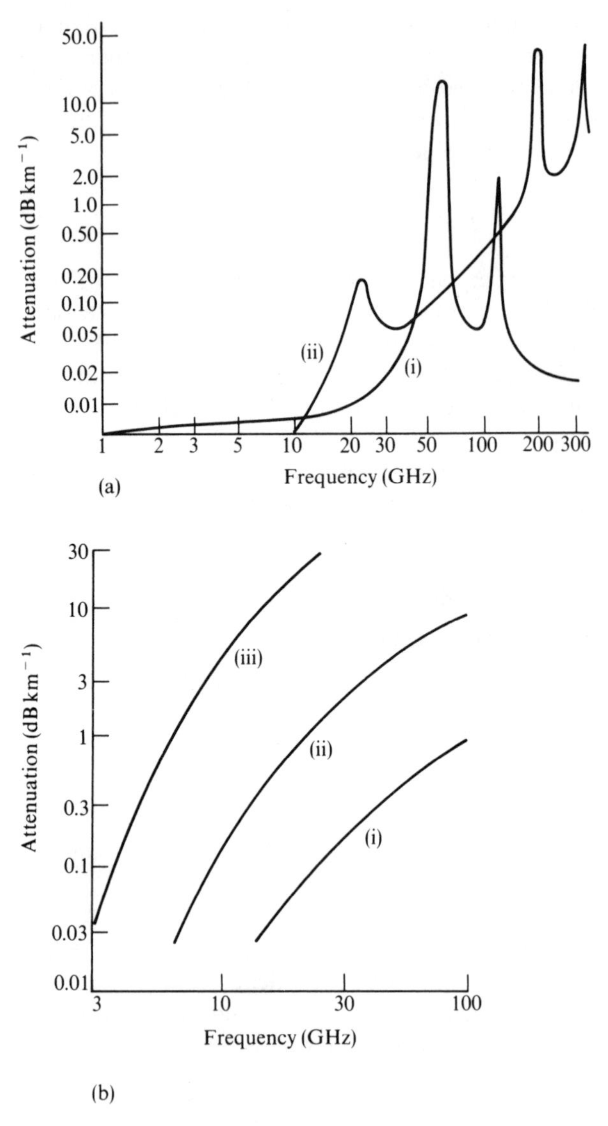

Figure 9.1 Frequency dependence of attenuation by atmospheric constituents. (a) Gases: (i) oxygen at one atmosphere, 20°C; (ii) water vapour, density 7.5 g m^{-3}, corresponding to the average at ground level in temperate climates. (b) Attenuation by rainfall: (i) 1 mm h^{-1}; (ii) 10 mm h^{-1}; (iii) 100 mm h^{-1}.

cent and 0.0001 per cent reliability mentioned above correspond to 52 minutes and 0.5 minutes per year respectively, so that statistical information on rate of rainfall over short times is needed. This matter is treated in works by Hall (1979) and Harden *et al.* (1978). The attenuation in a rain-storm can be very severe, particularly at the higher frequencies, and it is impractical to make good this loss by providing surplus gain in the link. It then becomes important to estimate the size of rainstorms, so that alternative routes can be planned to maintain communication.

9.4 Tropospheric propagation

Apart from attenuation, considered above, propagation through the troposphere is dominated by the small but significant refractive index of the atmosphere. This is usually expressed as the deviation from unity, measured in parts per million,

$$n = \varepsilon_r^{\frac{1}{2}} = 1 + N10^{-6}$$

or

$$N = (n - 1)10^6$$

A semi-empirical formula (e.g. Hall, 1979) relating N to atmospheric pressure and humidity is

$$N = (77.6P + 1720m)T^{-1}$$

in which P is pressure in millibars, m water concentration in $\mathrm{g\,m^{-3}}$ and T temperature in kelvin. The value of m lies typically between 2 and 20, so that the second term is significant in the overall value of N. This indicates the liability of N to fluctuation. The variation of N with height can be approximately expressed by

$$N = N_s \exp(-h/h_0)$$

An **average atmosphere** has been defined for which $N_s = 315$, $h_0 = 7.36\,\mathrm{km}$. For heights significant in terrestrial links, measured in kilometres,

$$N_{av} \cong 315 - 40h \tag{9.4}$$

The variations with height are of much greater significance than horizontal variations, so that the atmosphere may be to a good approximation considered as stratified. The propagation of microwave beams through the atmosphere is usually treated by consideration of optical ray paths.

9.4.1 Stratified atmosphere

A ray path through a spherically stratified medium is illustrated in Figure 9.2. Application of Snell's law ($n_s \sin\theta_s = n_{s+1} \sin\theta_s'$) and the geometry gives the equation

$$n_s r_s \sin\theta_s = n_{s+1} r_{s+1} \sin\theta_{s+1}$$

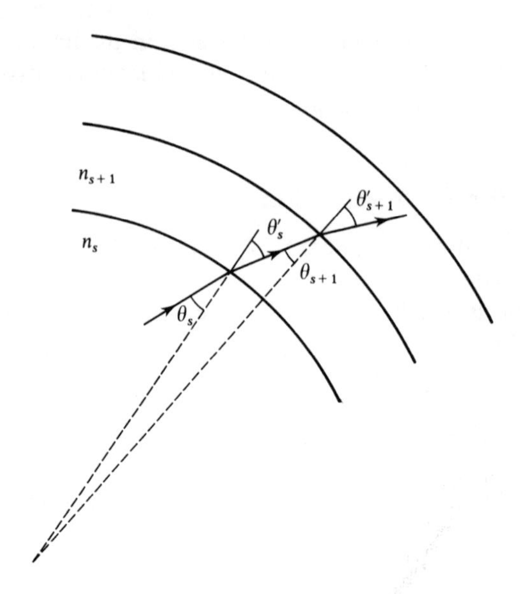

Figure 9.2 A ray in a spherically stratified medium.

or

$$n(r)r\sin\theta = n_0 a \sin\theta_0 \tag{9.5}$$

in which n_0 is the surface refractive index, a the radius of the Earth and θ_0 the angle of the ray to the vertical at launch. Equation (9.5) may be used to derive a differential equation by use of the relation

$$\tan\theta = r\frac{\mathrm{d}\phi}{\mathrm{d}r}$$

but when a linear equation such as (9.4) is a valid approximation (as for terrestrial links when the path is confined to a relatively short distance from the Earth's surface) an alternative approach is possible. We then have

$$n(r) = n_0(1 - kz)$$

If z is measured in metres, for an average atmosphere $k = 4 \times 10^{-8}\,\mathrm{m}^{-1}$. We may write $r = a + z$, in which a is the Earth radius equal to 6.38×10^6 m. Substituting both these expressions in equation (9.5) we have

$$n_0(1 - kz)(a + z)\sin\theta = n_0 a \sin\theta_0$$

With the range of values of z relevant to terrestrial links, both kz and z/a are very much less than unity, so that neglecting second order terms we find

$$n_0\left(1 + z\left(\frac{1}{a} - k\right)\right)a \sin\theta = n_0 a \sin\theta_0$$

This equation may be written in two ways, each with its own interpretation. Written in the form

$$n_0(a_e + z)\sin\theta = n_0 a_e \sin\theta_0 \tag{9.6}$$

with

$$\frac{1}{a_e} = \frac{1}{a} - k \tag{9.7}$$

it may be taken as relating to a spherical Earth of effective radius a_e with an atmosphere of uniform refractive index n_0. With the values of k and a given above, $a_e = 4a/3$.

Written in the form

$$n_e(z)\sin\theta = n_0 \sin\theta_0 \tag{9.8}$$

it describes a flat Earth with an atmosphere of effective refractive index

$$n_e(z) = n_0(1 + z/a_e)$$

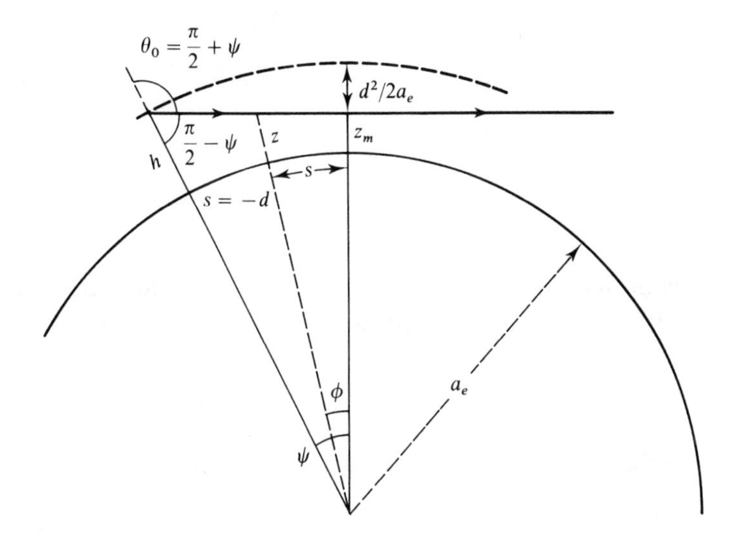

Figure 9.3 The path of a ray in a spherically stratified atmosphere with linear refractive index profile, referred to an Earth of effective radius a_e.

This 'flat Earth' interpretation can be extended by defining the effective refractive index as

$$n_e(z) = n(z)(1 + z/a)$$

in which $n(z)$ is the real refractive index. For the 'average' atmosphere, $n_e(z)$ increases linearly with increase in altitude, in contrast to the real refractive index. Under certain meteorological conditions, such as temperature inversion, the rate of decrease of $n(z)$ may be high enough to cause $n_e(z)$ to decrease at first, before eventually increasing linearly.

Ray paths on the effective Earth radius model will be straight lines, as shown in Figure 9.3. In this figure the angle θ_0 has been expressed as $\pi/2 + \psi$, with ψ a small angle. From the figure we then have

$$(a_e + z_m) = (a_e + h)\cos\psi \cong (a_e + h)(1 - \tfrac{1}{2}\psi^2)$$

whence

$$\begin{aligned} z_m &= h - \tfrac{1}{2}\psi^2 a_e \\ &= h - \frac{d^2}{2a_e} \end{aligned} \tag{9.9}$$

In this equation d is the arc distance between the antenna and the point of minimum altitude, height z_m. At a distance s from this latter point towards the antenna we have

$$a_e + z = (a_e + z_m)\sec\phi \cong (a_e + z_m)(1 + \tfrac{1}{2}\phi^2)$$

whence

$$\begin{aligned} z &= z_m + \tfrac{1}{2}a_e\phi^2 \\ &= z_m + \frac{s^2}{2a_e} \end{aligned} \tag{9.10}$$

Using the flat Earth model, a ray path will be a parabola described by equation (9.10), regarding (s, z) as Cartesian coordinates. This is portrayed in

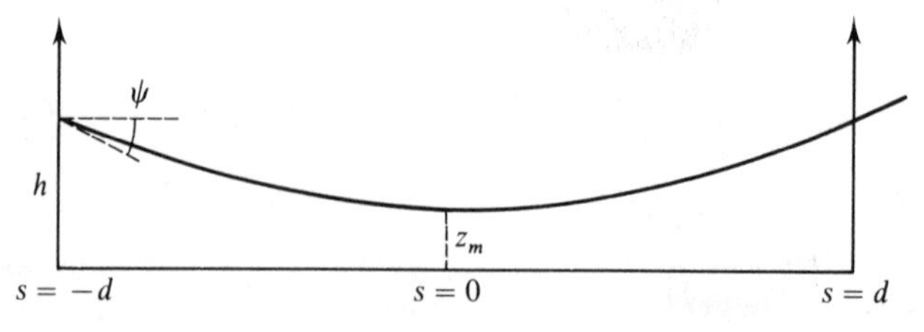

Figure 9.4 The path of the ray of Figure 9.3 referred to a flat earth surface.

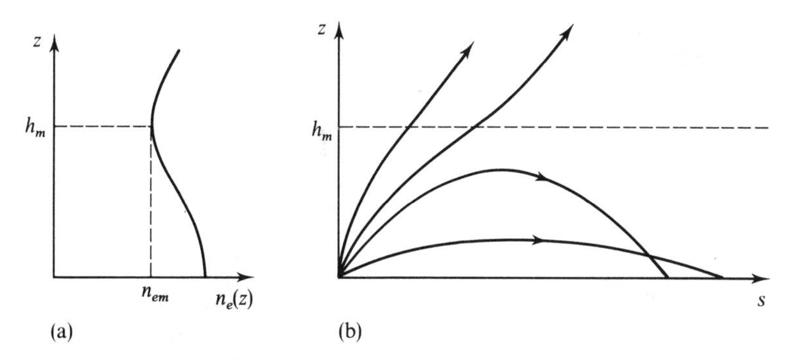

Figure 9.5 Ray paths associated with an effective refractive index initially decreasing with height.

Figure 9.4. In either form, the ray paths may be used for various purposes, such as planning for clearance of obstacles or to estimate path length, angle of arrival of a ray or delay.

The value of $4a/3$ for the effective Earth radius is consequent on the 'average' atmosphere. The value of k may be greater or less than 4×10^{-8}, depending on meteorological conditions. If k is negative, that is, if the refractive index increases with height, then $a_e < a$, and the clearance z_m of equation (9.9) decreases. It is found desirable to plan for $a_e = 0.7a$ to allow for varying conditions.

9.4.2 Ducting

As discussed above, in certain conditions the effective refractive index $n_e(z)$ can at first decrease with increase of altitude, and thus have the form indicated in Figure 9.5(a) in which a minimum n_{em} occurs at height h_m. Equation (9.8) then shows that if $\theta_0 > \sin^{-1}(n_{em}/n_0)$, a value of θ at h_m cannot be found, implying that the ray is reflected back towards the ground. This is shown on the flat Earth model in Figure 9.5(b). In such a situation, wave propagation takes place in two dimensions rather than in three, so that fields fall off more slowly with distance and waves may be propagated much further than normal. Such a formation is referred to as a **duct**. From these various considerations it is clear that the received signal may follow various paths in changing conditions, leading to change in strength, angle of arrival and delay, all of which have their effect on link performance.

9.4.3 Propagation over obstacles

Obstacles such as hills or buildings that obstruct the line of sight will clearly affect the signal received. The position of obstacles with respect to the line-of-sight path can be established by plots of the form of Figure 9.4. Because of the

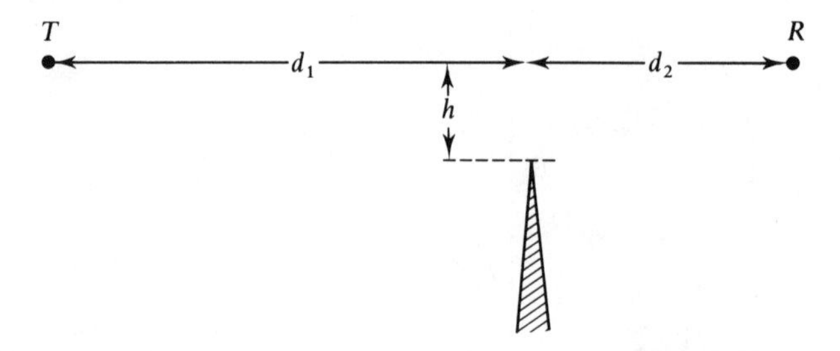

Figure 9.6 The geometry of a knife-edge obstacle near a transmission path TR.

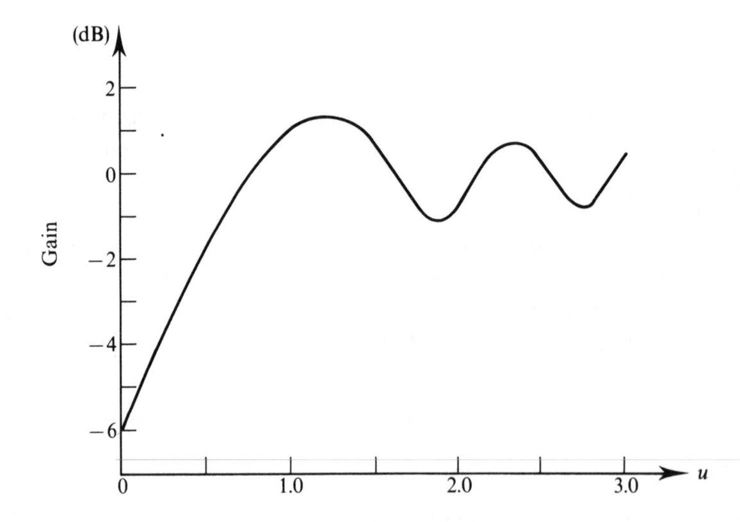

Figure 9.7 Showing the effect of the knife-edge obstacle of Figure 9.6 on received signal strength. Ordinate: gain above free space value. Abscissa: $u = h(2(d_1 + d_2)/\lambda d_1 d_2)^{\frac{1}{2}}$, in which λ is the wavelength and other variables are as in Figure 9.6.

wave nature of the beam, however, it is not sufficient to merely clear the line-of-sight path. A useful problem which can be solved (as in Brown and Clarke, 1980) concerns the effect of a conducting half-plane, or knife-edge, as illustrated in Figure 9.6. The power ratio of the signal received to that for a free space path can be shown to be

$$\tfrac{1}{2}[(C(u) + 0.5)^2 + (S(u) + 0.5)^2]$$

in which $C(u)$ and $S(u)$ are the standard Fresnel integrals (e.g. Abramowitz and Stegun, 1968) and

$$u = h[2(d_1 + d_2)/\lambda d_1 d_2]^{\frac{1}{2}}$$

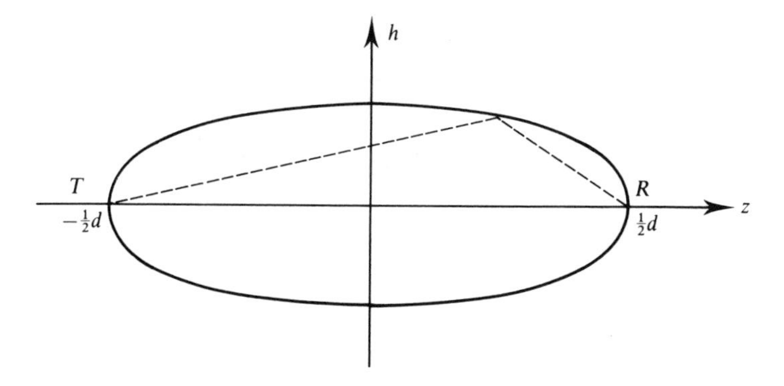

Figure 9.8 Showing the permissible proximity of an obstacle transmission path as a function of position for constant gain loss.

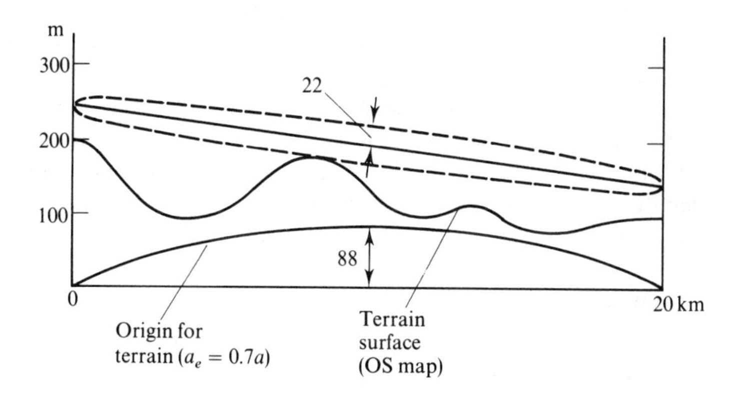

Figure 9.9 Path planning for a 20 km link. The format uses a rectilinear ray with terrain profile offset from an earth surface of equivalent radius $0.7a$. Dotted line indicates required clearance zone. Distances in metres.

This expression is illustrated graphically in Figure 9.7, from which it will be seen that the variable u should have a minimum value of about 0.8 in order to avoid significant loss beyond that for a free space path. For a given line of sight the condition that u shall be a given constant allows the permissible clearance at any point on the path to be established. In terms of the variables shown in Figure 9.8, u constant yields the relation

$$\frac{8h^2}{\lambda du^2} + \frac{4z^2}{d^2} = 1$$

which is the equation of an ellipse, as displayed in the diagram. The condition that the path length TPR is greater than TR by $\lambda/2$ defines the first Fresnel zone (e.g. Sander and Reed, 1986). which may be shown to correspond to

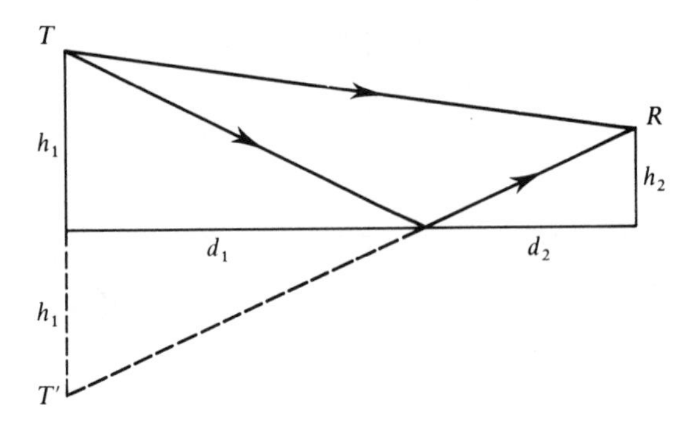

Figure 9.10 Direct and reflected rays between transmitter and receiver over a flat Earth.

$u = 2^{\frac{1}{2}}$. Thus a zone corresponding to about 60% of the first Fresnel zone around the ray path must be left free of obstacles. The way this can be used for path planning is illustrated in Figure 9.9, in which a straight ray format is used, with an exaggerated vertical scale. The terrain is mapped taking as the origin an Earth surface of radius a_e, for which the middle is raised up over the ends by an amount $d^2/8a_e$. Alternatively, a flat Earth presentation may be used when the path becomes parabolic as in Figure 9.4.

9.4.4 Ground reflections

In addition to the direct line-of-sight path it is possible for rays to arrive after reflection from the ground, as indicated for a flat Earth in Figure 9.10. Geometry shows that for the usual situation when heights are much less than the spacing, the path length difference is $2h_1h_2/d$. The amplitude of the received signal will be greater than the free space value by a factor

$$H = 1 + R\exp(-\mathrm{j}4\pi h_1 h_2/\lambda d) \tag{9.11}$$

The reflection coefficient R for a smooth surface may be calculated from the Fresnel formula (e.g. Sander and Reed, 1986):

vertical polarization

$$R_v(\alpha) = \frac{n\sin\alpha - (n^2 - \cos^2\alpha)^{\frac{1}{2}}}{n\sin\alpha + (n^2 - \cos^2\alpha)^{\frac{1}{2}}} \tag{9.12}$$

horizontal polarization

$$R_h(\alpha) = \frac{\sin\alpha - (n^2 - \cos^2\alpha)^{\frac{1}{2}}}{\sin\alpha + (n^2 - \cos^2\alpha)^{\frac{1}{2}}} \tag{9.13}$$

in which α is the angle between ray and reflecting surface (grazing angle) and n^2 is related to the relative permittivity ε_r, conductivity σ, and frequency by the expression

$$n^2 = \varepsilon_r - j\sigma/\omega\varepsilon_0$$

The application of these formulae to reflection from the Earth is complicated by the dependence of both ε_r and σ on frequency and on the type of ground or sea surface. Further complication arises because the reflecting surface may be curved, and may be rough rather than smooth.

For a smooth Earth and a ray at grazing incidence, R may be assumed equal to -1, giving from equation (9.11)

$$|H| = 2|\sin(2\pi h_1 h_2/\lambda d)| \tag{9.14}$$

If h_2 is varied for fixed h_1, then at ground level the signal will be zero, rising to a maximum of $|H| = 2$.

In general such a gain will not be realized, both because of the factors just mentioned and because no consideration has been given to the effect of antenna radiation pattern. With a narrow transmitting beam, the signal at the point of reflection will be smaller than in the direct ray; moreover, the receiving antenna will discriminate against the reflected ray. This will have the result of reducing the effective value of R in equation (9.11). Apart from specular reflection from the ground, there will also be some scattering of radiation from rough ground. The received signal is thus the result of a number of contributions, and variation of meteorological conditions will alter the phases and directions of arrival, leading to a variable signal.

9.4.5 Fading

In various places in the discussion of Section 9.4.1 on the determination of ray paths in the atmosphere, reference has been made to the variability of meteorological conditions and their effect on propagation. This variability results in variation of received signal strength on a microwave link, including deep fades for which allowance must be made in system design.

It is convenient to discuss this important topic in the context of a typical microwave link (Vigants, 1975) operating at about 5 GHz over a path length of perhaps 40 km and designed to suffer little ground reflection. In such a link the received signal is found to be liable to various fluctuations. Normal conditions are those usually found around midday, when atmospheric conditions are relatively stable. At such times rapid fluctuations are found of tenths of a decibel and slow variations of some one or two decibels. Towards sunrise and sunset when conditions are changing, much greater variations can be observed, when for seconds at a time the signal level can drop to very low values. Such occurrences are referred to as **deep fades**, and are a significant cause of link 'outage'. The rapid fluctuations are due to small scale inhomogeneities, and may be compared to the scintillations observed when viewing stars. Slow changes in a stable atmosphere produce small changes in the ray path and associated variation of signal. If two rays arrive at the antenna giving signals of similar amplitudes but with 180° phase difference, then cancellation will occur. It has been found that on such a link the fades can be accounted for by not more than three rays, explained by layers of decreasing refractive index as discussed in Section 9.4.2. The existence of these different rays has been demonstrated directly by Delange (1952) by transmitting very short pulses (~ 3 ns). On reception two or three pulses have been observed, with a differential delay of as much as 7 ns. A model exhibiting this effect is presented in Section A.11. The same effect can be demonstrated as by Crawford and Jakes (1952), by the transmission of a frequency-swept signal, in effect recording the frequency response of the channel. It has been found (Rummler, 1979; Lundgren and Rummler, 1979) that fading conditions can be simulated by a channel response of the form

$$(1 - \exp(\pm j(\omega - \omega_0)\tau)) \tag{9.15}$$

with appropriate choice of ω_0 and delay τ.

Such a response can be given circuit formulation and be used to simulate fading for laboratory tests. Observations on many links have provided statistical data on the occurrence and severity of fades. The data of interest concerns

(1) depth of fade, quantified by

$$L = \frac{\textit{amplitude during fade}}{\textit{normal amplitude}} < 1$$

(2) the time $t_f(L)$ for which the fade level of the signal is numerically smaller than a given value L as indicated in Figure 9.11;

(3) the number of fades in unit time for which L is smaller than some given value;

(4) the total fraction of time for which the received signal is less than L times the normal.

The average, $\bar{t}_f$, and the number per unit time are both found (Vigants, 1975) to be approximately proportional to L for small values of L.

An empirical relation for $\bar{t}_f$ was found:

$$\bar{t}_f \cong 410\,L\,\text{sec} \qquad L < 0.1 \tag{9.16}$$

Denoting the fraction in (4) by T/T_0, it was found possible to relate this

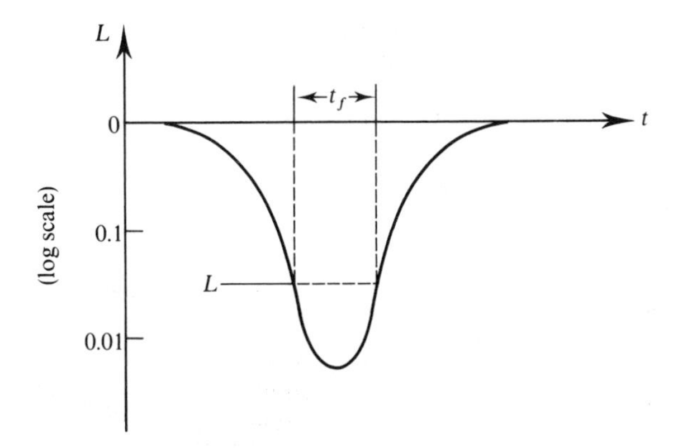

Figure 9.11 Showing the parameters characterizing a fade on a point-to-point link.

quantity to the length of link D (km) and frequency f (GHz) by the formula

$$T/T_0 \cong 6KfD^3\,10^{-7}\,L^2 \qquad L < 0.1 \tag{9.17}$$

in which K is a constant of the order of unity for ordinary undulating country, rising for paths across water, and falling for dry terrain. Equations (9.16) and (9.17) both apply to worst month periods. A whole year may be taken as equivalent to three of these periods.

The general type of these relations seems to be of wide application although the numerical factors will depend on local climate and terrain. It will readily be seen that equation (9.17) predicts very significant effects: consider the case when an extra propagation loss L_p in equation (9.2) has been allowed for at 40 dB, corresponding to a permissible depth of fade $L = 0.01$. For a 40 km link operating at 5 GHz equation (9.17), taking $K = 1$, gives a value for T/T_0 of 1.9×10^{-5}, or 0.002%. This is already of the order of the figures mentioned in Section 9.1, and, for n links in succession, must be multiplied by n. Thus for a 200 km path with 5 links, $T/T_0 \cong 0.01\%$. It is therefore necessary to find means of lessening the single link figures.

9.4.6 Channel equalization

Equation (9.15) describes a fading channel in terms of frequency characteristics, which are illustrated for a typical case in Figure 9.12. The position of the 'notch' may be in or out of the band of interest. It is possible to design a filter that would give an overall flat response, or 'equalize' the channel. Since the channel response varies slowly with time an adaptive circuit is used (Chamberlain and Price, 1983), relying on sampling at three different frequencies to derive the parameters for the compensating filter. Such a technique cannot compensate for the loss in amplitude but it can reduce the

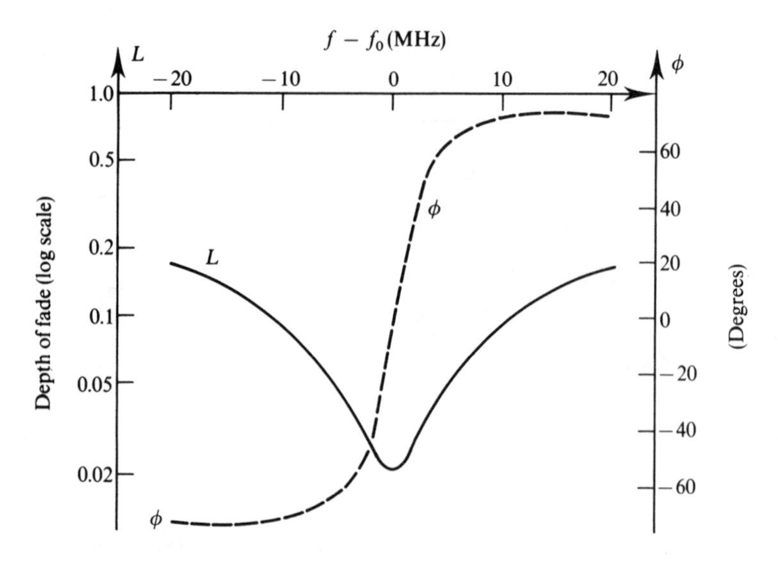

Figure 9.12 Attenuation and phase associated with a 2-path fade, calculated from $L\exp(j\phi) = a(1 - b\exp(-j(\omega - \omega_0)\tau))$ with $a = 0.51$, $b = 0.96$, $\tau = 2.8$ ns. The graphs are periodic with period 357 MHz.

effects of non-linear phase shifts and so reduce the rate of error in digital systems.

9.4.7 Diversity

The effects of a given fade are specific to a particular frequency and spatial position of the receiving antenna. It is found that the use of two frequencies, sufficiently separated, gives two received signals for which fades are not completely correlated. Thus by choice of the better frequency, reliability can be greatly increased. Such an arrangement is said to use **frequency diversity**. However, to employ two frequencies represents an undesirable use of the available frequency spectrum, and the same effect can be obtained by **space diversity**: the signals received on two vertically separated antennas are also not completely correlated, and the better signal can be used. The improvement can be quantified in terms of the ratio T/T_0 of equation (9.17). The value of T/T_0 for a link with diversity is reduced by a factor I below that for a single link, so that

$$(T/T_0)_{\text{div}} = (T/T_0)_s / I$$

With the type of link for which equation (9.17) was established, expressions for I were found (Vigants, 1975). Assuming two identical antennas, for space diversity

$$I \cong 1.2 \times 10^{-3} s^2 f/(DL^2) \tag{9.18}$$

in which s is the vertical separation in metres, with D (km) and f (GHz) as before. For frequency diversity

$$I \cong 80\,\Delta f/(f^2 DL^2) \tag{9.19}$$

The case previously evaluated used $D = 40\,\text{km}$, $f = 5\,\text{GHz}$, $L = 0.01$. A space diversity system using similar antennas 10 m apart vertically is found to give $I = 150$, leading to a value of 1.3×10^{-5} per cent for T/T_0. The same improvement would require frequency separation of about 0.2 GHz. It is to be noted that horizontal separation of the antennas is found to be far less efficacious than vertical separation.

To make use of diversity it is necessary to arrange switching between channels, which can be done at lower frequencies if two receivers are used. Alternatively, with space diversity the two RF signals may be combined using an adaptive phase shifter to ensure that they are in phase (Martin-Royle *et al.*, 1977).

Above 10 GHz, attenuation by rain becomes increasingly severe, and neither frequency nor space diversity alleviates the problem. Heavy rainfall occurs in cells of limited size, so that alternative transmission paths can avoid simultaneous severe fading. The spacing in **route diversity** depends on cell size: in the UK it is found that switched paths 4 km apart give satisfactory performance.

9.5 Troposcatter links

The microwave links described in the previous sections require a clear line-of-sight path between ends of a stage. They cannot therefore be used to span expanses of sea or desert, for example, where intermediate stages cannot be provided. Although radiation is transmitted beyond the horizon by diffraction, the attenuation is far too high to provide a feasible link. It is found, however, that adequate beyond-the-horizon signals are obtained through the phenomenon of tropospheric scatter, in which inhomogeneities in the troposphere act as scattering centres for incident radiation (Collin, 1985). Links of the order of a few hundred miles are found possible in this way. The principle of such a link is illustrated in Figure 9.13(a). Antennas at two points P and Q on the Earth's surface are beamed at the horizon seen from each station. Their intersecting radiation patterns define a volume S. Within this volume scattering centres illuminated by a beam from P cause radiation to be scattered within the radiation pattern of the antenna at Q. The geometrical variables important in the situation are shown in Figure 9.13(b).

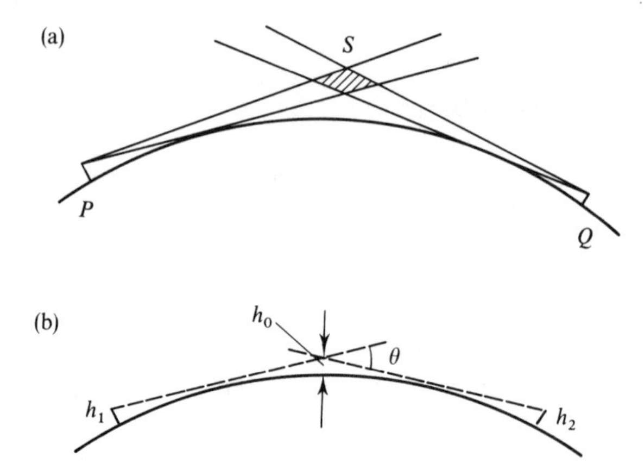

Figure 9.13 The geometry of over-the-horizon propagation by troposcatter. Forward scatter is strongly dependent on the angle θ.

The most important variables are the scatter angle, θ, and the altitude of the scattering volume. The scattered radiation is mainly in a narrow cone about the direction of the incident ray, so that the received signal decreases rapidly as θ increases. To give some idea of the orders of magnitude involved, Figure 9.14 (after Picquenard, 1974) shows the dependence on scattering angle of the attenuation over and above the free space loss, applied to the configuration of Figure 9.13(b). The scattering centres are thought to arise from turbulence, and are therefore dependent on climate. The figure applies to a frequency of 1 GHz, and at a frequency of f GHz attenuation should be increased by $10 \log f$ dB. The geometry of Figure 9.13(b) will only apply to a sea path: on land, terrain will limit the lowest angle at which beams may be launched, thus increasing θ and the height of the scattering volume. It is found that this increase in height also causes greater attenuation, presumably due to the distribution of scattering centres, and the values in Figure 9.14 should be further increased by $10 \log(h/h_0)$ dB. The large number of individual scatterers distributed over a large volume give rise to many contributions to the received signal, which will arrive at the antenna over a range of directions. This means that a highly directive antenna will pick up less than the full signal power, and it is found that an 'antenna-to-medium coupling loss' has to be included. This effect is illustrated in Figure 9.15 (after Picquenard, 1974).

9.5.1 Fading

The extra attenuation shown in Figure 9.14 represents an average value: it is liable both to slow and to fast variations. Slow variations may be on the scale of hours, days or a year, associated with climate and change of weather conditions. Rapid fading is also present, as might be expected from the large

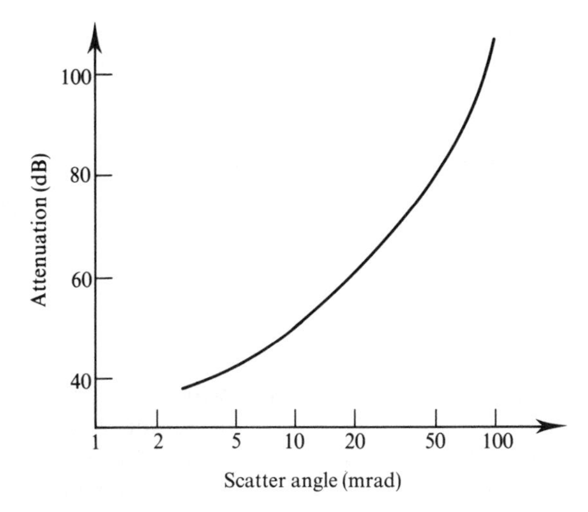

Figure 9.14 Showing the loss above free space loss on a troposcatter link as a function of scatter angle.

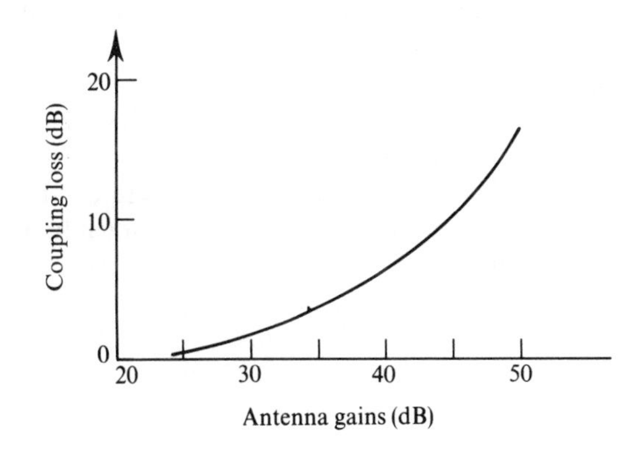

Figure 9.15 Showing the antenna-medium coupling loss between similar antennas on a troposcatter link.

number of randomly placed scatterers involved, which can give deep fades. As with line-of-sight paths, diversity can be used: as a rule of thumb one per cent in frequency or 100 wavelengths spatially are found to give satisfactory performance. An alternative is 'angle diversity' in which the receiver antenna has two feeds giving beams in slightly different directions, thus using different scattering volumes (Hall, 1979).

9.5.2 Bandwidth

It has been mentioned previously that the received signal is made up of components travelling by different paths, and hence with a range of delay times. The usable bandwidth of the link is given by

$$B \cong (\Delta\tau)^{-1} \tag{9.20}$$

in which $\Delta\tau$ is the maximum delay differential. A high gain antenna reduces the number of rays contributing to the signal and so increases the available bandwidth.

9.6 Cellular radio

The previous sections have been concerned with propagation in a relatively unobstructed medium. Quite different is propagation within an urban area that contains buildings of all shapes and heights. Propagation in such an environment is relevant to the needs of mobile radio telephony, at frequencies of the order of 1 GHz. In such an environment the signal at any point near ground level is the combination of signals arriving along many different paths and thus is liable to very severe fading. The mean signal amplitude decreases away from a base transmitter approximately proportional to d^{-2} rather than d^{-1} for free space transmission. Coverage is achieved by dividing the city area into cells. Adjacent cells operate on different frequencies, but frequency reuse is possible in cells further away. The cells can be some miles in radius using transmitter powers of a few watts both at the base and on the mobile units. The real complication is the automatic switching that must take place to keep a particular mobile set in contact with the base station best suited for signal strength while preserving continuity with the public telephone network. Further information can be found in Cooper and Nettleton (1983), Jakes (1974) and in the special issue of the Bell System Technical Journal on Advanced Mobile Phone Systems (1979).

9.7 Statistics of fading

In the literature concerning variations of signal strength in various environments, two types of statistical distribution frequently appear, and the purpose of this section is to set out the appropriate definitions.

9.7.1 Rayleigh distribution

Received signals may consist of many contributions of random phase and range of amplitudes. In such a situation the received carrier will appear to be modulated with a randomly varying envelope. A similar situation arises with the signal picked up from a mobile receiver in an urban environment. The probability that the amplitude lies between v and $v + dv$ is given by the

Rayleigh distribution

$$p(v)\,\mathrm{d}v = \frac{v}{a}\exp\left(-\frac{v^2}{2a}\right)\mathrm{d}v \tag{9.21}$$

It may be verified that the mean square value of v is equal to $2a$. (This distribution was encountered in Section 8.8.1.) The cumulative probability that $v > v_0$ is given by

$$\begin{aligned} P(v > v_0) &= \int_{v_0}^{\infty} p(v)\,\mathrm{d}v \\ &= \exp(-v_0^2/2a) \end{aligned} \tag{9.22}$$

The analysis of observations is often assisted by expressing equation (9.22) in the form

$$10\log(v_0^2) = 10\log(2a) + 10\log\log(P(v_0))^{-1}$$

As $P(v_0)$ goes from zero to unity, the right-hand side goes from infinity to negative infinity. Graph paper with an abscissa labelled on the log log scale then gives a straight line with signal strength in dB, as indicated in Figure 9.16.

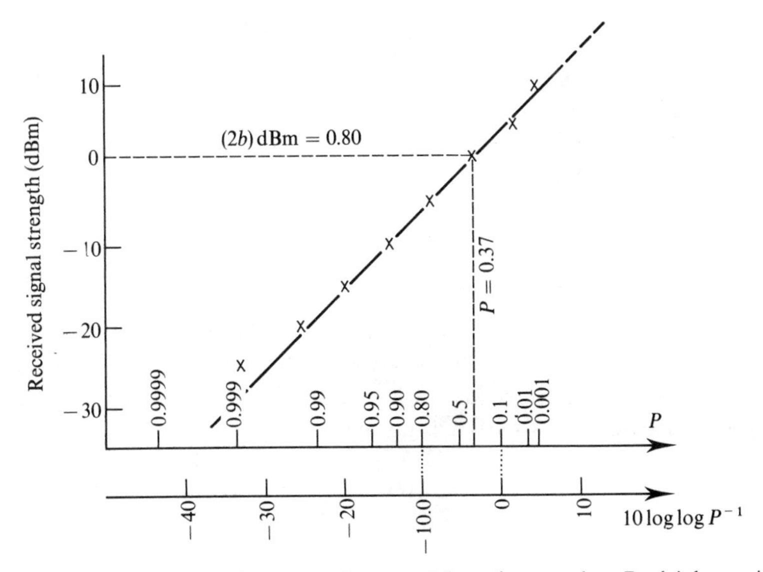

Figure 9.16 Illustrating the use of a special scale to plot Rayleigh statistics: $P(x > x_0) = \exp(-x_0^2/2a)$. Abscissa is equal to $10\log\log(P^{-1})$. The points may be regarded as experimental; the value of $2a$ can be read off against $P = e^{-1} = 0.37$. If Rayleigh statistics apply, the distance between $P = 0.1$ and $P = 0.8$ is the same as 10 dB on the ordinate.

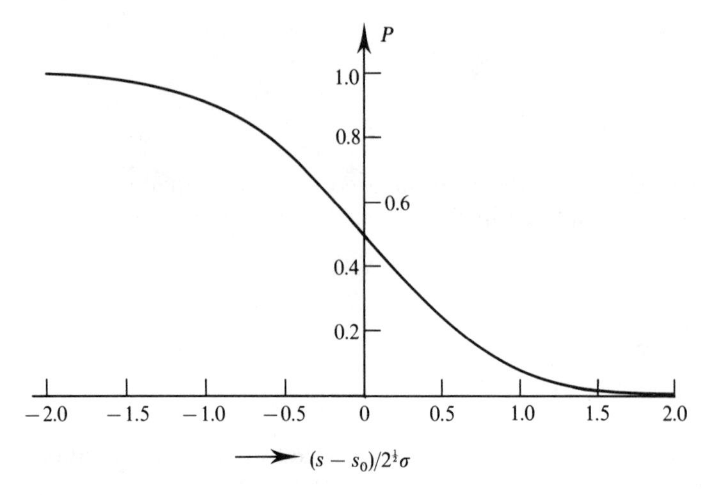

Figure 9.17 Cumulative probability P for a log-normal distribution. Signal strength s in dB. Mean signal s_0, standard deviation $P(x) = \frac{1}{2}(1 \pm \text{erf}(x)) \quad x \lessgtr 0$.

The probability of $v < v_0$ is given by

$$P(v < v_0) = 1 - P(v > v_0) = 1 - \exp(-v_0^2/2a)$$

For deep fades this may be expressed approximately by $(v_0^2/2a)$. Thus the probability of a deep fade is proportional to the power level. This may be seen in equation (9.17).

The Rayleigh distribution is found to apply to relatively fast fading situations, on a time scale of minutes.

9.7.2 Log-normal distribution

The variations in signal strength found after rapid fades are averaged out often to obey a normal or Gaussian distribution with signal strength measured on a logarithmic scale. Thus the probability of received power lying between s dB m and $(s + ds)$ dB m is given by

$$p(s)\,\text{d}s = \frac{1}{(2\pi)^{\frac{1}{2}}\sigma} \exp\left(-\frac{(s - \bar{s})^2}{2\sigma^2}\right) \tag{9.23}$$

in which σ is the standard deviation about $\bar{s}$. The cumulative probability is

$$P(s > s_0) = \tfrac{1}{2} - \tfrac{1}{2}\,\text{erf}\left(\frac{s_0 - \bar{s}}{2^{\frac{1}{2}}\sigma}\right) \tag{9.24}$$

in which erf(x) is the error function (Abramowitz and Stegun, 1968). This is illustrated graphically in Figure 9.17.

9.8 Some systems in service

In the earlier sections of this chapter, the considerations relevant to terrestrial links have been presented. This section gives brief descriptions of the salient features of three systems in current use in order to illustrate the way in which components and ideas are used to form equipment for service. The three systems are all operated by British Telecom, but are typical of their kind.

9.8.1 4/6 GHz analogue microwave link

The microwave link network installed by the British Post Office (now part of British Telecom) covers the UK, carrying much trunk telephony and distributing television. It was largely installed by the late 1970s, and is an analogue system using frequency modulation. The information in this section depends heavily on descriptions in the *Post Office Electrical Engineers Journal* (Martin-Royle *et al.*, 1976, pp. 162–8, 225–34 and 1977, pp 45–54) to which reference should be made for further details.

The average distance between adjacent stations is some 40 km, although the longest is about 100 km. Figure 9.18 shows a typical frequency assignment. The isolation of corresponding channels (e.g. 1, 1′) is achieved by both frequency separation and opposite polarizations on alternate channels. A schematic diagram of a unidirectional analogue channel is shown in Figure 9.19. As shown, the receive channels are $f_1, f_2, \ldots, f_8$, half with one

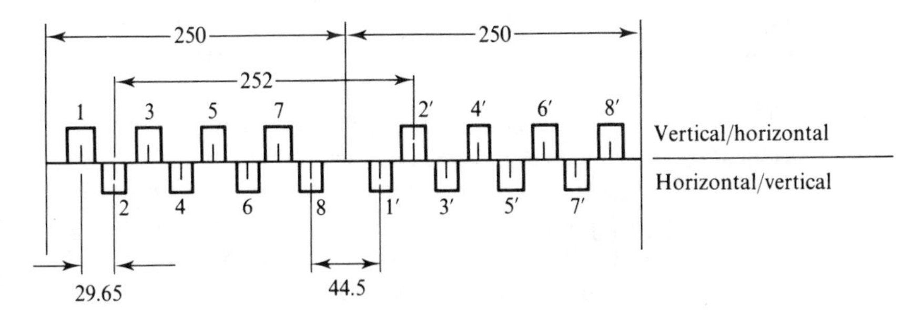

Alternate channels are on opposite polarization.
Frequencies in MHz.

Figure 9.18 Showing the frequency allocation of channels on the microwave link described in Section 9.8.1. Alternate channels are transmitted or received with opposite polarizations.

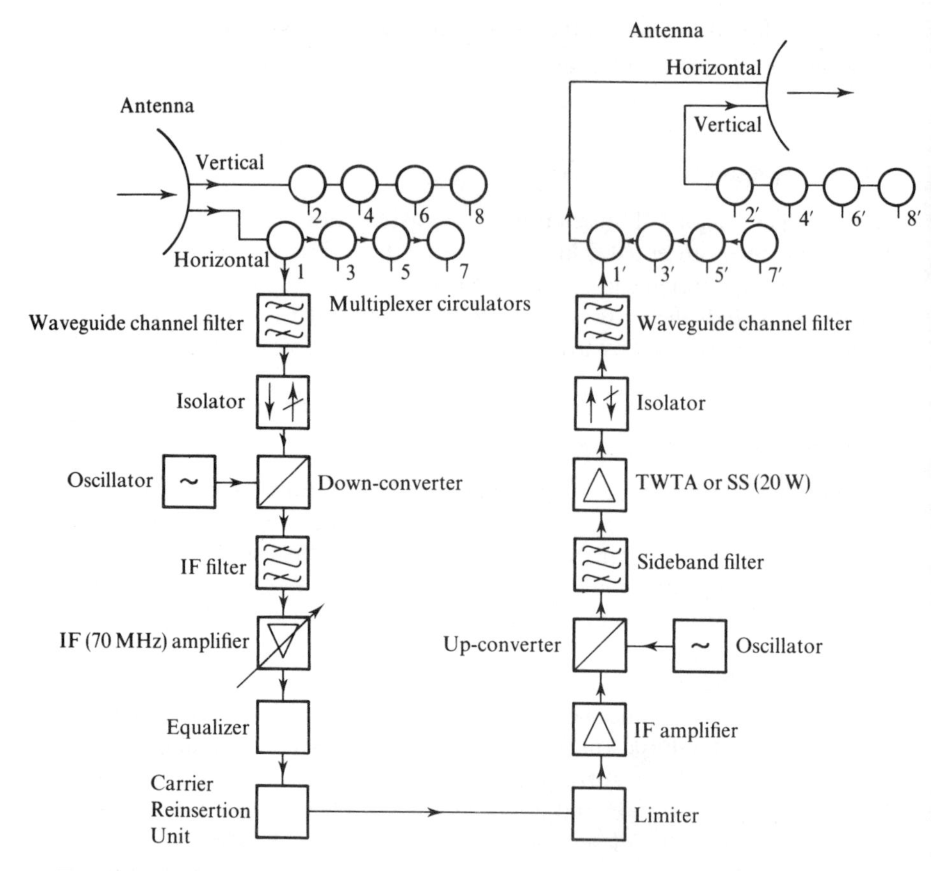

Figure 9.19 Schematic diagram showing layout of a single repeater stage of the link of Section 9.8.1.

polarization, half with the other. These signals are received simultaneously and are separated by passage along the chain of circulators: the channel filter will accept its design channel and reflect other frequencies, which are then sorted out by later filters. The isolator performs the obvious function of ensuring no local oscillator signal gets back to the antenna. A signal, such as one at f_1, is mixed with the correct frequency to give an intermediate frequency of 70 MHz, amplified to a standard level by a variable gain IF amplifier. If this station received the signal to retransmit by cable, a demodulator would recover the baseband signal from the frequency modulated IF carrier. For retransmission, the path is retraced: the IF carrier is frequency-changed in the up-converter to the new frequency f_1' and goes to the power amplifier (TWT or solid-state). This then goes to the antenna via isolator, filter and multiplexer. To insert baseband signals on the transmitter, an IF modulator would be used at the beginning of the transmit chain. Of the components, circulators and isolators have been discussed in Section 5.10. The following sections give notes on some of the other components.

WAVEGUIDE CHANNEL FILTERS

These consist of seven coupled, tuned waveguide cavities giving a Tschebychev response (e.g. Matthei *et al.*, 1980), with bandwidth 56 MHz to half-power.

MIXER

Conversion from microwave carrier to IF is accomplished using diode mixers, as discussed in Section 7.3. At 6 GHz a magic-T, as in Figure 7.6, would be appropriate, at 4 GHz the magic-T could be replaced by a coaxial line 'rat-race', as in Figure 5.22(b).

UP-CONVERTER

Any non-linear device may be used to produce the sum and difference of harmonic frequencies. It is more efficient in up-conversion to use a varactor diode as the non-linear element, as discussed in Section 7.6 on parametric amplification. Using a local oscillator as the pump and the input IF as the idler, the output from the remaining port has the desired frequency shift.

MICROWAVE OSCILLATOR

In order to obtain stability and low noise it is desirable that the microwave frequency used in the mixer and up-converter should be derived from a crystal oscillator. A scheme for doing this is indicated in Figure 9.20: a multiplication of × 10 is achieved by comparing a voltage controlled transistor oscillator at about 1 GHz with a crystal in the region of 100 MHz in a phase

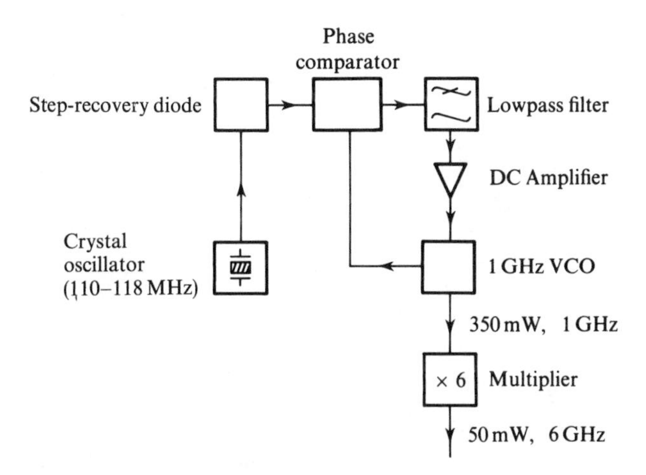

Figure 9.20 Circuit arrangement for production of a microwave frequency from a stable crystal controlled oscillator. The voltage control of the 1 GHz oscillator is effected through a varactor diode coupled to the oscillator resonant cavity.

Table 9.1 Typical microwave antennas used by the British Post Office (British Telecom)

Type of antenna	Polarization	Year of introduction	Typical power gain* (dB)	Typical aperture efficiency (%)	Remarks
3 m diameter paraboloid	Monopolar	Pre-1960	39.4 (4 GHz)	55 (4 GHz)	Heavy cast reflector, poor side lobe performance
Large horn (9 m high)	Bipolar	1961	45 (6 GHz)	52 (6 GHz)	Wideband (4–11 GHz), good electrical performance, mechanical construction creates maintenance problems
Early Cassegrain (3.7 m diameter)	Bipolar	1963	43 (6 GHz)	35 (6 GHz)	Single frequency band, good side lobe performance, low efficiency
Focal plane paraboloid (3.7 m diameter)	Bipolar	1965	44.5 (6 GHz)	45 (6 GHz)	Single frequency band, light spun-aluminium reflector, inexpensive to produce
Latest Cassegrain	Bipolar	1975	45.8 (6 GHz)	68 (6 GHz)	Single frequency band, good electrical performance, high aperture efficiency obtained by slotted taper feed and optimization of secondary reflector
Future	Bipolar	1979	Design objectives yet to be determined		Dual upper 6 GHz and 11 GHz frequency band, good electrical performance with high efficiency

* Measured relative to the gain of an isotropic antenna.
From Martin-Royle *et al.* (1977).

comparator. Voltage control is obtained by loading the cavity determining approximate oscillator frequency with a varactor diode, controlled in turn by the amplified output of the phase comparator. In this way some 350 mW at 1 GHz is obtained, subsequently multiplied using a step-recovery diode to give about 50 mW at 6 GHz.

MICROWAVE AMPLIFIER

Traditionally the travelling wave tube is used to provide wide band amplification in the microwave region. Power levels of the order of up to 20 W are required. In some cases, solid-state amplifiers are used, and as the obtainable power levels increase such usage will become more widespread. Development of travelling wave tubes has made lifetimes of the order of 20 000 hours possible.

An important feature of the power amplifier is linearity. A TWT is most efficient working near its maximum output, at which it becomes saturated. The distortion caused to the microwave signal causes the signals in one channel to affect other channels. The result is to introduce **intermodulation noise**, and to control this it is necessary to run the TWT at several decibels less than its saturated output.

CARRIER REINSERTION UNIT

The purpose of this unit is to ensure the presence of a carrier even if the input to the repeater falls to low levels: because the received signal is liable to variation on account of propagation changes, automatic gain control is provided at IF to keep the signal passed to the next section at the correct level. If the input signal were to fail, the automatic gain control would increase gain and produce a large noise signal. The carrier reinsertion unit operates at IF.

EQUALIZERS

In order to keep the signal free from distortion the amplitude and phase responses of the transmission path have to be carefully controlled, as discussed in Section 6.9. The equalizers provide the necessary corrections.

ANTENNAS

As an indication of the range of antennas in use in the system, Table 9.1 is reproduced from Martin-Royle *et al.*, 1977.

Table 9.2 Parameters of a troposcatter link.

Type of link	Analogue, FDM
Frequency (single way paths)	1.9–2.3 GHz 2.5–2.65 GHz
Spectrum width	3 MHz
Transmitter output power	1 kW
Feeder losses	5 dB
Onshore antenna gain	49 dB
Offshore antenna gain	43 dB
Scatter angle	16.2 mrad
Great circle distance	225 km
Transmitter antenna height above mean sea level	220 m
Receiver antenna height above mean sea level	40 m
Free space loss at 2 GHz	145.5 dB
Median scatter loss	62.2 dB
Scatter loss not exceeded for 99.99% of time	84.7 dB
84% confidence limit at 99.99% of time	6.1 dB
Receiver threshold	−126 dBW

9.8.2 A troposcatter link

The problem of providing telephone, telex and data communication between land and North Sea oil rigs was solved with a troposcatter link, fully described in Hill (1982), pp 42–8 and 70–9, which also outlines the various systems aspects of the project. Briefly, the distance in excess of 200 km is too far for a line-of-sight microwave link, and submarine cable would be both expensive and prone to damage in those waters. The general principles of troposcatter links were presented in Section 9.5, and this section will be restricted mainly to the system parameters appropriate to this particular case. In Table 9.2 the important parameters are displayed.

GENERAL CONSIDERATIONS

As indicated in Section 9.5, it is necessary to keep the scattering angle as low as possible, which implies that antennas should be as high as possible above sea level. The height of the shore-based antenna is dictated by terrain, and 220 m represents this factor. The off-shore height is limited by the size of the platform. The different structures that may be accommodated are reflected in the difference between antenna gains. To combat the severe fading experienced on such a link some form of diversity is used. In this case two antennas 100

Plate 3 British Telecom microwave links. (a) The Telecom Tower in London, which supports a variety of antennas. Horn-paraboloid antennas (Section 3.10.1, Figure 3.16(b)) are visible, together with parabolic reflector antennas incorporating feed systems of various patterns. Some use front horn-feeds supported in different ways; others are fed from the rear, using a subreflector. 'Drum skin' weather protecting radomes appear on some antennas. (b) 'Billboard' reflector antennas at the shore terminal of a troposcatter link to North Sea oilfields (Section 9.8.2). The offset feeds may be identified, separately mounted in front of the long focus reflectors. *(Photographs courtesy of British Telecom.)*

(a)

(b)

wavelengths apart are provided at each end of the link, all equipped to receive both vertically and horizontally polarized signals. An identical transmission signal feeds both antennas at the same end, except that the one transmits vertical polarization and the other horizontal. Thus, diversity is provided by four effectively independent paths. In this connection, Figure 9.21 illustrates typical variations of signal strength over a single path of 265 km: short term variations of the order of ±15 dB take place, superimposed on larger seasonal changes. The process by which the four paths are combined is considered later. The two directions on a link operate at different frequencies: in the 1.9–2.3 GHz band the separation is 213 MHz, in the 2.5–2.65 GHz band, it is 95 MHz. The general outline of this arrangement is illustrated in Figure 9.22.

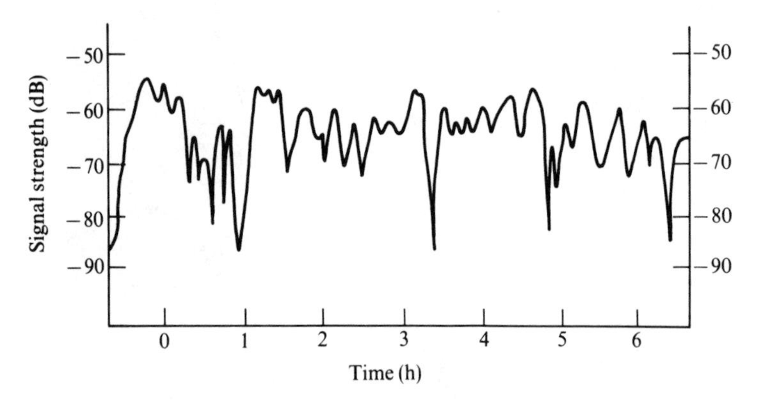

Figure 9.21 Typical record of received signal strength on a 250 km troposcatter link, showing magnitude of fading.

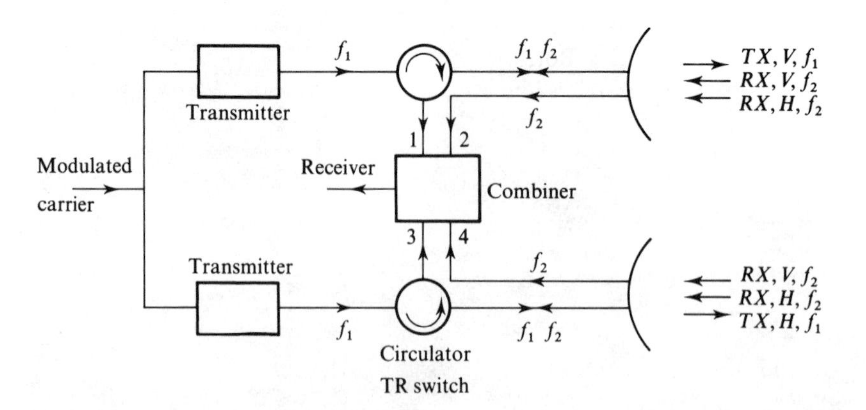

Figure 9.22 The layout of transmit/receive paths to provide 4-path diversity on the troposcatter link described in Section 9.8.2.

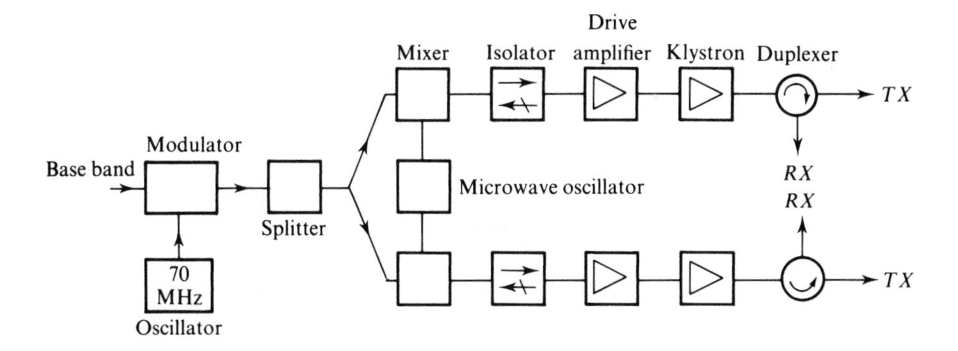

Figure 9.23 Configuration of the transmitter section of the troposcatter link described in Section 9.8.2.

ANTENNAS

The shore based antenna is an offset parabolic reflector some 18 m × 18 m, a so called billboard antenna. It gives a beam approximately 0.5° wide, with a gain of 49 dB. The off-shore antenna has to be smaller, and has a parabolic reflector some 9 m in diameter with centre feed. Mesh construction is used. The feeds for both types are responsive to both vertical and horizontal polarization, and so are 2-port devices. In Figure 9.22 the circulators act to separate the transmit and receive paths of similar polarization; the orthogonal polarization appears at its own port. The two transmitters in the diagram are fed from the same carrier and give coherent signals.

TRANSMISSION CHANNEL

The transmission channels are illustrated in Figure 9.23. The base band signal frequency modulates an IF carrier subsequently frequency translated to the microwave frequency. Both mixers are supplied from the same oscillator, ensuring coherent carriers in the two channels. Subsequently a drive amplifier feeds to the klystron power amplifiers. The klystron used has four tunable cavities giving a bandwidth of 13 MHz to half-power points and capable of delivering 1 kW of RF power. The magnetic field is provided by a large permanent magnet. The power gain is 37 dB.

RECEIVER

A single receiver chain is shown in Figure 9.24. Four such identical chains are used, one for each receive path. The outputs go to a combiner, discussed later. The RF filter has a bandwidth of 35 MHz; the IF filter bandwidth is 3 MHz and its design is critical to system performance. The RF amplifier is a low

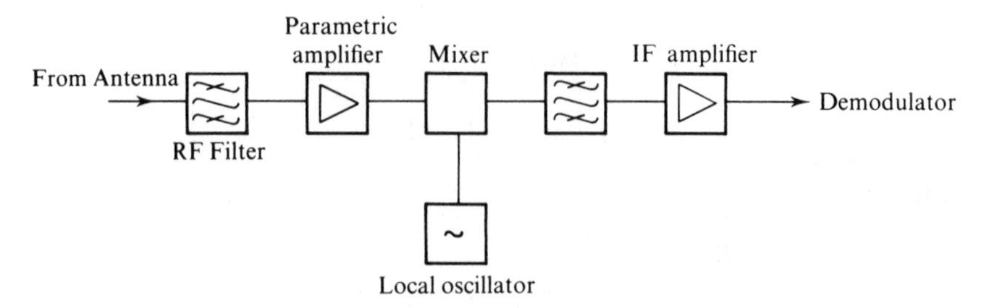

Figure 9.24 Configuration of the receiver section of the troposcatter link described in Section 9.8.2.

noise parametric amplifier. These typically have a gain of 18 dB over a bandwidth of 19 MHz, with noise figure of 1.8 dB.

DIVERSITY PATH COMBINER

The problem of combining the outputs from the different paths in diversity reception was encountered earlier. Before direct addition can take place at IF, phases have to be adjusted, since each path may have a different phase. In Figure 9.25 the method of phase adjustment is shown schematically. Each unit forms a self-regenerative phase shifter. The combiner produces its output at a design frequency ω_2 with a predetermined level F. Assuming the mixers to be product mixers and the filters F_1, F_2 to pass approximately $\omega_1 - \omega_2$ and ω_2 respectively, the signals shown in the diagram will be found. Thus, combining the four outputs will give a signal that depends on the sum of the squares of the four channel amplitudes. The input angular frequency ω_1 corresponds to 70 MHz and ω_2 to 59.3 MHz. This arrangement gives a 5 dB improvement in signal-to-noise ratio over a single channel, and in the presence of rapid fading an improvement of over 30 dB in signal strength.

The output from the combiner must then be demodulated and the base band signal recovered.

9.8.3 Digital microwave link

This section describes some of the features associated with a microwave link operating in a higher frequency range than that described in Section 9.8.1 and which uses digital signals. Such a link is in regular use at 11 GHz and higher frequencies, forming a constituent part of large networks. Barber and Godfrey (1985) and Hyamson *et al.* (1979) provide examples of two such systems, of which the latter, operated by British Telecom, will be used as a basis for discussion. The general concepts follow closely the lines of the analogue system of Section 9.8.1, and in most respects the layout of Figure 9.19. Differences lie partly in the details of the microwave circuitry for the

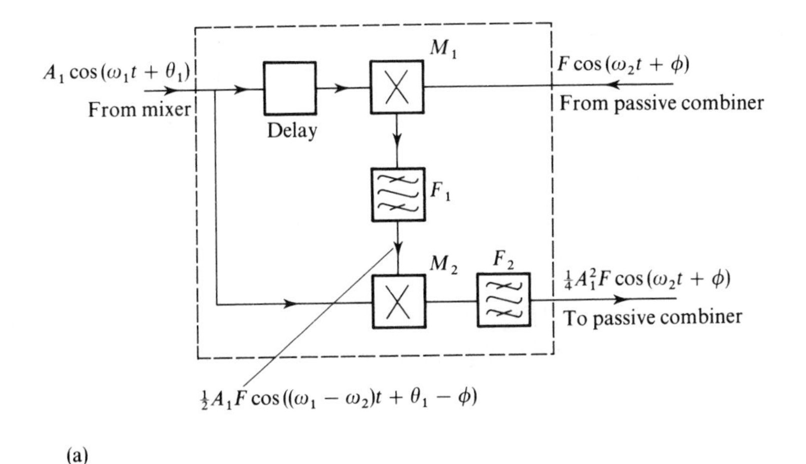

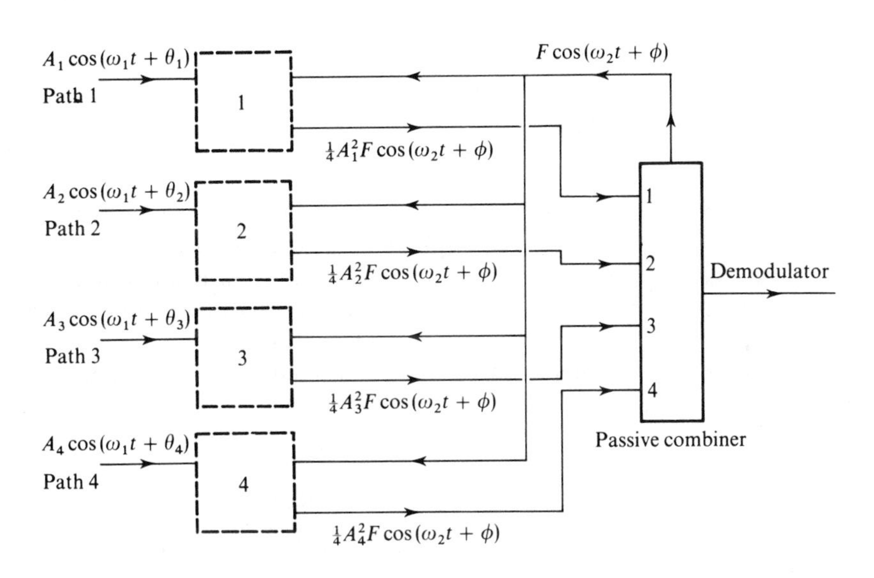

Figure 9.25 Illustrating the method of combining signals over the four paths of a diversity system: (a) single self-regenerative phase correcting circuit; (b) arrangement to combine four signals.

higher frequency and of antenna dimensions, and partly in the modulation methods employed. In the analogue system using frequency modulation, the modulation process occurred at intermediate frequency subsequently up-converted to the microwave carrier. This may also be the case with digital modulation, but digital modulation is also frequently performed directly on the carrier followed only by power amplification. This would be either a

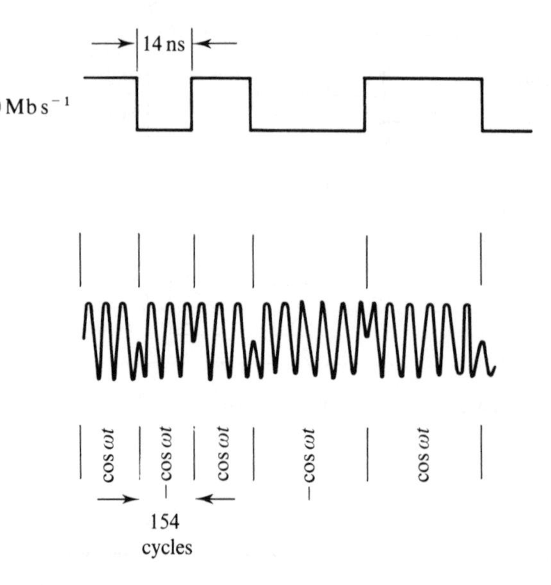

Figure 9.26 Waveform for PSK modulation.

travelling wave tube or solid-state amplifier typically delivering a few watts. For the various types of digital modulation, reference must be made to the literature, e.g. Stremler (1982). A type which has found considerable favour in the present application is 'quadrature phase shift keying' commonly referred to as QPSK. This form of modulation and its implementation at carrier frequency is described later in this section. It is convenient to consider first simple phase shift keying.

PSK

The waveform associated with simple phase shift keying is shown in Figure 9.26. In this method a binary input stream is used to change the phase of the carrier by 180° at the transition from digit value one to digit value zero, and back again on the reverse transition. The diagram is based on a binary data stream at 70 Mb s^{-1} and a carrier frequency of 11 GHz. The phase reversal can be implemented by the circuit configuration shown in Figure 9.27. The carrier at the input port one of the circulator is reflected at the termination of port two and then directed to port three. Thus if the reflection coefficient is $\Gamma = K \exp(j\phi)$, the amplitude is multiplied by K and phase advanced by ϕ (in addition to the effect of path length). If the termination is an ideal diode with zero forward and infinite reverse impedance, change of state of the diode will change the carrier phase by 180°. Thus, a binary stream controlling the diode will modulate the phase of the carrier as required. Demodulation is carried out at intermediate frequency, 140 MHz, and is accomplished by mixing with

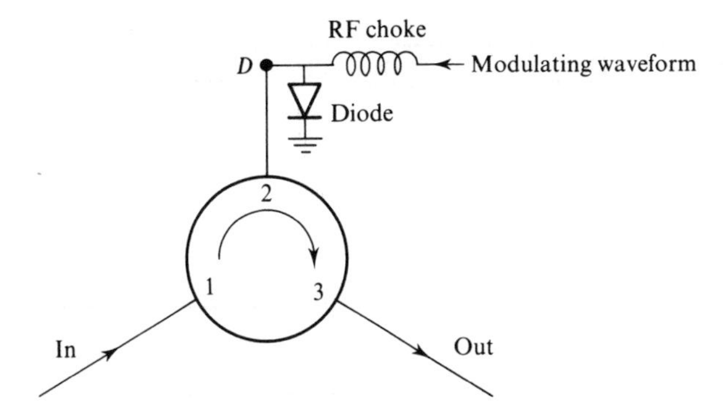

Figure 9.27 Diode modulator for PSK.

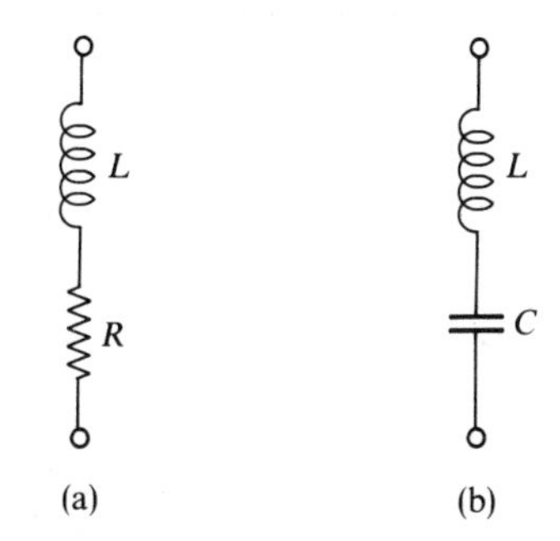

Figure 9.28 Equivalent circuit for diode used in modulation: (a) forward bias; (b) reverse bias. Diodes can be made for which at 11 GHz, $L = 0.15$ nH, $C = 0.1$ pF; R will be of the order of 0.5 Ω.

a local reference oscillator, when a positive output results from an in-phase signal, and negative from an out-of-phase signal. The local oscillator has to be phase locked to a signal recovered from the carrier. To obtain a spectral line to which the reference oscillator may be locked, it is necessary to pass part of the signal at IF to a non-linear device, such as a full wave rectifier circuit, from which the second harmonic is extracted. The reference oscillator at IF is then phase locked to this second harmonic.

This scheme of things depends on the diode being ideal. In practice the equivalent circuit of a diode is as shown in Figure 9.28, but with additional components, a phase reversal of 180° in reflection coefficient can be established.

QPSK

Although simple to implement, PSK is limited to a maximum data rate of 1 bit Hz^{-1}, as a result of only two possible states. In QPSK four possible states

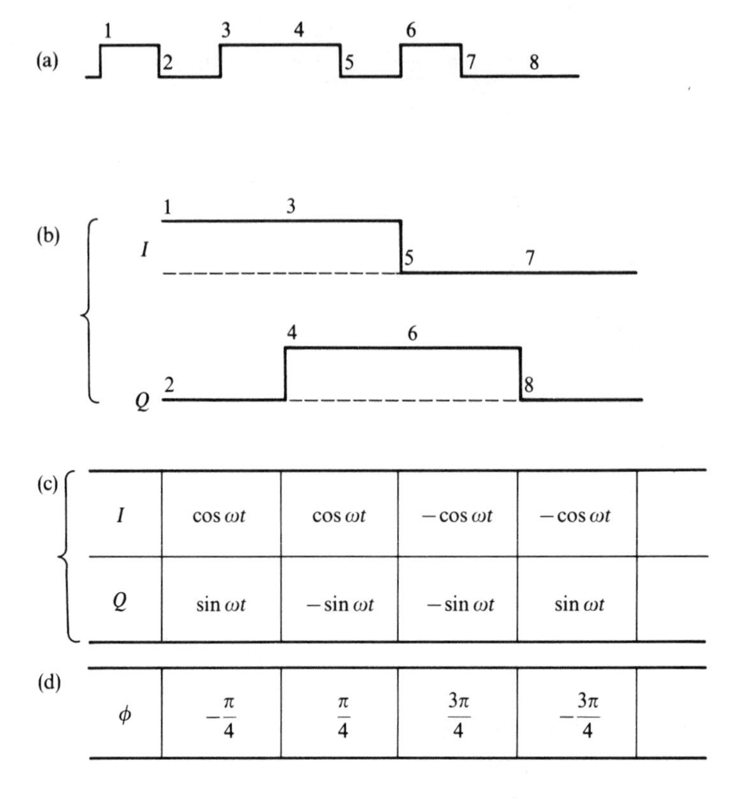

Figure 9.29 QPSK modulation: (a) data stream (140 $\mathrm{Mb\,s^{-1}}$); (b) I, Q data (70 $\mathrm{Mb\,s^{-1}}$); (c) carrier in symbol period for I, Q channels; (d) phase of resultant carrier.

occur, giving a maximum rate of 2 bit $\mathrm{Hz^{-1}}$. The states are four possible phases 90° apart. The method is most simply explained by division into in phase (I) and quadrature (Q) channels, as indicated in Figure 9.29. The incoming binary data stream (at 140 $\mathrm{Mb\,s^{-1}}$), Figure 9.29(a), is divided to produce two 70 $\mathrm{Mb\,s^{-1}}$ streams, Figure 9.29(b). Each of these streams modulates a carrier as in the PSK system outlined previously, with the carrier in the quadrature channel leading that in the in phase channel by 90°. This is shown in Figure 9.29(c). The two carriers are then summed, and the resultant transmitted carrier has in each symbol period the phase shown in Figure 9.29(d). Demodulation is effected by splitting the received carrier into two channels. In one channel, phase is compared with an in phase carrier, in the other with a quadrature carrier. It will then be found that the I and Q streams are obtained and can be recombined to form the initial 140 $\mathrm{Mb\,s^{-1}}$ stream. As in PSK the reference oscillator for demodulation must be derived from the incoming carrier. For QPSK this involves extracting the fourth harmonic. This is all done at intermediate frequency.

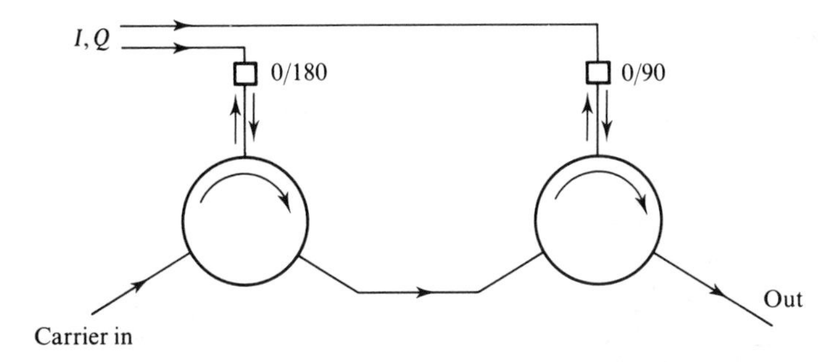

Figure 9.30 Serial switching for QPSK modulation.

Various alternatives exist for the implementation of the scheme of modulation. For example, the two PSK modulators could be fed from the same carrier. Use of a 3 dB hybrid junction to combine the two outputs would result in one carrier being in phase quadrature with the other. This gives the same output as shown in Figure 9.29. Another possibility, actually chosen for the British Telecom system, is to use serial switches. This scheme is shown in Figure 9.30. One diode is as in the PSK modulator, giving 180° phase shift between states. The other is arranged with suitable components to give only a 90° phase shift. The serial use of 0/180° and 0/90° switches gives the same four possible states for the output carrier. (It is necessary to recode the I and Q streams to the serial switches in order to realize the same carrier output as in Figure 9.29(d).) In a practical realization, allowance must be made for imperfections in the circulators, and for change in magnitude of the reflection coefficients because of the non-zero resistance of the diode in the forward direction.

SYSTEM CONFIGURATION

Figure 9.31 shows in simplified form the configuration of a repeater stage. The overall similarity to that of Figure 9.19 will be apparent. Apart from RF modulation, the other difference is the use of a regenerator after the

Table 9.3

Channel width	67 MHz
Transmitter power	10 W
Overall feeder and multiplex loss	15 dB
Antenna gain	49 dB
Noise figure	8 dB
Receiver noise bandwidth	80 MHz

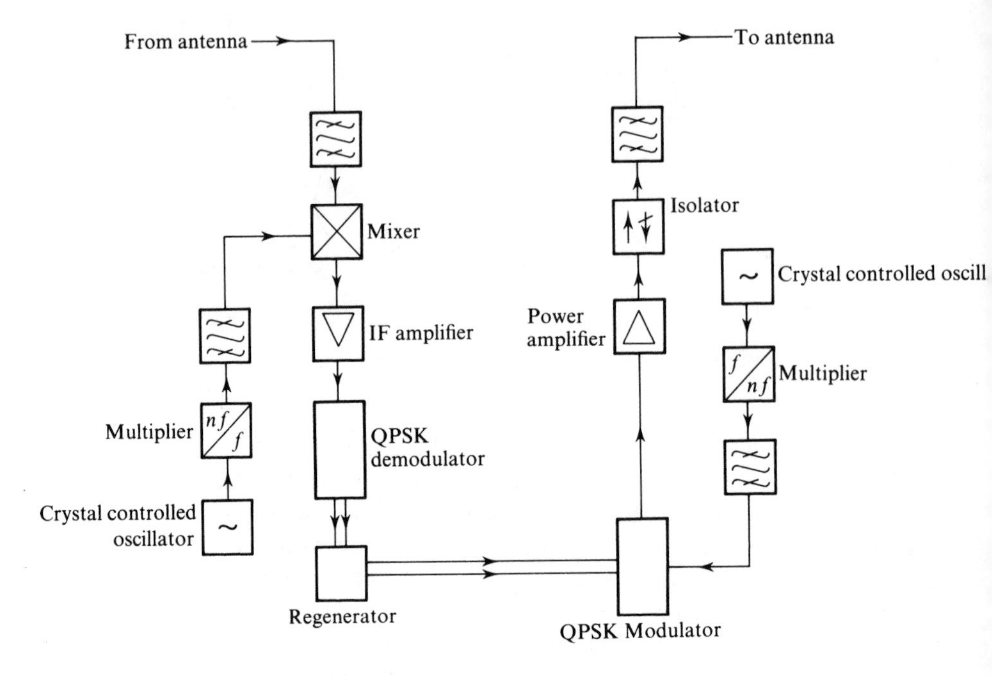

Figure 9.31 Schematic form of 11 GHz digital repeater.

demodulator, rather than an amplifier. At a terminal stage, the output from the regenerator would be recoded into a single data stream, and similarly the input data stream would be recoded and applied to the QPSK modulator. Supervisory circuits have not been shown. Some characteristics are given in Table 9.3.

EXERCISES

9.1 A system operating at 2 GHz uses antennas 3 m in diameter. The transmitter power is 2 W and the receiver has a bandwidth of 80 MHz with a noise figure of 5 dB. The signal-to-noise ratio at the demodulator is to be 23 dB to ensure the required error rate. Estimate the maximum length of a single link which will allow a fading margin of 35 dB. The effective area of an antenna may be assumed to be 65% of the physical aperture and the antenna noise temperature may be assumed to be 290 K. Ground reflections may be neglected.

9.2 Assuming that equation (9.17) is applicable to the situation described in question 9.1, estimate the time during a worst month that the system, operated with the calculated hop length, will fall below requirements due to fading. Use this result to obtain an estimate of the percentage of time out per year (taken as three worst months) for a connection involving 10 such links.

9.3 Assuming the data of Figure 9.1 applies for both oxygen and water vapour, estimate the attenuation between stations 40 km apart at frequencies of 6 GHz, 20 GHz and 100 GHz. Change of density with height should be neglected. Is this a plausible assumption?

9.4 A transmitting antenna is 130 m above ground level at a distance of 30 km from a receiving antenna 50 m above ground level. Estimate, assuming an effective Earth radius of 4300 km, the least clearance above a smooth Earth. Estimate also the angles made by the ray to the horizontal at both antennas.

9.5 The effective refractive index above a flat Earth decreases with height according to the formula

$$n^2(z) = n_0^2(1 - 2\kappa z)$$

Show that a ray that leaves the ground at a small grazing angle ε has the equation

$$z = -\tfrac{1}{2}\kappa x^2 + \varepsilon x$$

the origin of coordinates being taken at the launch point.

Such a ray is received back at ground, at a distance L from the launch point. Obtain expressions, in terms of L, for the launching angle and the maximum height reached. The time delay in traversing an element of path $\mathrm{d}s$ is $n\,\mathrm{d}s/c$. Show that, if terms in ε^4 are neglected, the total time delay is given by the expression

$$\frac{n_0 L}{c}\left(1 - \frac{\kappa^2 L^2}{24}\right)$$

Show that this last result is also obtained by using a straight-ray model with an effective Earth radius equal to $-\kappa^{-1}$.

9.6 A radio link at a frequency of 4 GHz is set up over flat terrain between two antennas 20 km apart on towers 40 m high. The reflection coefficient for the reflected ray is $0.3 \angle 0$. Show that the received signal strength is about 1.7 dB above that for free space transmission. Show also that as the receiving antenna is raised and lowered variations of about 5 dB in signal strength would be expected.

9.7 It is necessary to set up a microwave link at 2 GHz across some rolling country over a distance of some 50–60 km. Preliminary study of the map indicates two likely places for the ends of the link 55 km apart.

Both are hills, end A at 220 m, end B at 200 m. The major heights in between are at C, D and E: C is 18 km from A, height 120 m; D is 30 km from A, height 90 m; E is 10 km from B, height 140 m. Standard towers are 25 m high. Investigate path clearances by making a diagram as in Figure 9.9 and make proposals for the salient characteristics of the link, assuming an effective Earth radius of 4500 km. A ratio of 100 : 1 between horizontal and vertical scales is suggested.

9.8 Using the theory presented in earlier sections, investigate and discuss the system designs specified in Section 9.8.

References

Abramowitz, M. and Stegun, I. A. (1968). *Handbook of Mathematical Functions.* New York: Dover.

'Advanced mobile phone system'. (1979). Special Issue, *BSTJ*, **58**, January.

Barber, S. G. and Godfrey, B. W. (1985). 'Design considerations for a family of 64QAM digital radios'. *J. Elect. and Electronic Eng., Australia*, 468–71.

Chamberlain, J. K. and Price, A. J. (1983). 'Improved agile notch adaptive equalizer'. *Electronics Letters*, **19**, September.

Clarke, R. H. and Brown, J. (1980). *Diffraction Theory and Antennas.* Chichester: Ellis Horwood.

Collin, R. E. (1985). *Antennas and Radiowave Propagation.* New York: McGraw-Hill.

Cooper, G. R. and Nettleton, R. W. (1983). 'Cellular mobile technology – the great multiplier'. *IEEE Spectrum*, 30–7.

Crawford, A. B. and Jakes, W. C. (1952). 'Selective fading of microwaves'. *BSTJ*, **31**, 68–90.

Delange, O. E. (1952). 'Propagation studies at microwave frequencies by means of very short pulses'. *BSTJ*, **31**, 91–103.

Feher, K. (1981). *Digital Communications: Microwave Applications.*

Hall, M. P. M. (1979). *Effects of the Troposphere on Radio Communication.* Stevenage: Peter Peregrinus.

Harden, B. N., Norbury, J. R. and White, W. J. K. (1978). 'Estimation of attenuation by rain on terrestrial radio links in the UK at frequencies from 10–100 GHz'. *Microwaves, Optics and Acoustics*, **2**, 97–104.

Hill, S. J. (1982). 'British Telecom transhorizon radio services to offshore oil/gas production platforms'. *Brit. Tel. Comm. Eng.*, **1**, 42–8, 70–9.

Hyamson, H. D., Muir, A. W. and Robinson, J. M. (1979). 'An 11GHz high capacity digital radio system for overlaying existing microwave routes'. *IEEE Trans.*, **COM-27**, 1928–37.

Jakes, W. C., Ed. (1974). *Microwave Mobile Communications*. New York: Wiley.

Lundgren, C. W. and Rummler, W. D. (1979). 'Digital radio outage due to selective fading – observation vs prediction from laboratory simulation'. *BSTJ*, **58**, 1073–100.

Martin-Royle, R. D., Dudley, L. W. and Fevin, R. J. (1976). 'A review of the British Post Office microwave radio-relay network (Parts 1 and 2)'. *POEEJ*, **69**, 162–8 and 225–34.

Martin-Royle, R. D., Dudley, L. W. and Fevin, R. J. (1977). 'A review of the British Post Office microwave radio-relay network (Part 3)'. *POEEJ*, **70**, 45–54.

Matthei, G., Young, L. and Jones, E. M. T. (1980). *Microwave Filters, Impedance Matching Networks and Coupling Structures*. Norwood, Mass.: Artech House.

Picquenard, A. (1974). *Radio Wave Propagation*. London: Macmillan, Section 8.3.2.4.

Rummler, W. D. (1979). 'A new selective fading model: application to propagation data'. *BSTJ*, **58**, 1037–71.

Sander, K. F. and Reed, G. A. L. (1986). *Transmission and Propagation of Electromagnetic Waves* (2nd Edn.). Cambridge: Cambridge University Press.

Stremler, F. D. (1982). *Introduction to Communication Systems* (2nd Edn.). Reading, Mass.: Addison-Wesley.

Subramanian, M., O'Brien, K. C. and Puglis, P. J. (1973). 'Phase dispersion characteristics during fade in a line-of-sight radio channel'. *BSTJ*, **52**, 1877.

Vigants, A. (1975). 'Space-diversity engineering'. *BSTJ*, **54**, 103–42.

CHAPTER 10

SATELLITE SYSTEMS

OBJECTIVES

This chapter is concerned with satellite communication systems. It firstly discusses orbits and then describes the microwave configuration of a satellite transponder. The factors determining performance are analysed.

10.1 Introduction

The microwave links considered in the last chapter are restricted to line-of-sight paths, and are thereby restricted in length, if only because of the curvature of the Earth's surface. A communication satellite enables long paths to be set up between points on the Earth's surface from both of which the satellite is visible. The nature of the satellite has changed considerably over the few decades of development. In 1961 the 'Echo' project (see 'Project Echo', *BSTJ* **40**) involved a purely passive reflector, a balloon with a conducting skin. Within a few years the first active satellite was placed in orbit, acting as a relay station by receiving and retransmitting after amplification. These early satellites were spin stabilized. The antenna was part of the satellite body, rotated with it, and was designed to give a toroidal shaped radiation pattern. This inevitably meant that a large part of the radiated power, which totalled only about 10 W, went to waste as far as reception at a single Earth station was concerned. The very weak signals dictated large parabolic reflectors and very low noise receivers. Subsequent development has allowed an increase in RF power for transmission, but the biggest development has been in antenna design. The technique of 'despinning' the antenna platform (rotating it about the axis of spin in a direction opposite to the spin of the satellite) allowed the use of directional antennas: for example, Intelsat III mounted an Earth coverage horn. More recently, reflector antennas giving spot beams have been included together with configurations designed to produce a particular 'footprint' of coverage on the Earth's surface. This is accomplished by using a multiplicity of feeds to a single reflector, so that the assembly of beams creates the desired pattern. In this way the available power is directed in an optimum fashion. Apart from permitting Earth stations to use smaller reflectors, this technique has given greater flexibility in the way satellite circuits can be used.

The following sections describe in rather greater detail some microwave configurations appropriate to present satellite systems. For the considerations relating to launch and deployment reference should be made to Fthenakis (1984) and Wise (1978). A historical overview can be found in Rudge (1985).

10.2 Orbits

The simplest orbit to visualize is one which is circular, necessarily in a plane containing the centre of the Earth. Textbooks on dynamics show that the period in hours of such an orbit is given by

$$T = 0.0176\,R^{3/2}/a \tag{10.1}$$

in which R is the orbit radius in kilometres and a the Earth radius, 6380 km. Thus a relatively low-flying satellite with altitude 1500 km, suitable for investigation of the Earth's surface, has an orbit period just under 2 h.

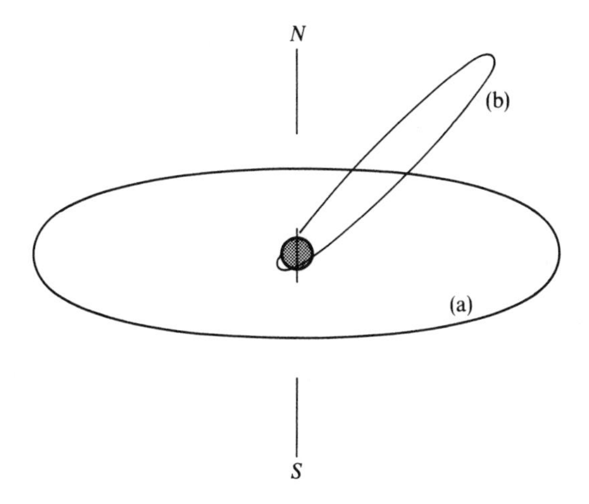

Figure 10.1 Orbits: (a) geostationary, radius 42 300 km; (b) typical Molniya orbit, greatest altitude 39 500 km, minimum 500 km.

The distance and bearing from a fixed earth station vary rapidly with time, a great disadvantage if the object is to set up a communication link between two points. To maintain such a link a number of satellites would be necessary to ensure that a line of sight was always available. Such an arrangement is used for a navigational satellite system.

It can be verified from equation (10.1) that an orbit with altitude of about 36 000 km has a period of 24 h, and thus a satellite in such an orbit on the Earth's equatorial plane will remain stationary with respect to the Earth. (More precisely, the required period is one sidereal day equal to 23 h 56 min 4 s.) This is the **geostationary orbit**: the parameters of this orbit will be considered in Section 10.2.1.

In high latitudes, a satellite in the geostationary orbit will be low over the horizon or below it. The USSR has made use of a highly elliptical orbit, as shown in Figure 10.1, for domestic communications. The period of such an orbit is given by equation (10.1) with R taken as the major semi-axis of the ellipse. One of Kepler's laws of planetary motion states that the radius vector from Earth centre to satellite sweeping out equal areas in equal times. Hence the satellite will remain visible for a relatively long time while passing through its most distant point. Several Molniya satellites in such orbits are used. The orbit shown has a period of about 12 h.

10.2.1 The geostationary orbit

The geometry of the geostationary orbit is shown in Figure 10.2. It will be seen that the angle subtended by the whole Earth at the satellite is about 17°, and that the line of sight grazes the Earth's surface at about latitude 81°. At very low angles of elevation, atmospheric path effects and noise radiated by

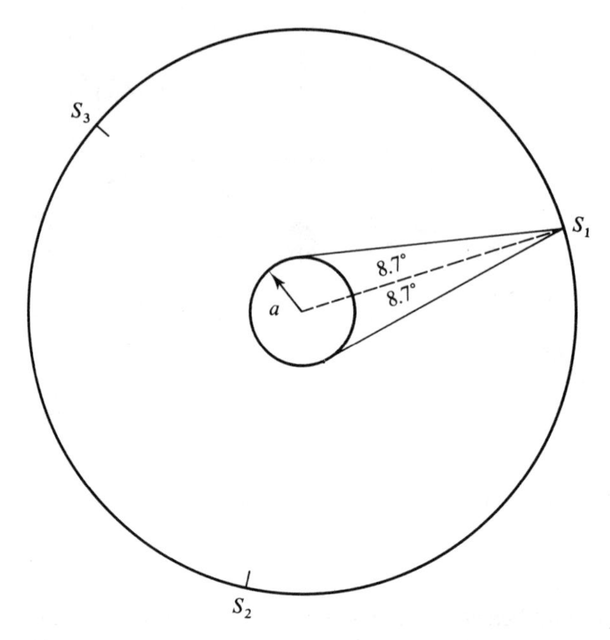

Figure 10.2 The geometry of the geostationary orbit and global coverage by three satellites. Earth radius, $a = 6380$ km, orbit radius 42 300 km.

the Earth render the satellite link unsatisfactory, so that it is not practical to site an Earth station above latitude 75°. Calculation of the angle of elevation (angle above horizon) for the general case leads to the result

$$\sin \alpha = \frac{\cos \phi \cos \theta - a/R}{(1 + (a/R)^2 - 2(a/R)\cos \phi \cos \theta)^{\frac{1}{2}}} \tag{10.2}$$

in which α is the angle of elevation, θ the latitude of the Earth station and ϕ the difference in longitude between Earth station and satellite.

Apart from high latitudes it can be seen from Figure 10.2 that three equally spaced satellites will give global coverage. Geography dictates that two of these positions would correspond to mid-Atlantic and mid-Indian ocean, both approximately 60° of longitude away from the UK. Thus, putting $\phi = 60°$, $\theta = 52°$, $a/R = 0.151$ in equation (10.2) it is found that $\alpha \cong 10°$. For a satellite in the same longitude as an Earth station, the elevation would be 30°.

The long path length between satellite and Earth station has two effects: a long time delay, and increased free space loss. The two-way distance is about 75 000 km, giving a delay of 250 ms. The one-way free space loss may be calculated as 196 dB at 4 GHz and 205 dB at 11 GHz.

The above treatment relates to an ideal state of affairs, treating the Earth as effectively a point mass and ignoring external perturbations. In reality, a perfect geostationary orbit does not exist: the Earth's gravitational field is asymmetric, other gravitational pulls have an effect and radiation pressure causes perturbations. A satellite in the nominally geostationary orbit will tend to drift in the east–west plane, and it must be kept on station by thrust jets at appropriate intervals. Regulations require station keeping to be held to $\pm 0.1°$. A deviation in the north–south plane leads to diurnal variation of the satellite position as seen by an Earth station, a factor which influences the size that an Earth station antenna may profitably have, unless tracking facilities are provided. At present stabilities of the order of $\pm 0.1°$ can be achieved. It may be noted that diurnal variations will give rise to Doppler shifts which may be significant in high bit-rate digital systems. At 12 GHz, a variation of $\pm 0.1°$ will give a Doppler shift of ± 110 Hz.

Another factor of the geostationary orbit is that at certain periods of the year a satellite will fall into the Earth's shadow for some time during each day, when the solar cell array will no longer provide primary power. Depending on the power level required, operation may be carried on using storage cells. It can also happen around the equinoxes that the sun as well as the satellite lies within the beam of the earth station antenna. Because of noise radiated by the sun, severe degradation of the signal-to-noise ratio may occur.

The geostationary orbit is limited in capacity. A satellite is acting as a relay to a particular Earth station, and this station acts as interference to an adjacent satellite. At present a separation of 2° is aimed at, representing 180 satellites at uniform spacing. The number of satellites in orbit is of this order.

10.3 Physical parameters

From the point of view of establishing a communication link, the factors of most interest are antenna dimensions and primary power available. Both of these are influenced by the ability to deploy structures after launch as well as

Table 10.1 Communication satellites

Satellite	Intelsat I	Intelsat V
Year of launch	1965	1979
Mass (kg)	38	967
Primary power (W)	33	1200
RF power (W)	10	200
Effective bandwidth (MHz)	50	2200*
Telephone circuits	240	12 000
Deployed size (m)	0.72 × 0.59	15.8 × 6.8

* Achieved by frequency reuse within a bandwidth of 500 MHz.

by the permissible weight. They are further influenced by the ability to stabilize the satellite with respect to the Earth and the Sun. Whereas early satellites were stabilized by spinning about an axis, modern satellites can be stabilized with respect to three axes in space. Coupled with the ability to deploy antenna structures and large arrays of solar cells, performance has increased dramatically, and continues to do so. Some figures relating to the Intelsat series of communication satellites are shown in Table 10.1.

The projected Olympus series of communications satellites will have launch weights in the range 2300–3500 kg and primary power capabilities of 3.5–7.5 kW.

10.4 RF configuration

In its simplest conception, a commmunication satellite receives a transmission from an Earth station, converts it to a lower frequency, amplifies and then retransmits. A schematic diagram of such a **transponder** is shown in Figure 10.3. The change in frequency eases the need for isolation between receive and transmit antennas which would be required were the same frequencies to be used. The frequency pairs 6/4 GHz and 14/11 GHz are in current use. The power amplifier mainly used to date has been the travelling wave tube, in spite of the obvious problems of lifetime associated with an electron tube in the space environment. It has been found possible to design tubes with adequate lifetime and that are sufficiently robust. These are used on account of attainable power and efficiency. Some solid-state amplifiers have been used, and more will be developed in the future. To cover the seven to ten year lifespan normally given to a satellite, considerable redundancy is provided, which can be brought into use by remotely controlled switching. One property of the TWT power amplifier is non-linearity at maximum output; this causes intermodulation with frequency division multiplex signals, and to limit intermodulation noise the amplifier will be operated at less than maximum output power. Efficiency of the order of 50% may be achieved. The transponders in a communications satellite might well each have a bandwidth of 36 MHz, each separated by 40 MHz from the adjacent channels, and provide RF output powers of the order of 20–40 W. Such a

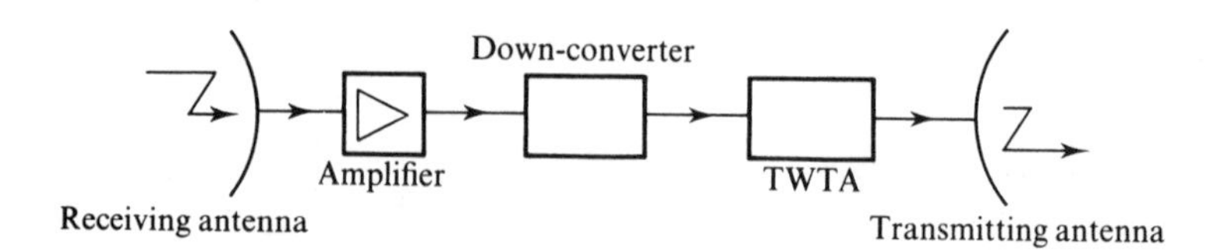

Figure 10.3 Schematic diagram of a satellite transponder. TWTA, travelling wave tube amplifier.

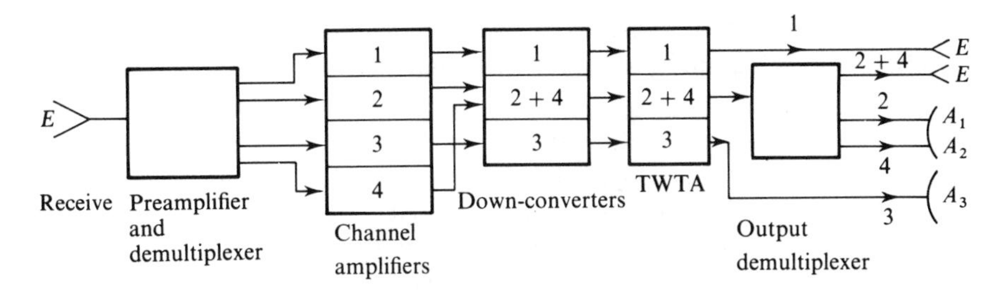

Figure 10.4 Illustrating a possible configuration of a multi-channel transponder. Antennas: E denotes Earth coverage horn; A_1, A_2, A_3 directional antennas of varying coverages, including a spot beam.

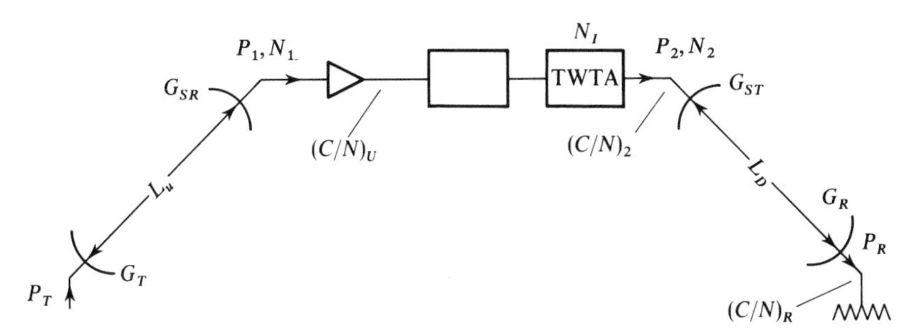

Figure 10.5 Showing the various stages of an Earth-satellite link and the effect on carrier-to-noise ratio. Symbols are defined in Section 10.5.

transponder will handle some 300 (two-way) telephone circuits or one TV channel.

Figure 10.3 represents a single channel in its simplest form. In practice there will be many transponders all fed from the same antenna, the outputs of which will be recombined and distributed among different transmitting antennas. A more detailed example is portrayed in Figure 10.4.

10.5 System performance

As in the case of terrestrial links outlined in Chapter 9, satisfactory performance depends on maintaining an adequate carrier-to-noise ratio at the receiving end of a link, although in a satellite link there are far more limitations on the component parameters than in the terrestrial case. Figure 10.5 shows in schematic form the parameters involved in a satellite link from Earth station transmitter through a transponder back to an Earth station receiver. The symbols in the diagram have the following significance:

P_T	Transmitter power
G_T	Antenna gain, Earth transmitter
G_{SR}	Antenna gain, satellite receiver
G_{ST}	Antenna gain, satellite transmitter
G_R	Antenna gain, Earth receiver
L_U	Uplink loss, $(4\pi R/\lambda)^2$
L_D	Downlink loss $(4\pi R/\lambda)^2$
P_R	Received power
P_1, P_2	Powers at points indicated in Figure 10.5
N_1, N_2, N_R	Noise powers at points indicated in Figure 10.5
(C/N)	Carrier-to-noise ratio at indicated points
N_I	Intermodulation noise power added by TWTA

In addition, analysis must introduce

B	Bandwidth
T_A	Temperature of Earth antenna and receiver
T_S	Effective temperature of satellite antenna and receiver
G_S	Gain of transducer (P_2/P_1)

We have the following relations:

$$P_1 = P_T G_T G_{SR}/L_U \tag{10.3}$$

$$N_1 = kT_0 B \tag{10.4}$$

in which T_0 may be taken as Earth temperature since the Earth largely fills the radiation pattern of the satellite receiving antenna. It is sometimes convenient to denote by P_T^* the effective isotropic radiated power (Section 3.3) of the transmitter, $P_T G_T$.

By definition of effective noise temperature

$$\left(\frac{C}{N}\right)_U = \frac{P_1}{kT_S B} = \frac{P_T^*}{L_U} \cdot \frac{G_{SR}}{kT_S B} \tag{10.5}$$

At the transponder output

$$P_2 = G_S P_1 \tag{10.6}$$

$$N_2 = kT_S B G_S + N_I \tag{10.7}$$

Finally

$$P_R = P_2 G_{ST} G_R / L_D = P_2^* G_R / L_D \tag{10.8}$$

$$N_R = N_2 G_{ST} G_R / L_D + kT_A B \tag{10.9}$$

The final carrier-to-noise ratio is most simply expressed in the form

$$\begin{aligned}\left(\frac{C}{N}\right)_R^{-1} &= \frac{N_2}{P_2} + \frac{kT_A B}{P_R} \\ &= \frac{kT_S B}{P_1} + \frac{N_I}{P_2} + \frac{kT_A B}{P_R}\end{aligned} \tag{10.10}$$

The first term has been associated with the uplink through equation (10.5), and the last may similarly be related to the downlink by defining

$$\left(\frac{C}{N}\right)_D = \frac{P_R}{kT_A B} = \frac{P_2^*}{L_D} \cdot \frac{G_R}{kT_A B} \tag{10.11}$$

With this notation, we may write

$$\left(\frac{C}{N}\right)_R^{-1} = \left(\frac{C}{N}\right)_U^{-1} + \frac{N_I}{P_2} + \left(\frac{C}{N}\right)_D^{-1} \tag{10.12}$$

Note that in both equations (10.5) and (10.11) the ratio G/T occurs: this is taken as a figure of merit of the antenna and its receiving system. It is commonly quoted in the form $10 \log (G/T)$, and given units dB K^{-1}.

To investigate the significance of equation (10.12) it is convenient to consider a particular case, using practically relevant figures. We first note that to obtain global coverage from the satellite a conical beam of about 8.5° semi-angle is required. A direct calculation using the idealized result in Section 3.8.3 gives $G = 22$ dB. In practice a figure somewhat less than this would be realized, say 19 dB. Assume that two such antennas are present on the satellite, that the uplink operates at 6 GHz, and that the downlink operates at 4 GHz. Consider as the Earth station a parabolic reflector antenna 10 m in diameter. Allowing an aperture efficiency of 55%, the effective receiving area would be 4.3 m^2, corresponding to a gain of 53 dB at 6 GHz and 49.5 dB at 4 GHz. A realistic transmitter power is 1 kW, and satellite transmitter power 20 W. The noise temperature of the satellite receiving antenna will be close to 290 K, and the amplifier will be assumed to have a noise figure of 3 dB. The only remaining parameter to be specified, apart from P_2/N_I is the noise temperature of the Earth receiving system. An antenna at these frequencies pointing to the zenith in a clear sky will have an effective temperature of only a few degrees, which will be increased by the effect of antenna and feeder losses. To this of course has to be added the equivalent noise temperature of the receiver. For an antenna with low elevation, side lobes will receive Earth noise. A plausible value for the antenna temperature would be 10 K, and for a low noise amplifier, 80 K. (This corresponds to a noise figure of 1.05 dB, and would be a parametric amplifier.) A bandwidth of 36 MHz will be assumed, and a path length of 36 000 km. These values are displayed below:

$$
\begin{aligned}
P_T &= 1\,\text{kW} \\
P_2 &= 20\,\text{W} \\
G_T &= 53\,\text{dB} \\
G_R &= 49.5\,\text{dB} \\
G_{SR} = G_{ST} &= 19\,\text{dB} \\
L_U &= 199\,\text{dB} \\
L_D &= 196\,\text{dB} \\
T_S &= 580\,\text{K} \\
T_A &= 10 + 80 = 90\,\text{K} \\
B &= 36\,\text{MHz}
\end{aligned}
$$

From these figures we find

EIRP Earth transmitter, $P_T^* = 83\,\text{dBW}$
EIRP satellite transmitter, $P_2^* = 32\,\text{dBW}$
satellite receiver figure of merit, $(G_{SR}/T_S) = -8.6\,\text{dB}\,\text{K}^{-1}$
Earth station figure of merit, $(G_R/T_A) = 30\,\text{dB}\,\text{K}^{-1}$
satellite gain, $G_S = 110\,\text{dB}$.

Using the result $10\log(kB) = -153$, the carrier-to-noise ratio is evaluated as follows:

$$(C/N)_U = 83 - 199 - 8.6 + 153 = 28.4\,\text{dB}$$
$$(C/N)_D = 32 - 196 + 30 + 153 = 19\,\text{dB}$$

Ignoring intermodulation noise

$$(C/N)_R = 18.5\,\text{dB}$$

An allowance of 30 dB for P_2/N_I would reduce the value of $(C/N)_R$ to 18.2 dB. (The factor P_2/N_I is controlled by the output level of the TWTA. It will be necessary to run it at several decibels below saturation.) It is clear from this analysis that it is the downlink which dominates the overall performance. The value of 18 dB is too low for a high quality link, and no allowance has been made in the calculation for additional path losses incurred by the tropospheric segment of the path.

The original Intelsat I radiated 5 W into a toroidal beam of width 11°, corresponding to a loss of some 20 dB compared with the above calculations. The Earth stations, now referred to as Intelsat-A standards, used parabolic reflectors 30 m in diameter. The receivers originally used cryogenic maser amplifiers, subsequently replaced by parametric amplifiers. Reduction in Earth station complexity is a result of improvement in downlink performance, basically requiring greater EIRP from the satellite. Such improvement permits smaller reflectors and simpler front ends to the receivers, thus reducing costs. The output powers from travelling wave tubes has remained around the 20–40 W level, although higher powers have been used for special purposes, and are envisaged for direct broadcasting satellites. The great improvements which have taken place are in the design of antennas, and in the ability to launch and deploy them. Antennas are now used which restrict coverage to specific areas, thus making much better use of the RF power available. Further, the use of the 11/14 GHz band has increased antenna gains for the same size structures. As an example, the Intelsat V satellite has the following antenna 'farm' (e.g. Neusten and Marchant, 1978):

1 global beam	18° × 18°′	6/4 GHz link
2 hemispheric beams	14° × 5°	
2 zonal beams	9° × 3°	
2 steerable spot beams		
East spot	3.2° × 1.8°	11/14 GHz
West spot	1.6° × 1.6°	

Frequency reuse is possible in two ways: use of orthogonal polarizations, and isolation of receive/transmit antennas. The use of such beams is dependent on the ability to keep the satellite antennas correctly pointed which advances in satellite technology have made possible.

The Standard-A Intelsat receiving system referred to above not only calls for a reflector diameter of 30 m but requires the figure of merit (G/T) to be better than 40.7 dB K^{-1}. A Standard-B system which has a more limited capability uses a reflector 11 m in diameter with a figure of merit of 31 dB K^{-1}. Smaller earth terminals can be used for smaller traffic requirements.

10.6 Use of higher frequencies

The numerical investigation set out in the previous section related to a link using 6/4 GHz. Operation at the higher frequencies of 11/14 GHz is increasingly taking place. The additional factor that needs to be taken into account is the possibility of attenuation in the tropospheric segment of the downlink path. This was discussed in Section 9.3 with reference to terrestrial links, and is relevant to satellite links.

Measurements outlined by O'Neill and Hayter (1982) suggest that moderate rainfall seldom gives rise to attenuation of more than 2 dB, although this can rise to 7 dB in severe rain. The earlier discussion on diversity is relevant.

It must also be noted that the presence of an attenuating medium in the down path will, in addition to decreasing the signal received, also increase the noise received by the antenna. This effect is similar to that caused by attenuation in a feeder after the antenna.

10.7 Satellite antennas

The purpose of this section is to draw attention to the design requirements for satellite antennas, and not to present design methods, which are discussed more fully by Mittra *et al.* (1983). As the sections have shown, whilst early communication satellites had only one receive/transmit antenna, present day satellites operate many RF channels, employ orthogonal polarizations for frequency reuse, and employ beams shaped to suit the service areas.

10.7.1 Reflector antennas

Various configurations have been developed for different commercial satellites. The approach towards producing a shaped beam has been to use an offset parabolic reflector (Figure 3.16(a)) with a cluster of feed horns appropriately placed so that the overlapping beams give the correct shape of contour area. An offset reflector is used since the arrangement allows a large feed array to be used without causing aperture blockage. As an example Intelsat-A carries two transmit antennas, approximately 1.3 m × 1.3 m, one fed by an array of 37 horns and the other by an array of 19 horns. These all transmit right-hand circular polarization, at 4 GHz. The receive reflector is slightly smaller, and covers eastern and western hemispheres separately, using left-hand circular polarization. Switching within the feed horn arrays allows various beam shapes to be selected. This design uses different antennas for the two orthogonal polarizations. In other designs, the same geometrical surface can be used to support two effectively independent reflectors, one for vertical polarization, one for horizontal polarization. This is done by making the reflector of parallel strips which will reflect when the electric vector is parallel to the strips but not when it is transverse.

The mechanical construction of reflectors is usually of a fine metallic mesh mounted on an open rib structure to give shape. This reduces the pressure exerted by the solar wind. Associated with multiple feed antennas are beam forming networks, accomplished with the aid of power dividers and phase shifting networks.

10.7.2 Phased array antennas

Adjustable service areas can also be obtained by use of phased arrays, as discussed in Sections 3.6.5, 10.4 and 8.7.3. Depending on requirements, such an array can be used to produce a steerable beam (for example, to track a movable user) or to place a null to reduce the effect of a source of interference. Realization depends, as with radar arrays, on the availability of suitable RF components, bearing in mind the additional problems of weight and complexity in the satellite environment.

10.8 Direct broadcasting satellites

The previous discussion has centred on communication systems of the type run by public utilities, with large antennas and connected to a public network. Quite different is the concept of an individual homestead receiving television signals from a satellite using suitably cheap and robust equipment. Direct broadcasting has been given room in the range 11.7–12.5 GHz. In Europe, 40 channels of bandwidth 27 MHz with channel separation of 19 MHz have been allocated to the various European countries.

The service assumes that for reception a reflector of diameter 0.9 m will suffice to give a figure of merit of 8 dB K^{-1} when used with a GaAs low noise front end. The proposed power flux density at the edge of the coverage area is -103 dB W m^{-1}. The satellite transmit beam requires a beam width of the order of $1° \times 1°$. It is envisaged that a single satellite in geostationary orbit will provide five channels, with a total RF power in excess of 1 kW. Each transmit antenna will be tailored to provide the appropriate coverage area for the service it is to provide. A pointing accuracy of about 0.1° is assumed. The uplink channels are contained in the band 17.3–17.8 GHz, using an antenna 5 m in diameter with 1 kW input power total.

Summary

This chapter has outlined the features of a microwave communication link using a satellite. The constraints on system performance have been considered and expressions derived showing the dependence of carrier-to-noise ratio on the parameters involved.

EXERCISES

10.1 The transmitter of a satellite in geostationary orbit radiates on 7.5 GHz with an EIRP of 39 dBW. Estimate the G/T figure of merit for the Earth receiving station if the carrier-to-noise ratio is not to be less than

15 dB, using a bandwidth of 36 MHz. Discuss the realization of this Earth receiving system.

10.2 The transmitter power in the satellite referred to in question 10.1 is 20 W. The antenna has an axially symmetric radiation pattern of the form $\exp(-\alpha\theta)^2$, θ being the angle off bore-sight. Estimate the earth coverage to 3 dB down on the maximum signal strength when the antenna is directed vertically downward.

10.3 The following proposals have been made for a direct broadcast satellite system using the geostationary orbit:

Downlink

Frequency	12.2–12.5 GHz
Maximum path attenuation	9 dB
Channel width	24 MHz
Transmission beam	$0.8^\circ \times 1.2^\circ$
C/N ratio greater than	14 dB
Power flux density at edge of coverage	-107 dB(W m^{-2})
Receiver figure of merit	10 dB K^{-1}
Receiving antenna beamwidth	1.7°

Uplink

Based on antenna 5 m in diameter with efficiency 65% operating at 17 GHz, maximum power 1 kW.

Discuss as fully as possible the realization of the overall system.

References

General

Useful information concerning some commercial systems may be found in

Feher, K. (1981). *Satellite/Earth Station Engineering*. Englewood Cliffs, New Jersey: Prentice Hall.

Specific studies

'Project Echo'. (1961). *BSTJ*, **40**, pp. 975–1233.

Fthenakis, E. (1984). *Manual of Satellite Communication.* New York: McGraw-Hill.

Mittra, R., Imbriale, W. A. and Maanders, E. J. (1983). *Satellite Communication Antenna Technology.* Amsterdam: North-Holland, Ch. 4.

Neusten, M. and Marchant, P. (1978). 'Satellite relays and distribution.' *IBA Tech. Rev.* **11**, 47–54.

O'Neill, H. J. and Hayter, D. (1982). 'Propagation tests.' *IBA Tech. Rev.* **18**, 28–34.

Rudge, A. W. (1985). 'Sky-hooks, fish-warmers and hub-caps: milestones in satellite communications'. *IEEE Proc. Pt. F*, (Feb), 1–12.

Wise, F. H. (1978). 'Fundamentals of satellite broadcasting.' *IBA Tech. Rev.*, **11**, 18–26.

CHAPTER 11

MICROWAVE MEASUREMENTS

OBJECTIVES

This chapter considers methods of measurement of the various parameters that have been used to characterize microwave systems, including frequency, power, noise and scattering parameters.

11.1 Introduction

The designed performance of a system can only be achieved if the component parts are behaving according to the assumed specification. At various stages such specifications have to be checked and appropriate adjustments made. This chapter will be restricted to some of the microwave aspects of measurements, although measurements in other frequency ranges may of course be equally important to the overall system, as at intermediate and video frequencies. The specifically microwave aspects were considered in Chapter 1, where transmitters, receivers and propagation links were identified. The propagation link is partly concerned with the properties of antennas, the measurement of which was considered in Section 3.11, and partly with the measurement of power. The transmitter is required to produce a suitably modulated microwave carrier, free of spurious signals, at the required power. Similar comments apply to local oscillators used in receivers. Measurements thus concern power and frequency spectrum. The function of the receiver is to amplify the small input signals, introducing as little noise as possible. The major characteristics affecting the receiver are noise performance, bandwidth and dynamic range. The bandwidth and behaviour at large signal levels are determined mainly in the IF amplifier and will not be considered further here. The measurement of noise necessarily requires a signal source or noise source of known power. Implicit in all these considerations is the need to know, and therefore measure, frequency. Apart from these 'block' items, various microwave components have to be characterized (usually by scattering parameters) as the matching of a source and a load to connecting transmission lines investigated by measurement of a reflection coefficient and non-linear devices such as transistors mentioned earlier. These various measurements can be carried out with the aid of a number of measuring instruments in different configurations. We may identify

(1) signal sources of determinable power and frequency;

(2) power measuring devices;

(3) noise measurement;

(4) measurement of reflection coefficients and scattering parameters;

(5) measurement of frequency and spectral content.

These categories will be briefly examined in the following sections.

11.2 Signal sources

The traditional signal source consists of an oscillator of which the frequency and output can be adjusted by appropriate controls, such as size of a cavity resonator or DC power supplied. Given sufficient stability the frequency control can be calibrated; the power output can be monitored by a power measuring device and hence controlled by a feedback loop. A variable low power output can then be obtained by the use of a calibrated attenuator. To get a reliably known output, screening must ensure that the final output comes only through the attenuator. Such an instrument can be made at any frequency for which a suitable oscillator is available. The output signal is ideally a pure sine wave; it may contain spurious signals in the form of sine waves at other frequencies, or a random content near to the signal frequency, the **phase noise** (Robins, 1982). Limits have to be placed on these unwanted signals.

Among the possible sources and amplifiers those described in earlier chapters may be listed, namely:

- transistor at lower microwave frequencies;
- Gunn diodes;
- Impatt diodes;
- backward wave oscillator, as a swept frequency source;
- travelling wave tube.

For a number of purposes a switched precision frequency source is needed, as in automatic testing of equipment. This is provided by the frequency synthesizer.

11.2.1 Frequency synthesis

A frequency synthesizer effectively produces a frequency which is a high multiple of a low reference frequency, and by selecting the multiplier various output frequencies can be chosen. Figure 11.1 shows a circuit using a phase-locked loop which produces the harmonic of order N from a reference frequency. In Figure 11.2 a multiple loop synthesizer circuit is shown. As illustrated in the diagram, the output frequency is given by

$$f_o = \left(N + \frac{P}{R}\right) f_R$$

If P is a programmable divider, f_o can be increased in steps of f_R/R rather than in steps of f_R. Frequencies that can be synthesized by either of these

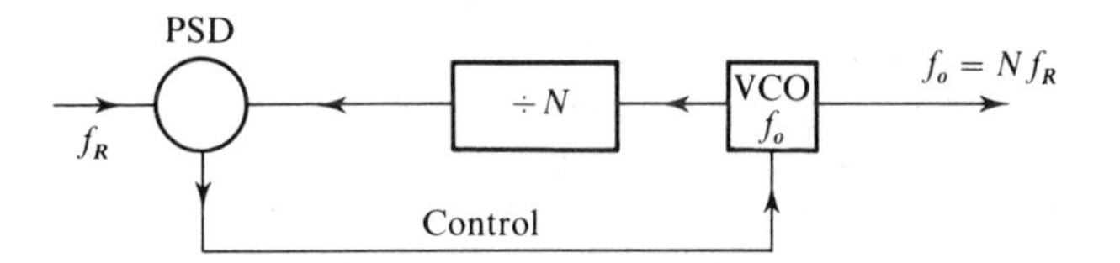

Figure 11.1 Circuit configuration to give a specific harmonic of a stable reference frequency. The output is coherent with the reference and adjustable by altering the ratio of division. VCO, voltage controlled oscillator; PSD, phase sensitive detector.

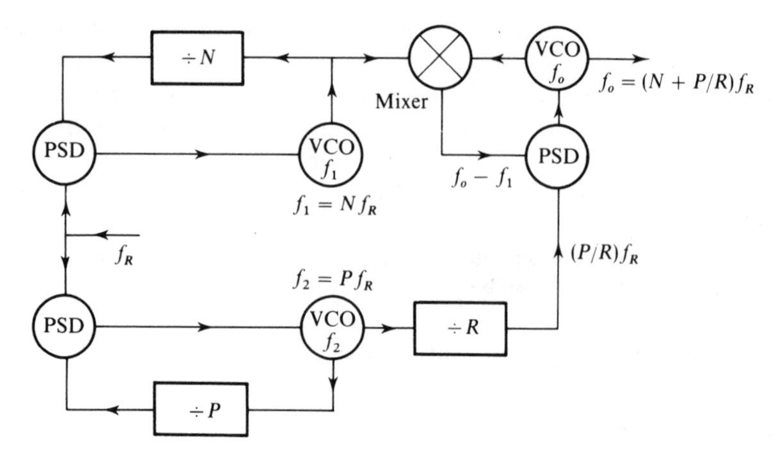

Figure 11.2 A multi-loop synthesizing circuit. This circuit produces a coherent output adjustable in steps equal to some fraction of the reference frequency. N, R fixed division ratios; P, programmable divider.

circuits are restricted by frequency limits of the programmable dividers and by operation of phase locked loops to less than some tens of megahertz. Fixed dividers work at considerably higher frequencies, so the circuit of Figure 11.1 can be used to multiply a lower frequency synthesized signal by N. This of course multiplies the step size by N also. Frequency shifting can also be employed, as shown in Figure 11.3. In this circuit the local oscillator must also be synthesized from the reference frequency: this could be done by firstly deriving a high multiple of the reference frequency, using the circuit of Figure 11.1, followed by harmonic generation and selection of a suitable harmonic by filtering. In this way the low frequency synthesized signal can be translated to a higher frequency. This discussion shows how, in principle, a microwave signal can be derived from a low frequency reference source: practical realization is a demanding exercise. For further information the references McAllister (1980; 1983) and Manassewitsch (1980) may be consulted.

11.3 Power measurement

It is important to characterize a source that delivers power through a transmission line or waveguide to a load. If the source and load have zero reflection coefficients with respect to the connecting line, no ambiguity of definition exists: the power delivered to the load is the available power from the source. Usually, the source will not be perfectly matched and it is necessary to distinguish available power delivered on conjugate matching from the power delivered to the transmission line terminated in its characteristic impedance. The latter may be termed the Z_0-available power. (This difference was illustrated in the example of Section 2.11.3.) Power

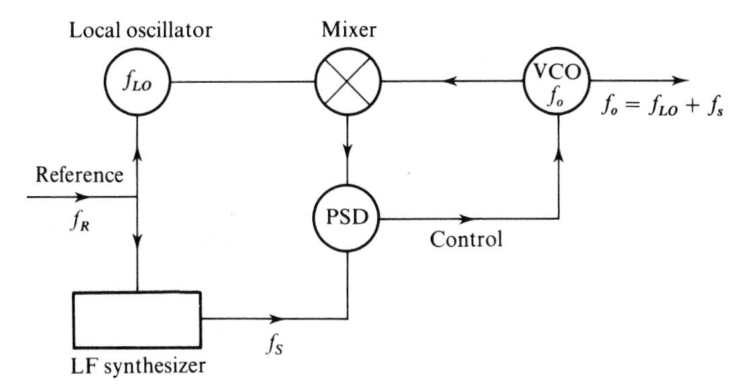

Figure 11.3 Circuit arrangement for translating the output of a low frequency synthesizer to a high frequency.

measurements actually refer to power dissipated in the load, and will in general differ from both of the above characteristic powers. As a result, there is uncertainty in relating the power recorded by a power meter to the power characterizing the source. An error of plus or minus 10 per cent may easily arise on account of mismatch. The uncertainty can be reduced if measured values for the reflection coefficients are available.

An obvious way of measuring power is to measure the heat produced on absorption. At a high level, the power can be absorbed in a fluid cooled load: measurement of temperature rise and flow rate will then give the required result. Errors arise when all the microwave energy is not dissipated in the load, and also by convective and conductive losses of heat. This technique is useful above a few watts.

Another calorimetric technique is to measure the rise in temperature of an absorbing load which can also be heated with DC power. The DC power can then be substituted for microwave power to give the same temperature rise. This method is relatively insensitive to environmental conditions, but the different distributions of heat production in the two cases can cause error.

A related technique uses a **bolometer**: an absorber which has a large temperature coefficient of resistance. The absorbers are typically thin wires **(barretters)** or thermistors, mounted in the waveguide or transmission line to give a good VSWR, with electrical connections to make the bolometer one arm of a Wheatstone bridge, as indicated in Figure 11.4. The supply voltage V can be adjusted successively to balance the bridge with and without the microwave power. The difference in DC power supplied then gives the microwave power. The bridge can be made to balance automatically. Thermistors are more sensitive than barretters, and they are electrically and mechanically more robust. Being semiconductors they are sensitive to changes in ambient temperature, but this can be compensated for by, for example, using a balancing thermistor in the bridge.

Thermocouples can be used as absorbing elements; they can be

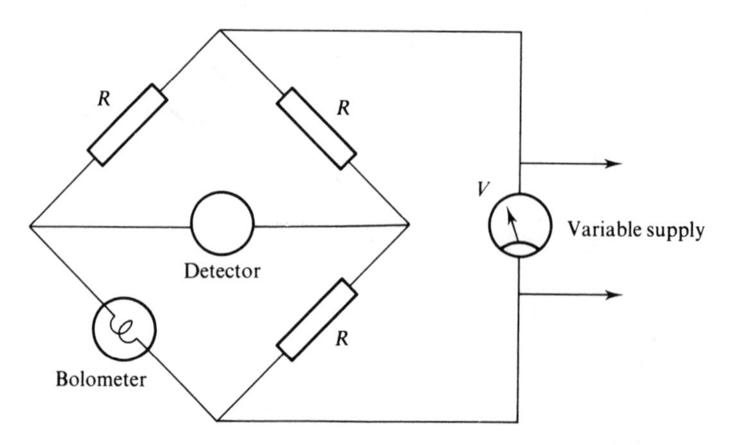

Figure 11.4 Power measurement with a bolometer. At balance, the bolometer resistance is equal to R, and the DC power is $V^2/4R$. When the bolometer is dissipating RF power, the DC power must be reduced by the same amount to maintain balance.

junctions of dissimilar metals or semiconductors, the latter giving greater sensitivity. Sensors based on thermojunctions may use several junctions in series.

The above sensors all depend on the conversion of microwave power to heat: a crystal diode or Schottky barrier diode gives a rectified output voltage depending on microwave power and can therefore be used as a sensor. A sensor based on such elements is rapid acting, but must be calibrated. In a microprocessor based instrument the calibration data can be stored and readings automatically corrected (see, for example, Spenley and Foster, 1982).

The use of modern solid-state technology permits the design of stable sensors with good VSWR over a wide range and sensitive down to -70 dB m (Hewlett-Packard, 1977).

A general discussion on power measurement techniques will be found in Bailey (1985).

11.4 Noise measurement

From the definition of noise figure in Section 6.7.2, its measurement is seen to require determination of signal-to-noise ratio at input and output. Further, the definition presumes a linear amplifier. Signal and noise at output can be measured on a true power meter. The input signal must come from a known source, and calculation of input noise power requires a measurement of noise bandwidth. However, we are frequently concerned with a complete receiver, which is only linear as far as the detector, and in which it is not possible to attach power measuring devices to the IF amplifier. This problem, and that of

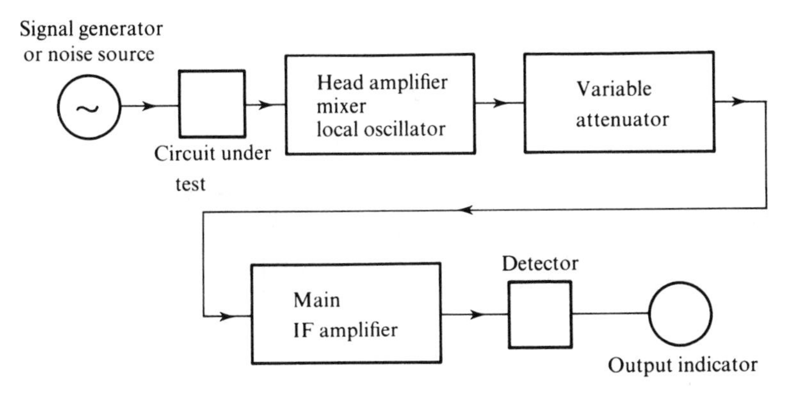

Figure 11.5 Circuit configuration for measurement of the noise figure of an RF amplifier. All significant sources of noise occur prior to the attenuator.

measuring noise bandwidth, are overcome if a noise source is used instead of a source generating a sine wave carrier. The usual configuration is shown in Figure 11.5. The receiver is configured with a variable attenuator in the IF amplifier chain, at a point where all the noise is generated on the input side. (This is practically convenient, since a separate head amplifier is commonly used.) The output monitor can conveniently be diode current (DC). A base reading of output current is taken with no excess noise from the noise generator: extra noise is then provided, and the attenuator adjusted to give the same diode current reading. Since the statistical properties of the 'signal' are unchanged, the noise power at the output of the attenuator is the same in both cases. The extra noise contributing to the output noise will be equal to p_nB_n in which p_n is the excess noise spectral density and B_n the noise bandwidth. If the attenuation is increased by power ratio A we have

$$FkT_0B_nA = FkT_0B_n + p_nB_n$$

Hence

$$F = p_n/(kT_0(A - 1))$$

11.4.1 Noise sources

A number of noise sources are available:

- thermal noise from resistor or matched load;
- thermionic diodes;
- gas discharge tubes;
- avalanche junction diodes.

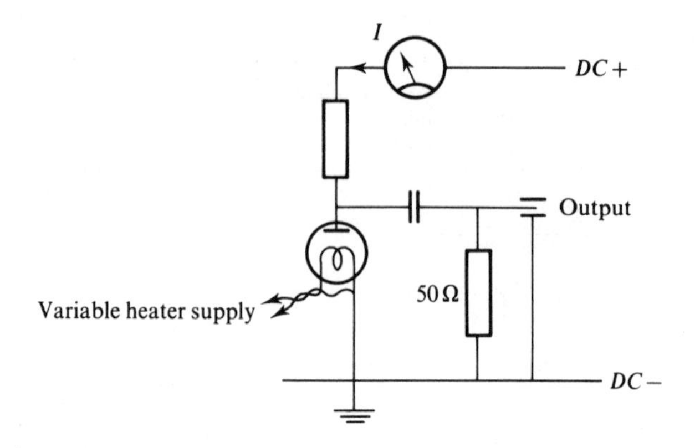

Figure 11.6 An absolute noise generator using a thermionic noise diode.

Of these, the first two are absolute. The available power from a matched load is kT W Hz^{-1} and the current through a temperature limited thermionic diode carries a mean-square noise current of $2eI$ A^2 Hz^{-1}, in which e is the electronic charge and I the mean current. The matched load at different temperatures is used at microwave frequencies, especially for very low noise amplifiers, when the load is either at liquid nitrogen temperature, 77 K, or room temperature. Heated loads can also be made. The thermionic diode cannot be used at microwave frequencies, because the transit time is too long, but is used for measurement of IF amplifiers, in a circuit of the type shown in Figure 11.6. The resistor is chosen to match the line used, either 50 Ω or 75 Ω.

Gas discharge tubes have effective, repeatable, noise temperatures of the order of 10^4 K, but require calibration. Avalanche breakdown in junction diodes provides a convenient solid-state noise source, which again requires calibration. Further information on these sources and their calibration will be found in Bailey (1985).

11.5 Measurement of scattering parameters

In general, scattering parameters are complex numbers varying with frequency, so that complete specification involves phase information as well as amplitude. Phase measurements are much the more demanding, and frequently knowledge of magnitudes is sufficient. Systems that determine magnitudes only are called **scalar analysers**; a system giving phase information as well is termed a **vector analyser**. Scattering parameters further fall into two categories: parameters like S_{11} and S_{22} which are reflection coefficients, involving measurements at one port only, and those like S_{21} which involve transmission measurements. Methods used for measuring scattering parameters range

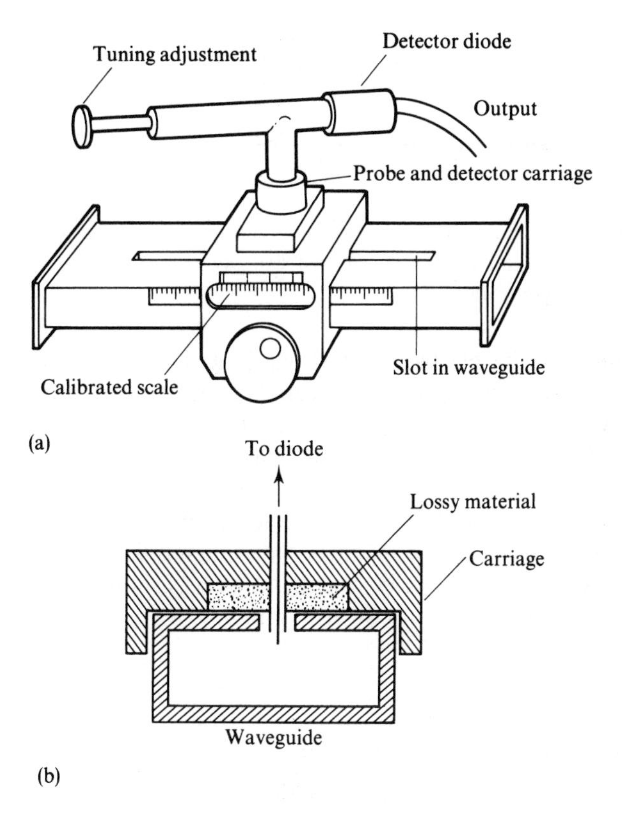

Figure 11.7 Standing wave detector in rectangular waveguide. (a) General form of instrument. (b) Section through the probe. The lossy material acts to reduce variations arising through leakage of radiation between waveguide and carriage. The probe penetration adjustment is not shown.

between simple manual measurements of standing wave ratios and highly sophisticated microcomputer controlled network analysers.

11.5.1 Measurement of standing wave pattern

As discussed in Section 2.10.2, the pattern of the standing wave on a transmission line or waveguide can be used to determine the reflection coefficient both in magnitude and in phase. The standing wave pattern can be explored experimentally if it is possible to put a longitudinal slot in the side of the transmission line or waveguide which does not disturb the wall currents and through which a probe can be inserted to sense the electric field inside. A centre line slot in the broad face of a rectangular waveguide satisfies the conditions, and an instrument based on this is shown in Figure 11.7. The probe is usually connected directly to a crystal diode, the DC output of which

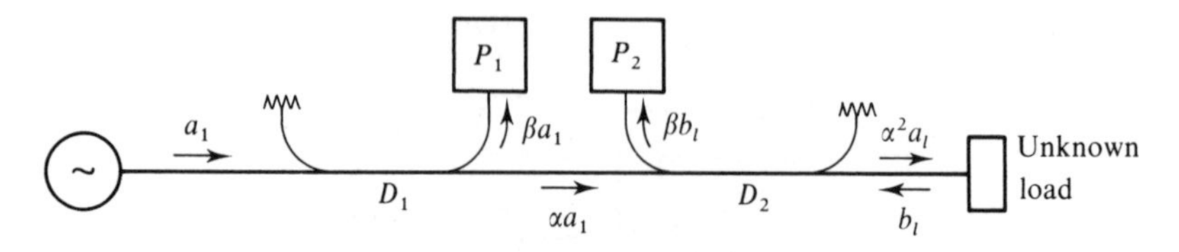

Figure 11.8 Measurement of the magnitude of an unknown reflection coefficient. D_1, D_2 are directional couplers; P_1, P_2 are matched power meters. The symbol MWW indicates matched termination.

is a monitor of the electric field strength within the guide. It is often advantageous to modulate the source supplying the waveguide assembly with square waves, and thus to observe the modulation output from the crystal detector rather than a small direct current. This technique avoids troubles from DC drift and reduces the effect of flicker noise. Once the probe position is located with respect to the chosen reference place, direct observation of the standing wave pattern can be made, both in magnitude and in position. The reflection coefficient can then be calculated, as was shown in Section 2.10.2. The problems associated with such a standing wave detector are mainly mechanical, concerning

- probe position accuracy;
- maintaining a constant insertion;
- perturbation of the waveguide field.

In addition, the detector output will not bear a simple relationship to wave amplitude: it is found to be a square law at low levels, tending to linear at high. The detector law can be calibrated by setting up a known standing wave pattern, for example by a short circuit. The mechanical problems get worse as frequency increases and all dimensions get smaller. Very great care is needed to reach the one per cent level of error (Ginzton, 1957).

Measurements using a moving probe standing wave detector are inevitably slow. An alternative that will give the magnitude of any reflection coefficient, but not the argument, is indicated in Figure 11.8. Directional couplers enable quantities proportional to forward and backward waves to be measured directly, giving $|\Gamma|^2 = P_2/P_1$. Several sources of error exist: the coupling factor and directivity of each coupler, as well as the individual measurements of power. (The definitions of coupling factor and directivity are illustrated in Figure 11.9. The importance of matching output arms can be seen.)

11.5.2 Network analysers

The basic arrangement for measuring a reflection coefficient is illustrated in Figure 11.10(a). Directional couplers take samples of incident and reflected

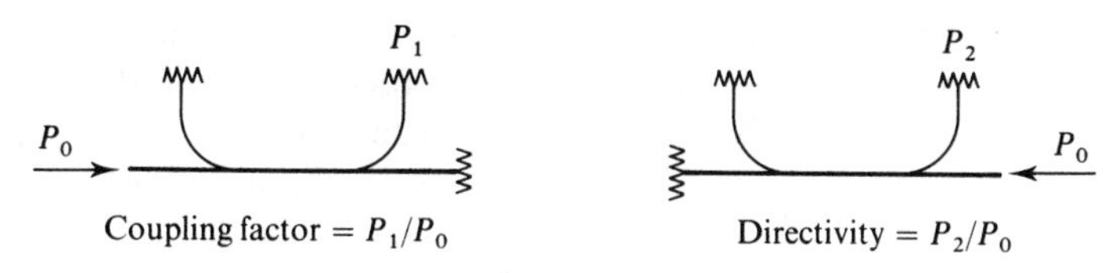

Figure 11.9 Illustrating the definition of coupling factor and directivity of a directional coupler.

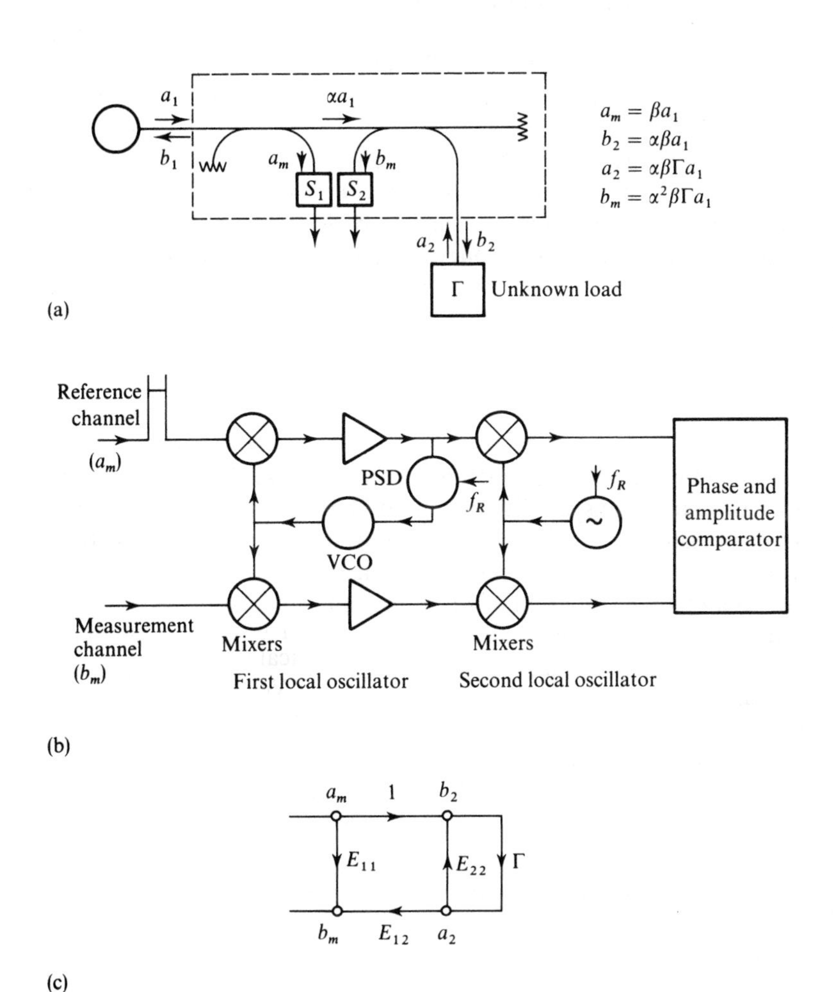

Figure 11.10 The measurement of an unknown reflection coefficient in magnitude and phase. (a) Circuit configuration; S_1, S_2 indicate matched sensors or connectors. (b) Possible form of circuitry for the comparison of two microwave signals by frequency changing; the two channels are as nearly identical as possible, and both local oscillators are derived from the same reference source. (c) Signal flow graph showing a three-term error model, characterizing imperfections in the circuit realization.

waves, so that the ratio b_m/a_m is a measure of the reflection coefficient of the unknown load. The circuit arrangement for extracting magnitude and phase is outlined in Figure 11.10(b). In this arrangement each microwave signal is down-converted to an intermediate frequency of the order of 20 MHz in identical mixers using the same local oscillator. The local oscillator is controlled so that the intermediate frequency is locked to a stable reference. By using identical channels and including a line lengthener in the microwave reference channel, phase relations are preserved. After a second down-conversion to a frequency of the order of 200 kHz, phase sensitive detectors can be used to give in-phase and quadrature components, and hence the complex value of the reflection coefficient. It is essential to keep the signal inputs to the mixers small compared to the local oscillator so as to ensure linearity.

A scalar analyser may be made at significantly lower cost by measuring power instead of complex amplitude at the two measurement ports. Suitably sensitive power meters (such as calibrated Schottky diodes) can be used directly without additional circuitry. Logarithmic amplifiers for the power meters in conjunction with a swept source permit a display of $|\Gamma|$ against frequency.

ERROR CORRECTION

The circuit of Figure 11.10(a) operates in the way described if the couplers are perfect. In practice this condition will not be realized, and errors will result. Errors will also result from imperfect connectors, the effect of which will be change on reuse. These errors can be regarded as equivalent to the action of a linear circuit between the actual reflection coefficient and the measured value. This may be seen as follows: the microwave circuit within the dotted lines in Figures 11.10(a) is a linear 2-port, and hence has a scattering matrix description

$$b_1 = S_{11}a_1 + S_{12}a_2$$

$$b_2 = S_{21}a_1 + S_{22}a_2$$

Further a_m, b_m must depend linearly on a_1, a_2 so that

$$a_m = P_1a_1 + P_2a_2$$

$$b_m = Q_1a_1 + Q_2a_2$$

Using the relations $a_2 = \Gamma b_2$, imposed by the unknown load, we find

$$a_1 = S_{21}^{-1}(1 - \Gamma S_{22})b_2$$

and hence we find that the ratio b_m/a_m, the nominal measured reflection coefficient, can be expressed in terms of Γ in the form

$$\frac{b_m}{a_m} = \frac{E_{11} - \Gamma E}{1 - \Gamma E_{22}}$$

If we write $E = E_{11}E_{22} - E_{12}E_{21}, \ldots$, the coefficients E_{ij} may be regarded as the scattering parameters of a 2-port circuit. Since only three independent coefficients appear in the expression for b_m/a_m it is customary to take $E_{21} = 1$. The values of E_{11}, E_{22} and E can be obtained from the measured results obtained with three known loads, usually taken as a matched load, short in the reference plane ('direct' short) and offset short. Once these quantities have been found, measured values for the unknown load can be corrected. It is convenient to represent the error correcting circuit by a signal-flow graph as shown in Figure 11.10(c). (The theory of signal-flow graphs is dealt with, for example, in Somlo and Hunter, 1985).

Clearly, this operation involves a substantial amount of computing, especially since it must be repeated for all frequencies of interest. It is normally a feature of an automatic analyser, in which the frequency of the

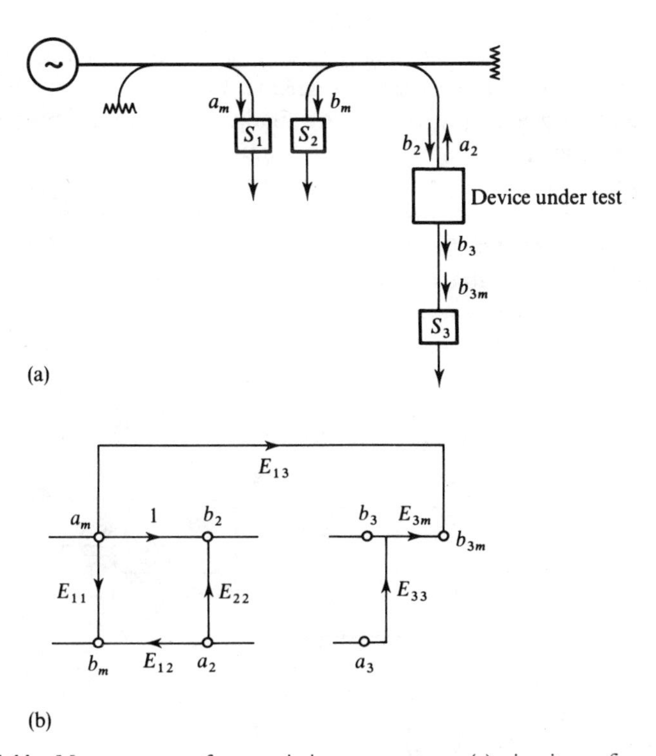

Figure 11.11 Measurement of transmission parameters: (a) circuit configuration; (b) signal flow graph for six-term error model.

microwave source is computer controlled and all measured values are automatically translated into digital form.

TRANSMISSION MEASUREMENTS

This discussion relates to reflection measurements. The system configurations for transmission measurements (S_{21}) is indicated in Figure 11.11(a). Additional error correcting terms occur, as in Figure 11.11(b). These have to be determined by further calibration measurements. Thus measurement in one direction involves six error terms, said to constitute a **6-term error model**. Measurements in the reverse direction can be carried out by physically reversing the device under test, or by using a switch mechanism. In the latter case the calibration has to be carried out for both switch positions.

A review of the development of automatic microwave network measurements is given in Adam (1978), and error correction is reviewed by Fitzpatrick (1978).

11.5.3 6–port analysers

The method outlined in the previous section for the measurement of reflection coefficient demands measurement of phase as well as amplitude, and thus demands complicated equipment. An alternative method exists whereby only power measurements are made. This uses a 6-port network, shown schematically in Figure 11.12. Without specifying the nature of the 6-port, apart from linearity, the possibility of the method can be seen from the following considerations. Ports three, four, five and six are terminated so that ports one and two form effectively a 2-port linear network. Of the complex amplitudes a_1, a_2, b_1 and b_2, two only will be independent, and we may choose these two to be a_2 and b_2.

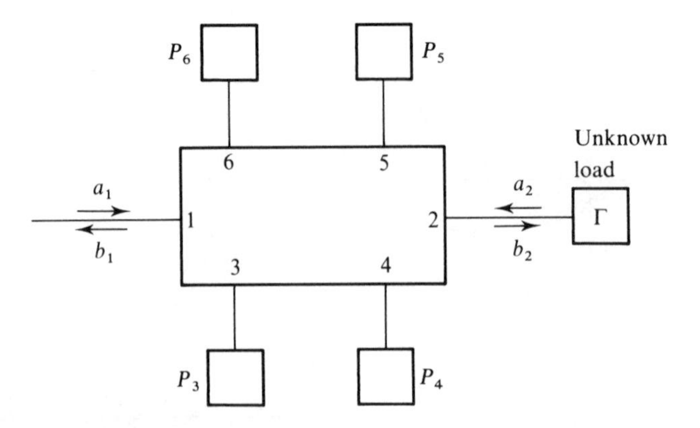

Figure 11.12 Configuration for 6-port measuring technique for an unknown reflection coefficient.

The complex amplitudes at the termination of ports three, four, five and six will each be linear functions of these two variables, and the power in the load will be proportional to the square of the modulus of the complex amplitude. We may therefore write

$$P_i = |M_i a_2 + N_i b_2|^2 \qquad i = 3, 4, 5, 6$$

These four equations we reduce to three by taking port six as reference, giving

$$\frac{P_i}{P_6} = \left|\frac{M_i a_2 + N_i b_2}{M_6 a_2 + N_6 b_2}\right|^2 \qquad i = 3, 4, 5$$

The reflection coefficient of the unknown load, Γ, determines $a_2 = \Gamma b_2$ so that we finally have three equations

$$p_i^2 = \frac{P_i}{P_6}\left|\frac{M_6}{M_i}\right|^2 = \left|\frac{\Gamma + c_i}{\Gamma + c_6}\right|^2 \qquad c_i = N_i/M_i$$

Each of these equations defines a locus in the plane of the Smith Chart, which is a circle of radius r_i, cente Γ_i given by

$$\Gamma_i = (p_i c_6 - c_i)/(1 - p_i^2)$$

$$r_i = p_i |c_6 - c_i| /(1 - p_i^2)$$

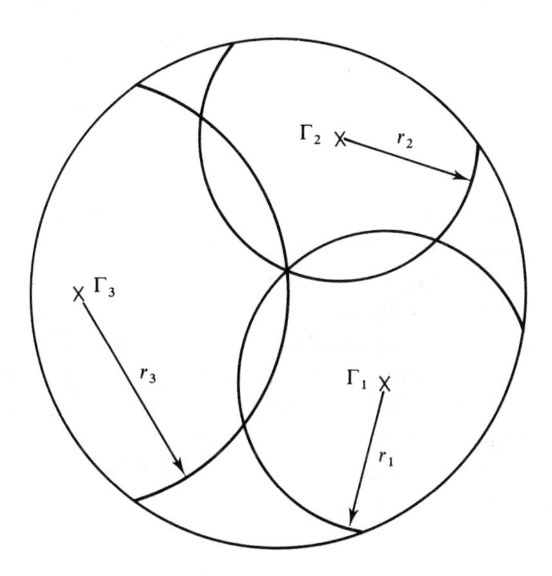

Figure 11.13 Loci derived from 6-port measurements, displayed in the reflection coefficient plane. The unknown reflection coefficient is determined by the intersection of the three circles.

in which all the terms are known either by measurement or from knowledge of the network. Plotting these three loci would ideally give a result like that shown in Figure 11.13. In a real measurement the intersection of the three circles giving Γ would not be perfect, and the properties of the network should be chosen to give a well conditioned intersection.

An example of a 6-port network is shown in Figure 11.14 (Engen, 1978) which uses 3 dB hybrid couplers and power dividers. (The hybrids are numbered in accordance with equation (5.5).) Assuming ideal couplers and perfect matching we find

$$4P_3/P_6 = |\Gamma + 1|^2$$
$$4P_4/P_6 = |\Gamma - 1|^2$$
$$4P_5/P_6 = |\Gamma - \mathrm{j}|^2$$
$$4P_7/P_6 = |\Gamma + \mathrm{j}|^2$$

Thus for this circuit the centre of the circles all lie on the unit circle at the points $(\pm 1, 0)$, $(0, \pm \mathrm{j})$. In practice, a calibration procedure will be needed, as described in the previous section, for the vector network analyser. Possible procedures are discussed in Engen (1978) and Somlo and Hunter (1985).

11.6 Frequency measurement

There are two aspects of frequency measurement which have to be considered: firstly, the measurement of the frequency of a source, and secondly, measurement of the total frequency content, including, for example, noise and side-bands from modulation.

11.6.1 Source frequency

In principle, the measurement of frequency is a matter of counting, and if counters are available, frequency can be determined by counting cycles in a prescribed time. However, in the microwave range above the frequency at which available counters will work, other methods have to be adopted. The standard method involves the use of a transfer oscillator. The arrangement is indicated in Figure 11.15, in which it is assumed that f and f_i are known. The mixer, being a non-linear device, will produce the range of frequencies $f_s \pm nf$. By use of a sampling mixer or a harmonic generator after the transfer oscillator, very high harmonics can be used. If f is adjusted to give a beat frequency of f_i then

$$f_s - nf = f_i$$

If f is then altered to $f + \delta f$, f_i changes by $n\delta f$, whence the value of n can be determined and f_s found. Another possibility is to increase f from f_1, at

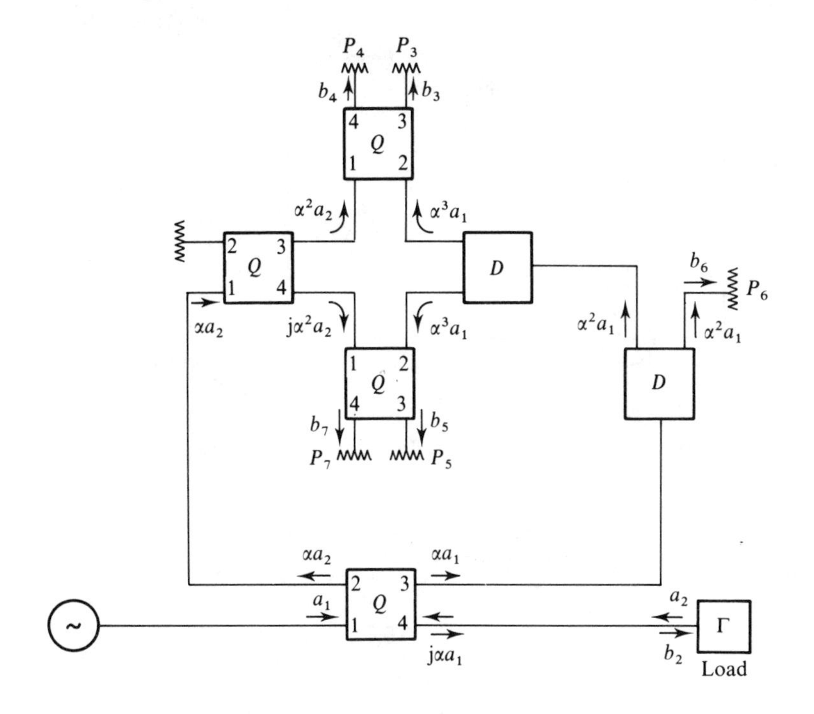

Q 3 dB hybrid

$$S = \alpha \begin{bmatrix} 0 & 0 & 1 & \mathrm{j} \\ 0 & 0 & \mathrm{j} & 1 \\ 1 & \mathrm{j} & 0 & 0 \\ \mathrm{j} & 1 & 0 & 0 \end{bmatrix}$$

$\alpha = 1/\sqrt{2}$

D power divider

$b_2 = \mathrm{j}\alpha a_1;\ b_3 = \alpha^3(b_2 + a_2);$

$b_4 = \mathrm{j}\alpha^3(a_2 - b_2);\ b_5 = \alpha^3(b_2 + \mathrm{j}\alpha a_2)$

$b_6 = -\mathrm{j}\alpha b_2;\ b_7 = -\alpha^3(\mathrm{j}b_2 + a_2)$

Figure 11.14 A possible 6-port measurement system, using 3 dB hybrid couplers and power dividers.

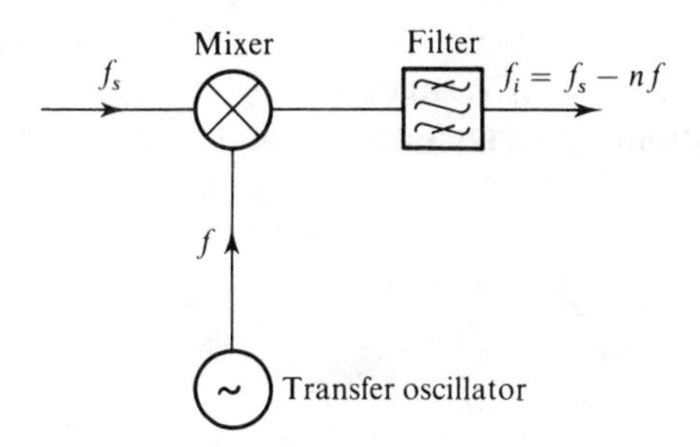

Figure 11.15 Illustrating the use of a transfer oscillator in frequency measurement. One particular harmonic of the transfer oscillator output will give rise to an output from the mixer which is in the range of the filter. If n can be determined and f_i measured, f_s will be known.

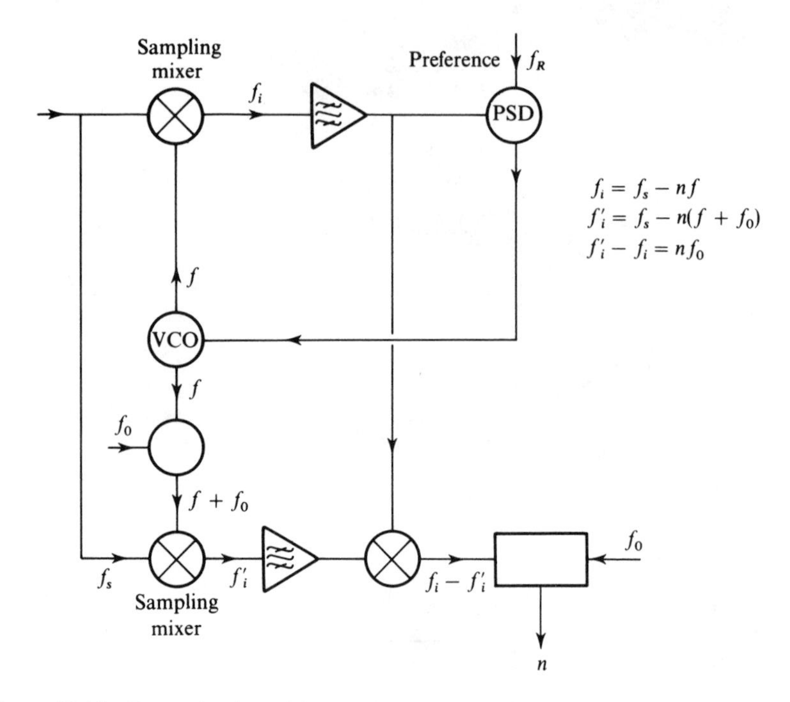

Figure 11.16 Determination of harmonic number. The transfer oscillator is the VCO, frequency f; f_i is made equal to f_R by controlling f; the harmonic number is obtained by measuring $f_i - f_i'$ in terms of f_o, a small frequency offset. The counter measuring n can be used to determine the gating time for the counter measuring f so that an automatic read out can be obtained.

which the first observed output at f_i occurs, to the frequency f_2 at which f_i is again observed. Then

$$nf_1 = (n - 1)f_2$$

and again n may be determined. One implementation of these ideas is shown schematically in Figure 11.16. The transfer oscillator is a VCO controlled by a loop to give an output at f_i. A small offset f_o is added to f, sufficiently small that the same harmonic gives an output f'_i in the same range as f_i. Thus

$$f_s - nf = f_i$$
$$f_2 - n(f + f_o) = f'_i$$

Hence

$$nf_o = f_i - f_i'$$

allowing n to be determined.

In this discussion it has been assumed that the input to be determined

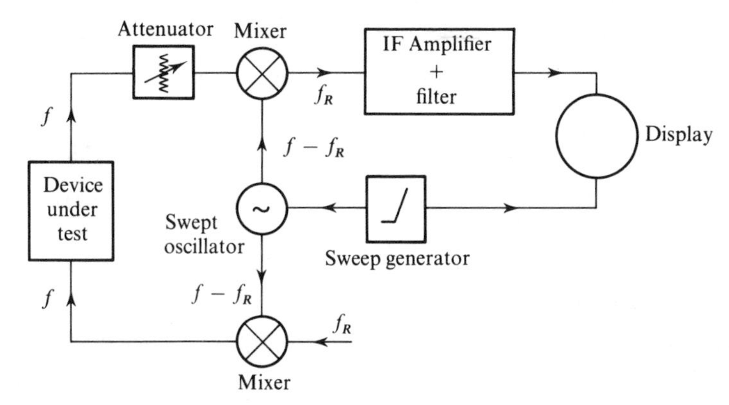

Figure 11.17 Schematic form for a spectrum analyser. The arrangement ensures that the intermediate frequency is equal to a reference frequency.

is a good sine wave: particular configurations for a measurement system must take into account the effect of amplitude and frequency modulation on the accuracy of the measurement.

11.6.2 Spectrum analysers

The complete frequency spectrum is investigated using a spectrum analyser. This is in principle a narrow filter whose mid-band frequency is scanned over the range to be investigated. In fact the filter is fixed in frequency, and the input signal down-converted using a swept local oscillator. The schematic form of such an instrument is shown in Figure 11.17. Although simple in principle, considerable understanding of the system is required to obtain optimum performance. Nominally steady state, the swept signal is in fact a transient, so that the choice of filter width, which determines resolution, is dependent on choice of sweep speed.

It is also possible to use digital techniques. The signal is sampled at regular intervals over a sufficient time to allow the spectral properties of the waveform to be determined by computation using the Fast Fourier Transform. In this way phase information is obtained as well as magnitude. Sampling can now be carried out at frequencies well into the microwave spectrum.

Further information will be found in manufacturers' literature connected with specific instruments.

References

Adam, S. F. (1978). 'Automatic microwave network measurements'. *Proc. IEEE*, **66**, 384–91.

Bailey, A. E., Ed. (1985). *Microwave Measurements*. Stevenage: Peter Peregrinus.

Engen, M. J. (1978). 'The 6-port measurement technique'. *Microwave J.*, **21**, No. 5, 18.

Fitzpatrick, J. (1978). 'Error models for systems measurement'. *Microwave J.*, **21**, No. 5, 63.

Ginzton, E. L. (1957). *Microwave Measurements*. New York: McGraw-Hill.

Hewlett-Packard (1977). 'Fundamentals of RF and microwave power measurements'. Application Note **64-1**.

McAllister, P. A. (1980). 'Phaselock techniques for synthesis of microwave frequencies'. *Proc. IEEE*, **H 127**, 112–15.

McAllister, P. A. (1983). 'Synthesised signal generator: a design example'. *Proc. IEEE*, **H**, 451–5.

Manassewitsch, V. (1980). *Frequency Synthesisers*. New York: Wiley.

Page, D. (1984). 'Developments in microwave counter design'. *New Electronics*, 17 April, 66.

Robins, W. P. (1982). *Phase noise in signal sources*. Stevenage: Peter Peregrinus.

Somlo, P. I. and Hunter, J. D. (1985). *Microwave Impedance Measurement*. Stevenage: Peter Peregrinus.

Spenley, P. and Foster, W. (1982). 'Automatic scalar analyser uses modern technology'. *Microwave J.*, **25**, No. 4, 83.

APPENDIX

A.1 International band designations

It is to be noted from Table A.1 that the alphabetical designations given to the 'old' bands are not strictly speaking correct. The nomenclature L, S, C, X, etc. properly relates to bands specifically used for radar within the wider limits shown in the table. This is shown by comparison with the waveguide designations given in Section A.6.

A.2 Guided waves in systems of linear conductors

Maxwell's equations in a homogeneous dielectric medium, with phasor representation of time-harmonic fields of angular frequency ω, are

$$\text{curl}\,\boldsymbol{E} = -\mathrm{j}\omega\mu\boldsymbol{H}$$
$$\text{curl}\,\boldsymbol{H} = \mathrm{j}\omega\varepsilon\boldsymbol{E}$$

Consider propagation in the z direction with propagation factor $\exp(-\mathrm{j}\beta z)$ and define

$$\boldsymbol{E}_t = E_x\boldsymbol{a}_x + E_y\boldsymbol{a}_y$$
$$\boldsymbol{H}_t = H_x\boldsymbol{a}_x + H_y\boldsymbol{a}_y$$

In Cartesian coordinates we find

$$\frac{\partial E_z}{\partial y} + \mathrm{j}\beta E_y = -\mathrm{j}\omega\mu H_x \tag{A.1}$$

$$-\mathrm{j}\beta E_x - \frac{\partial E_z}{\partial x} = -\mathrm{j}\omega\mu H_y \tag{A.2}$$

$$\frac{\partial E_y}{\partial x} - \frac{\partial E_x}{\partial y} = -\mathrm{j}\omega\mu H_z \tag{A.3}$$

$$\frac{\partial H_z}{\partial y} + \mathrm{j}\beta H_y = \mathrm{j}\omega\varepsilon E_x \tag{A.4}$$

$$-\mathrm{j}\beta H_x - \frac{\partial H_z}{\partial x} = \mathrm{j}\omega\varepsilon E_y \tag{A.5}$$

$$\frac{\partial H_y}{\partial x} - \frac{\partial H_x}{\partial y} = \mathrm{j}\omega\varepsilon E_z \tag{A.6}$$

A.2.1 TEM waves

Putting $E_z = 0$ and $H_z = 0$, we find

$$\frac{E_x}{H_y} = -\frac{E_y}{H_x} = \frac{\omega\mu}{\beta} = \frac{\beta}{\omega\varepsilon}$$

Table A.1 International band designations

<table>
<tr><th>New (NATO)</th><th>Old</th></tr>
<tr><td rowspan="2">A
0–250 MHz</td><td>HF
3–30 MHz</td></tr>
<tr><td>VHF
30–300 MHz</td></tr>
<tr><td>B
250–500 MHz</td><td rowspan="2">UHF
300–1000 MHz</td></tr>
<tr><td>C
500–1000 MHz</td></tr>
<tr><td>D
1–2 GHz</td><td>L
1–2 GHz</td></tr>
<tr><td>E
2–3 GHz</td><td rowspan="2">S
2–4 GHz</td></tr>
<tr><td>F
3–4 GHz</td></tr>
<tr><td>G
4–6 GHz</td><td rowspan="2">C
4–8 GHz</td></tr>
<tr><td>H
6–8 GHz</td></tr>
<tr><td>I
8–10 GHz</td><td rowspan="2">X
8–12 GHz</td></tr>
<tr><td rowspan="2">J
10–20 GHz</td></tr>
<tr><td>J Ku
12–18 GHz</td></tr>
<tr><td rowspan="2">K
20–40 GHz</td><td>K
18–27 GHz</td></tr>
<tr><td>Q Ka
27–40 GHz</td></tr>
<tr><td>L
40–60 GHz</td><td>40–60 GHz</td></tr>
<tr><td>M
60–100 GHz</td><td>O E
60–90 GHz</td></tr>
</table>

whence

$$\beta = \omega(\mu\varepsilon)^{\frac{1}{2}}$$

Further

$$\boldsymbol{E}_t \cdot \boldsymbol{H}_t = 0 \qquad \frac{E_t}{H_t} = Z_g = \left(\frac{\mu}{\varepsilon}\right)^{\frac{1}{2}} \quad \text{and} \quad \boldsymbol{H}_t = Z_g^{-1}\boldsymbol{a}_z \times \boldsymbol{E}_t$$

To satisfy equation (A.3) when $H_z = 0$ we introduce a scalar function $V(x, y)$ such that

$$E_x = -\frac{\partial V}{\partial x} \qquad E_y = -\frac{\partial V}{\partial y}$$

Substitution into equation (A.6) with $E_z = 0$ gives

$$\frac{\partial^2 V}{\partial x^2} + \frac{\partial^2 V}{\partial y^2} = 0$$

or in general

$$\nabla_t^2 V = 0 \tag{A.7}$$

On a perfectly conducting surface, the tangential component of the electric field must vanish, so that on such a surface V must be constant. Non-trivial solutions to equation (A.7) exist only when two separate surfaces take on different values of V. Supposing V is a specific solution of equation (A.7) having different values on the two conductors forming the transmission line, then defining the modal function $\boldsymbol{e}(x, y)$ by

$$\boldsymbol{e}(x, y) = -\nabla_t V$$

the general expression associated with a forward wave takes the form

$$\boldsymbol{E}_t = A\,\boldsymbol{e}\exp(-\mathrm{j}\beta z)$$
$$\boldsymbol{H}_t = AZ_g^{-1}\boldsymbol{a}_z \times \boldsymbol{e}\exp(-\mathrm{j}\beta z)$$

For a backward wave the sign of β is changed throughout, giving

$$\boldsymbol{E}_t = B\,\boldsymbol{e}\exp(\mathrm{j}\beta z)$$
$$\boldsymbol{H}_t = -BZ_g^{-1}\boldsymbol{a}_z \times \boldsymbol{e}\exp(\mathrm{j}\beta z)$$

Thus in general

$$\boldsymbol{E}_t = \boldsymbol{e}[A\exp(-\mathrm{j}\beta z) + B\exp(\mathrm{j}\beta z)]$$
$$\boldsymbol{H}_t = Z_g^{-1}\boldsymbol{a}_z \times \boldsymbol{e}[A\exp(-\mathrm{j}\beta z) - B\exp(\mathrm{j}\beta z)]$$

A.2.2 TE modes

Putting $E_z = 0$ in equations (A.1) and (A.2)

$$\frac{E_x}{H_y} = -\frac{E_y}{H_x} = \frac{\omega\mu}{\beta}$$

whence

$$Z_g = \frac{\omega\mu}{\beta} \qquad \boldsymbol{E}_t \cdot \boldsymbol{H}_t = 0 \qquad \boldsymbol{H}_t = Z_g^{-1}\boldsymbol{a}_z \times \boldsymbol{E}_t$$

Equations (A.4) and (A.5) then give

$$\boldsymbol{H}_t = \left(\frac{-\mathrm{j}\beta}{\sigma^2}\right)\nabla_t H_z$$

in which $\sigma^2 = (\omega^2\mu\varepsilon - \beta^2)$.

Finally, substitution into equation (A.3) yields

$$\nabla_t^2 H_z + \sigma^2 H_z = 0 \tag{A.8}$$

At a perfectly conducting surface the component of magnetic field normal to the surface must vanish, requiring that on the surface of each conductor

$$\frac{\partial H_z}{\partial n} = 0$$

This boundary condition in association with equation (A.8) constitutes an eigenvalue problem, to which solutions exist only for specific values of the constant σ. Let σ_c be an eigenvalue and ψ the associated solution. We then have

$$\omega^2\mu\varepsilon - \beta^2 = \sigma_c^2$$

or

$$\beta^2 = \omega^2\mu\varepsilon - \sigma_c^2$$

The cut-off wavelength for the mode is then given by $\lambda_c = 2\pi/\sigma_c$, and in terms of guide wavelength, λ_g, and free space wavelength, λ, the equation takes the form

$$\frac{1}{\lambda_g^2} = \frac{1}{\lambda^2} - \frac{1}{\lambda_c^2} \tag{A.9}$$

Defining a modal function by

$$\boldsymbol{e} = \sigma_c^{-1}\boldsymbol{a}_z \times \nabla_t\psi$$

we find that the field components associated with a forward wave can be expressed in the form

$$\boldsymbol{E}_t = A\,\boldsymbol{e}\exp(-\mathrm{j}\beta z)$$
$$\boldsymbol{H}_t = AZ_g^{-1}\boldsymbol{a}_z \times \boldsymbol{e}\exp(-\mathrm{j}\beta z)$$
$$\boldsymbol{H}_z = \left(\frac{-\mathrm{j}\lambda}{\eta\lambda_c}\right)A\,\psi\exp(-\mathrm{j}\beta z)$$

For the backward wave the sign of β must be changed throughout, giving finally

$$\boldsymbol{E}_t = \boldsymbol{e}[A\exp(-\mathrm{j}\beta z) + B\exp(\mathrm{j}\beta z)]$$
$$\boldsymbol{H}_t = Z_g^{-1}\boldsymbol{a}_z \times \boldsymbol{e}[A\exp(-\mathrm{j}\beta z) - B\exp(\mathrm{j}\beta z)]$$
$$H_z = \left(\frac{-\mathrm{j}\lambda}{\eta\lambda_c}\right)\psi[A\exp(-\mathrm{j}\beta z) + B\exp(\mathrm{j}\beta z)]$$

A.2.3 TM modes

An analysis similar to that of the last section but with $H_z = 0$ yields

$$Z_g = \frac{\beta}{\omega\varepsilon} \qquad \boldsymbol{H}_t = Z_g^{-1}\boldsymbol{a}_z \times \boldsymbol{E}_t$$

and

$$\boldsymbol{E}_t = \left(\frac{-\mathrm{j}\beta}{\sigma^2}\right)\nabla_t E_z$$
$$\nabla_t^2 E_z + \sigma^2 E_z = 0 \tag{A.10}$$

where, as before, $\sigma^2 = \omega^2\mu\varepsilon - \beta^2$.

The boundary condition is now $E_z = 0$ on the conducting surfaces, constituting a different eigenvalue problem. Take σ_c and ψ_c, to be specific solutions. The cut-off wavelength is, as before, given by $\lambda_c = 2\pi/\sigma_c$, and equation (A.9) holds. Defining

$$\boldsymbol{e} = -\sigma_c^{-1}\nabla_t\psi_c$$

we now find general expressions in the form

$$\begin{aligned}
\boldsymbol{E}_t &= \boldsymbol{e}[A\exp(-\mathrm{j}\beta z) + B\exp(\mathrm{j}\beta z)]\\
\boldsymbol{H}_t &= Z_g^{-1}\boldsymbol{a}_z \times \boldsymbol{e}[A\exp(-\mathrm{j}\beta z) + B\exp(\mathrm{j}\beta z)]\\
E_z &= \left(\frac{-\mathrm{j}\lambda_g}{\lambda_c}\right)\psi[A\exp(-\mathrm{j}\beta z) - B\exp(\mathrm{j}\beta z)]
\end{aligned}$$

A.3 Coaxial line

In this case the relevant solution of equation (A.7) may be taken as $V = -\ln\rho$, giving

$$\boldsymbol{e} = \rho^{-1}\boldsymbol{a}_\rho$$

This is the result used in Section 2.6. To calculate the attenuation constant arising from wall loss we need

$$\begin{aligned}
\oint \boldsymbol{H}\cdot\boldsymbol{H}^*\,\mathrm{d}l &= 2\pi a[\boldsymbol{H}\cdot\boldsymbol{H}^*]_{\rho=a} + 2\pi b[\boldsymbol{H}\cdot\boldsymbol{H}^*]_{\rho=b}\\
&= 2\pi(A/\eta)^2(a^{-1} + b^{-1})
\end{aligned}$$

The result in Section 2.6 follows.

A.4 Rectangular waveguide

TE MODES

Equation (A.9) in Cartesian form is

$$\frac{\partial^2 H_z}{\partial x^2} + \frac{\partial^2 H_z}{\partial y^2} + \sigma^2 H_z = 0$$

The boundary conditions, in terms of the notation of Figure 2.5, are

$$\frac{\partial H_z}{\partial x} = 0 \quad \text{on} \quad x = 0, a$$

$$\frac{\partial H_z}{\partial y} = 0 \quad \text{on} \quad y = 0, b$$

The appropriate solution is

$$\psi = \cos\left(\frac{m\pi x}{a}\right)\cos\left(\frac{n\pi y}{b}\right)$$

with

$$\sigma^2 = \left(\frac{m\pi}{a}\right)^2 + \left(\frac{n\pi}{b}\right)^2$$

The formulae of Section 2.8.1 follow.

TM MODES

The appropriate solution of equation (A.9) for which E_z vanishes on $x = 0, a$ and $y = 0, b$ is

$$\psi = \sin\left(\frac{m\pi x}{a}\right)\sin\left(\frac{n\pi y}{b}\right)$$

with

$$\sigma^2 = \left(\frac{m\pi}{a}\right)^2 + \left(\frac{n\pi}{b}\right)^2$$

The formulae of Section 2.8.2 then follow.

The attenuation constant in both cases has to be evaluated by the method of Section 2.5.2. Details may be found in texts on electromagnetic theory. The note in Section 2.5.2 concerning degenerate modes applies.

A.5 Circular waveguide

The correct form for equations (A.8) and (A.10) in cylindrical coordinates is

$$\frac{\partial^2 \psi}{\partial \rho^2} + \frac{1}{\rho}\frac{\partial \psi}{\partial \rho} + \frac{1}{\rho^2}\frac{\partial^2 \psi}{\partial \phi^2} + \sigma^2 \psi = 0$$

The appropriate solutions have the form

$$\psi = J_m(\sigma\rho)\cos(m\phi) \tag{A.11}$$

TE MODES

In the case of TE modes the boundary condition is $\partial\psi/\partial\rho = 0$ on $\rho = a$. Thus we must have

$$\sigma a = s_{mn}$$

in which s_{mn} are the zeros of $J'_m(u)$ introduced in Section 2.9. The cut-off wavelength, λ_{mn}, is given by

$$\lambda_{mn} = \frac{2\pi a}{s_{mn}}$$

and the appropriate form for ψ is

$$\psi = J_m\left(\frac{s_{mn}\rho}{a}\right)\cos(m\phi)$$

The results of Section 2.9.1 follow.

TM MODES

In this case $\psi = 0$ on $\rho = a$, so that the cut-off wavelength and corresponding solution are given by

$$\sigma a = t_{mn}$$

and

$$\psi = J_m\left(\frac{\rho t_{mn}}{a}\right)\cos(m\phi)$$

The results of Section 2.9.2 follow.

ATTENUATION

Evaluation of the expression for the attenuation constant is more straightforward than for the rectangular guide since only integration in ϕ is required. For a forward wave in the dominant (TE_{11}) mode, with $A = 1$

$$\boldsymbol{H}\cdot\boldsymbol{H}^* = \left(\frac{\lambda}{\eta\lambda_{11}}\right)^2\left(\cos^2\phi + \left(\frac{\lambda_{11}}{s_{11}\lambda_g}\right)^2\sin^2\phi\right)J_1^2(s_{11})$$

Hence

$$\oint \boldsymbol{H}\cdot\boldsymbol{H}^*\,\mathrm{d}l = \left(\frac{\lambda}{\eta\lambda_{11}}\right)^2 \pi a\left(1+\left(\frac{\lambda_{11}}{s_{11}\lambda_g}\right)^2\right) J_1^2(s_{11})$$

we have

$$P = \left(\frac{1}{2Z_g}\right)\langle \boldsymbol{e}\cdot\boldsymbol{e}^*\rangle$$

$$= \left(\frac{\pi a^2}{4Z_g}\right) J_1^2(s_{11})\left(1-\left(\frac{1}{s_{11}}\right)^2\right)$$

Using

$$\alpha = \left(\frac{R_s}{4P}\right)\oint \boldsymbol{H}\cdot\boldsymbol{H}^*\,\mathrm{d}l$$

we find after manipulation

$$\alpha = \left(\frac{R_s}{a\eta}\right)\left(\frac{1}{s_{11}^2-1}+\left(\frac{\lambda}{\lambda_{11}}\right)^2\right)\left(1-\left(\frac{\lambda}{\lambda_{11}}\right)^2\right)^{-\frac{1}{2}}$$

Evaluation of the attenuation constant for the TE_{01} mode leads to the expression

$$\alpha = \frac{R_s}{a\eta}\left(\frac{\lambda}{\lambda_{01}}\right)^2\left(1-\left(\frac{\lambda}{\lambda_{01}}\right)^2\right)^{-\frac{1}{2}}$$

This expression shows that in this mode attenuation decreases indefinitely as the frequency increases. This property was behind the development of the 'long haul' waveguide for use in trunk communications. A waveguide of 50 mm diameter was used, operating at 50 GHz. The project was rendered obsolete by the rapid development of optical fibres.

A.6 Standard rectangular waveguide

Table A.2 gives designations and dimensions of a selection of standard rectangular waveguides operating above 1 GHz, together with theoretical values for attenuation for copper and the maximum power before breakdown. The useful frequency range is calculated as $1.25 f_c$ to $1.9 f_c$, and the attenuation at $1.5 f_c$. The original sizes were specified in inch units, with a ratio of approximately 2:1 between wide and narrow sides. There have been many systems used to designate waveguide sizes: those given here are the WG notation commonly used in the UK and the WR notation used in the USA. The WR notation relates to the inch measure of the wide dimension, e.g. WR 90 is, in inches, 0.9 × 0.4.

Table A.2 Standard rectangular waveguides over 1 GHz

UK	USA	Useful frequency range		Dimensions		Attenuation	Power capacity
				a	*b*		
			GHz	mm	mm	dB m^{-1}	MW
WG 6	WR 650	*L*	1.14–1.73	165.1	82.6	0.0052	13.5
WG 7	WR 510		1.45–2.20	129.5	64.8	0.0075	8.3
WG 8	WR 430		1.72–2.61	109.2	54.6	0.0097	5.9
WG 9A	WR 340		2.17–3.30	86.4	43.2	0.014	3.8
WG 10	WR 284	*S*	2.60–3.95	72.1	34.0	0.019	2.4
WG 11A	WR 229		3.22–4.90	58.2	29.1	0.025	1.6
WG 12	WR 187	*C*	3.94–5.99	47.6	22.1	0.036	1.0
WG 13	WR 159		4.64–7.05	40.4	20.2	0.043	0.81
WG 14	WR 137		5.38–8.18	34.85	15.80	0.058	0.54
WG 15	WR 112		6.58–10.0	28.50	12.62	0.079	0.36
WG 16	WR 90	*X*	8.20–12.5	22.86	10.16	0.11	0.23
WG 17	WR 75		9.84–15.0	19.05	9.53	0.13	0.18
WG 18	WR 62	*J*	11.9–18.0	15.80	7.90	0.18	0.12
WG 19	WR 51		14.5–22.0	12.95	6.48	0.24	0.083
WG 20	WR 42		17.6–26.7	10.67	4.32	0.37	0.048
WG 21	WR 34		21.7–33.0	8.64	4.32	0.44	0.037
WG 22	WR 28	*Q*	26.4–40.1	7.11	3.56	0.58	0.025
WG 23	WR 22		33.0–50.1	5.69	2.85	0.81	0.016
WG 24	WR 19		39.3–59.7	4.78	2.39	1.1	0.010
WG 25	WR 15		49.9–75.8	3.76	1.88	1.5	0.007
WG 26	WR 12	*O*	60.5–92.0	3.10	1.55	2.0	0.005
WG 27	WR 10		73.8–112	2.54	1.27	2.7	0.003
WG 28	WR 8		92.3–140	2.03	1.02	3.8	0.002
WG 29	WR 7		110–170	1.65	0.83	5.2	
WG 30	WR 5		145–220	1.30	0.65	7.5	
WG 31	WR 4		172–261	1.09	0.55	9.7	
WG 32	WR 3		217–330	0.86	0.43	13.8	

A.7 The Smith Chart

This section shows the derivation of the constant resistance and constant reactance circles used in Section 2.10.1. The reflection coefficient corresponding to the normalized impedance $r + jx$ is given by the expression

$$\Gamma = \frac{r + jx - 1}{r + jx + 1}$$

Hence

$$r + jx = \frac{1 + \Gamma}{1 - \Gamma}$$

The contour for r constant is given by

$$2r = \frac{1 + \Gamma}{1 - \Gamma} + \left(\frac{1 + \Gamma}{1 - \Gamma}\right)^*$$

Writing $\Gamma = u + jv$ we find

$$u^2 - \frac{2ur}{r + 1} + v^2 = \frac{1 - r}{1 + r}$$

or

$$\left(u - \frac{r}{r + 1}\right)^2 + v^2 = \frac{1}{(r + 1)^2}$$

This is the equation of a circle, centre $(r/(r + 1), 0)$ and radius $(r + 1)^{-1}$. Similarly, a contour for constant x is given by

$$2jx = \frac{1 + \Gamma}{1 - \Gamma} - \left(\frac{1 + \Gamma}{1 - \Gamma}\right)^*$$

This leads to

$$(u - 1)^2 + \left(v - \frac{1}{x}\right)^2 = \frac{1}{x^2}$$

which is a circle centre $(1, x^{-1})$, and radius x^{-1}.

A.8 Coupled lines

Consider the case of two identical conductors symmetrically placed with respect to a ground plane, as shown in Figure A.1. Such a configuration is

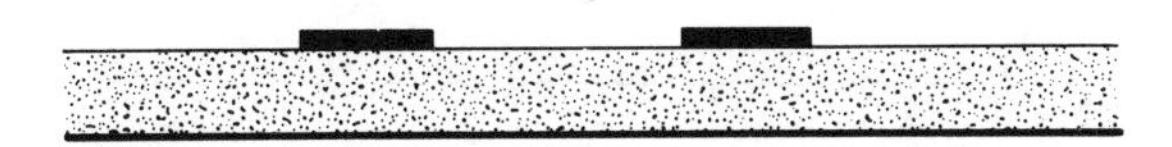

Figure A.1 Coupled lines consisting of two similar strips mounted on a dielectric over an earth plane.

most simply analysed in terms of even and odd modes of excitation, which are illustrated in Figure A.2, (a) and (b). Because of the coupling between the lines, both magnetic and electric, the two modes of excitation will yield different characteristic impedances Z_{0e}, Z_{0o}. If the dielectric medium is homogeneous, waves will travel in both cases with the same velocity (given by $(\mu\varepsilon)^{-\frac{1}{2}}$); if the medium is inhomogeneous, as illustrated in Figure A.1, consideration of the distribution of electric field lines makes it clear that the effective capacity is different in the two modes, so that the wave velocities will also be different. A given physical length of coupled line will therefore in general have different electrical lengths in the two modes. These are indicated by values for θ_e, θ_o in the even and odd modes respectively.

Figure A.3(a) shows the voltages and currents in the case of general excitation, and Figure A.3 (b) and (c) shows the same variables for even and odd modes respectively. The general case of Figure A.3(a) may be regarded as the superposition of Figure A.3, parts (b) and (c). Hence

$$\left.\begin{aligned} V_1 &= V_{1e} + V_{1o} & I_1 &= I_{1e} + I_{1o} \\ V_2 &= V_{1e} - V_{1o} & I_2 &= I_{1e} - I_{1o} \\ V_3 &= V_{4e} - V_{4o} & I_3 &= I_{4o} - I_{4e} \\ V_4 &= V_{4e} + V_{4o} & I_4 &= -I_{4o} - I_{4e} \end{aligned}\right\} \tag{A.12}$$

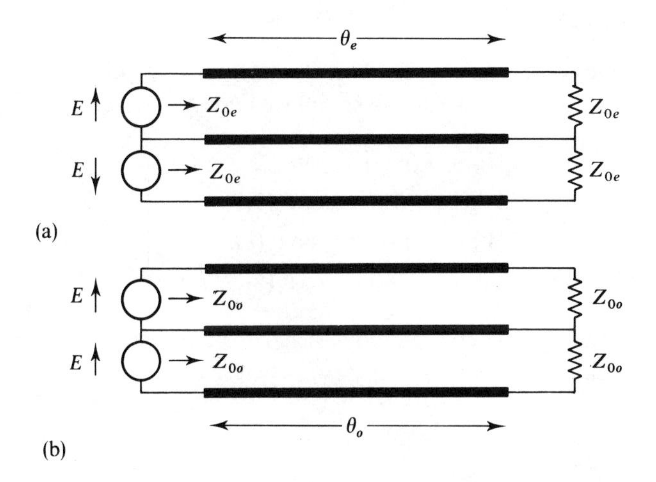

Figure A.2 Illustrating definitions of characteristic impedances: (a) even; (b) odd.

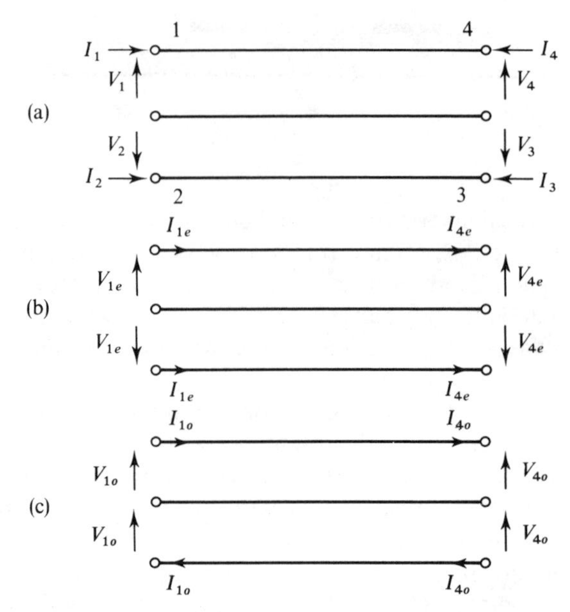

Figure A.3 Coupled lines regarded as a 4-port network:(a) arbitrary excitations; (b) even excitation; (c) odd excitation.

The variables in the even and odd modes of excitation are in each case related by the transmission matrix equation:

$$\begin{bmatrix} V_{1e} \\ I_{1e} \end{bmatrix} = \begin{bmatrix} \cos\theta_e & \mathrm{j}Z_{0e}\sin\theta_e \\ \mathrm{j}Y_{0e}\sin\theta_e & \cos\theta_e \end{bmatrix} \begin{bmatrix} V_{4e} \\ I_{4e} \end{bmatrix} \tag{A.13}$$

$$\begin{bmatrix} V_{1o} \\ I_{1o} \end{bmatrix} = \begin{bmatrix} \cos\theta_o & \mathrm{j}Z_{0o}\sin\theta_o \\ \mathrm{j}Y_{0o}\sin\theta_o & \cos\theta_o \end{bmatrix} \begin{bmatrix} V_{4o} \\ I_{4o} \end{bmatrix} \tag{A.14}$$

Using equations (A.12) to express V_{1e}, V_{2e}, etc. in terms of V_1, V_2, etc., and substituting in equations (A.13) and (A.14) we find

$$\left.\begin{aligned} V_1 + V_2 &= \cos\theta_e(V_3 + V_4) - \mathrm{j}Z_{0e}\sin\theta_e(I_3 + I_4) \\ I_1 + I_2 &= \mathrm{j}Y_{0e}\sin\theta_e(V_3 + V_4) - \cos\theta_e(I_3 + I_4) \\ V_1 - V_2 &= \cos\theta_o(V_4 - V_3) + \mathrm{j}Z_{0o}\sin\theta_o(I_3 - I_4) \\ I_1 - I_2 &= \mathrm{j}Y_{0o}\sin\theta_o(V_4 - V_3) + \cos\theta_o(I_3 - I_4) \end{aligned}\right\} \tag{A.15}$$

When in use as a coupler each port is fed from a line of impedance Z_0. Consider the case for which $Z_0^2 = Z_{0e}Z_{0o}$, when

$$Z_{0e} = \gamma Z_0 \quad Z_{0o} = \gamma^{-1}Z_0$$

Consider the coupler to be fed from port one, and terminated in Z_0 at the other ports, shown in Figure A.4. We may write the currents and voltages in

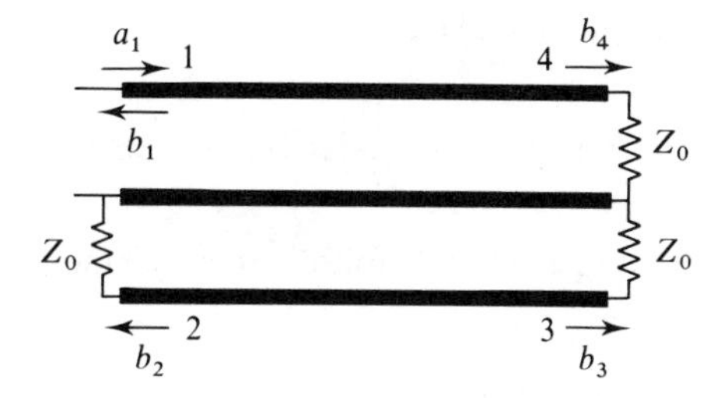

Figure A.4 Complex wave amplitudes at the various ports.

terms of the complex wave amplitudes as follows:

$$V_1 = (a_1 + b_1)Z_0^{\frac{1}{2}} \qquad I_1 = (a_1 - b_1)Z_0^{-\frac{1}{2}}$$
$$V_2 = b_2 Z_0^{\frac{1}{2}} \qquad I_2 = -b_2 Z_0^{-\frac{1}{2}}$$
$$V_3 = b_3 Z_0^{\frac{1}{2}} \qquad I_3 = -b_3 Z_0^{-\frac{1}{2}}$$
$$V_4 = b_4 Z_0^{\frac{1}{2}} \qquad I_4 = -b_4 Z_0^{-\frac{1}{2}}$$

Substituting in equations (A.15) we find

$$a_1 + b_1 + b_2 = (\cos\theta_e + j\gamma\sin\theta_e)(b_3 + b_4) \tag{A.16}$$
$$a_1 - b_1 - b_2 = (\cos\theta_e + j\gamma^{-1}\sin\theta_e)(b_3 + b_4) \tag{A.17}$$
$$a_1 + b_1 - b_2 = (\cos\theta_o + j\gamma^{-1}\sin\theta_o)(b_4 - b_3) \tag{A.18}$$
$$a_1 - b_1 + b_2 = (\cos\theta_o + j\gamma\sin\theta_o)(b_4 - b_3) \tag{A.19}$$

Adding equations (A.16) and (A.17) we find

$$b_3 + b_4 = 2a_1/D_e$$

where

$$D_e = 2\cos\theta_e + j(\gamma + \gamma^{-1})\sin\theta_e$$

Similarly equations (A.18) and (A.19) yield

$$b_4 - b_3 = \frac{2a_1}{D_o}$$

in which D_o is the same expression as D_e but with θ_o replacing θ_e. Thus

$$S_{31} = \left(\frac{1}{D_e} - \frac{1}{D_o}\right)$$

$$S_{41} = \left(\frac{1}{D_e} + \frac{1}{D_o}\right)$$

Subtracting equation (A.18) from equation (A.16)

$$S_{21} = \frac{1}{D_e}(\cos\theta_e + j\gamma\sin\theta_e) - \frac{1}{D_o}(\cos\theta_o + j\gamma^{-1}\sin\theta_o)$$

Finally from equations (A.16) and (A.19)

$$S_{11} = \frac{1}{D_e}(\cos\theta_e + j\gamma\sin\theta_e) - \frac{1}{D_o}(\cos\theta_o + j\gamma\sin\theta_o)$$

The remaining scattering parameters follow from symmetry.

A.8.1 Homogeneous medium

In this case $\theta_o = \theta_e$, $D_e = D_o$ and both S_{11} and S_{31} vanish: all ports are matched and some power is reversed in flow.

Explicitly

$$S_{21} = jD^{-1}(\gamma - \gamma^{-1})\sin\theta$$
$$S_{41} = 2D^{-1}$$

Clearly, increasing the proportion of power flowing to port two requires an increase of γ, and consequent close coupling, which may be difficult to achieve.

A.8.2 Mode interference coupler

If γ diverges by only a small amount from unity, in the simplest approximation $D_o \simeq \frac{1}{2}\exp(+j\theta_o)$ and $D_e \simeq \frac{1}{2}\exp(+j\theta_e)$. It can be seen that

$$\begin{aligned} S_{11} &\simeq 0 \\ S_{21} &\simeq 0 \\ S_{31} &\simeq \tfrac{1}{2}(\exp(-j\theta_e) - \exp(-j\theta_o)) \\ &= \exp(-\tfrac{1}{2}j(\theta_o + \theta_e))j\sin\tfrac{1}{2}(\theta_o - \theta_e) \\ S_{41} &\simeq \tfrac{1}{2}(\exp(-j\theta_o) + \exp(-j\theta_e)) \\ &= \exp(-j\tfrac{1}{2}(\theta_o + \theta_e))\cos\tfrac{1}{2}(\theta_o - \theta_e) \end{aligned}$$

The arrangement then provides a broad-band quadrature coupler, at the expense of a small mismatch at the input port.

A.9 Dependence of noise figure on source impedance

The internal noise sources in a linear 2-port device can be represented in the form of voltage and current generators at the input, followed by a noiseless 2-port. The various noise generators are shown in Figure A.5.

It is considered that only a narrow band of frequencies is transmitted, in which case the noise generators may be specified in terms of spectral densities. These spectral densities have to be regarded as statistical averages, related to the corresponding noise voltage.

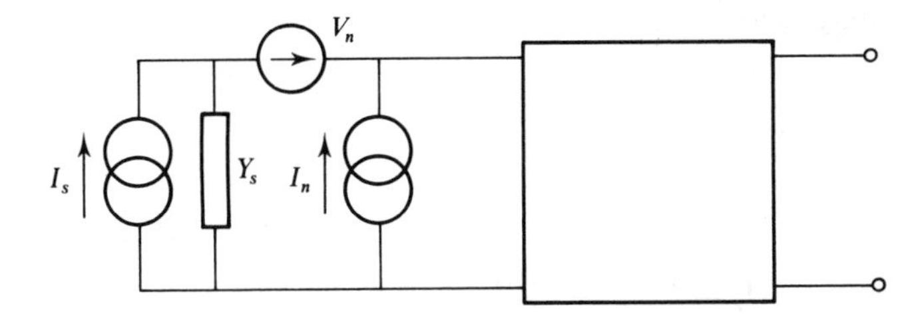

Figure A.5 The noise generators associated with a noisy 2-port network.

Two quantities are of particular interest: the time average of the square of one noise voltage or current, and the time average of the product of two different noise quantities. We express

$$\overline{v^2(t)} = 2\langle |V^2|\rangle \Delta f$$

in which Δf is the bandwidth and $\langle\,\rangle$ indicate the ensemble average. The spectral density is double sided in frequency. For the product

$$\overline{v(t)i(t)} = \langle (VI^* + V^*I)\rangle \Delta f$$

If v and i are uncorrelated $\langle VI^*\rangle \equiv 0$; if they are correlated we may write

$$\langle VI^*\rangle = Y_n^*\langle |V|^2\rangle$$

in which Y_n has dimensions of admittance.

Since the 2-port in Figure A.5 is noiseless, the composite noise source may be characterized in any convenient manner: consider the short-circuit current, given by

$$I_s + I_n + Y_s V_n$$

The mean square value is given by

$$\begin{aligned} 2\Delta f\langle (I_s + I_n + Y_s V_n)(I_s + I_n + Y_s V_n)^*\rangle \\ = 2\Delta f\langle (|I_s|^2 + |I_n|^2 + |Y_s|^2|V_n|^2 + I_s I_n^* + I_s^* I_n \\ + I_s Y_s^* V_n^* + I_s^* V_s V_n + Y_s^* V_n^* I_n + Y_s V_n I_n^*)\rangle \end{aligned}$$

Since the noise associated with the source may be assumed to be uncorrelated with the other noise sources $\langle V_n I_s^*\rangle = 0$ and $\langle I_s I_n^*\rangle = 0$. The equivalent sources V_n and I_n may be correlated. Let

$$I_n = I_n' + I_n''$$

where I_n' is completely uncorrelated with V_n.

We may then write

$$\langle V_n I_n^* \rangle = \langle V_n I_n''^* \rangle = Y_n^* \langle |V_n|^2 \rangle$$

Further,

$$\begin{aligned}\langle |I_n|^2 \rangle &= \langle (I_n' + I_n'')(I_n' + I_n'')^* \rangle \\ &= \langle |I_n'|^2 \rangle + \langle |I_n''|^2 \rangle \\ &= \langle |I_n'|^2 \rangle + |Y_n|^2 \langle |V_n|^2 \rangle\end{aligned}$$

Collecting terms it can be seen that for the mean square of the short-circuit noise current

$$2\Delta f \langle (|I_s|^2 + |I_n'|^2 + |V_n|^2 |Y_s + Y_n|^2) \rangle$$

In the absence of extra noise sources this expression reduces to $2\Delta f \langle |I_s|^2 \rangle$, so that the noise figure is given by

$$F = 1 + \frac{1}{\langle |I_s|^2 \rangle} (\langle |I_n'|^2 \rangle + \langle |V_n|^2 \rangle |Y_s + Y_n|^2)$$

Representing the noise sources by appropriate resistances

$$2\langle |I_s|^2 \rangle = 4kTG_s \qquad 2\langle |V_n|^2 \rangle = 4kTR_n$$

The expression for noise figure then takes the form given in Section 6.7.5:

$$F = F_0 + \frac{R_n}{G_s} |Y_s + Y_n|^2$$

in which F_0, R_n, Y_n are parameters of the 2-port, to be determined experimentally. (For further information, refer to Haus *et al.* (1960), ‘Representation of Noise in Linear Two-ports’ *Proc. I.R.E.* **48**, 69–73.)

A.10 Matched filters

It is convenient to describe a pulse in the form $As(t)$ in which the ‘shape’ function $s(t)$ is chosen so that

$$\int_{-\infty}^{\infty} (s(t))^2 \, dt = 1 \tag{A.20}$$

The pulse will be associated with a spectral density function given by

$$S(f) = \int_{-\infty}^{\infty} s(t)\exp(-j2\pi f t)\,dt \tag{A.21}$$

By Parseval's theorem

$$\int_{-\infty}^{\infty} |S(f)|^2\,df = \int_{-\infty}^{\infty} (s(t))^2\,dt = 1$$

Using the inverse Fourier Transform relation, we have

$$s(t) = \int_{-\infty}^{\infty} S(f)\exp(j2\pi f t)\,df \tag{A.22}$$

The matched filter is that whose transfer function $H(f)$ is related to the pulse spectral density by

$$H(f) = S^*(f) \tag{A.23}$$

It can be shown that this filter gives optimum performance in that the output rises to a maximum at $t = 0$, and gives the best signal-to-noise ratio in the presence of white noise.

The impulse response of the filter will be

$$h(t) = \int_{-\infty}^{\infty} H(f)\exp(j2\pi f t)\,df$$

Comparison with equation (A.22) shows that since $S^*(f) = S(-f)$,

$$h(t) = s(-t) \tag{A.24}$$

From this result it can be seen that a filter with the transfer function given by equation (A.23) is physically unrealizable. To derive a practically possible approximation a delay is allowed, so that

$$H(f) = S^*(f)\exp(-j2\pi f T_d)$$

The impulse response then becomes

$$h(t) = s(T_d - t)$$

The delay is chosen so that $s(T_d)$ is small. The response of the filter to the signal pulse can be written

$$\begin{aligned} v(t) &= A\int_{-\infty}^{\infty} H(f)S(f)\exp(j2\pi f t)\,df \\ &= A\int_{-\infty}^{\infty} |S(f)|^2\exp(j2\pi f(t - T_d))\,df \end{aligned} \tag{A.25}$$

or alternatively

$$v(t) = A \int_{-\infty}^{\infty} s(\tau)h(t - \tau)\,d\tau$$

$$= A \int_{-\infty}^{\infty} s(\tau)s(T_d - t + \tau)\,d\tau \qquad (A.26)$$

The maximum occurs at $t = T_d$, with

$$v(T_d) = A \int_{-\infty}^{\infty} (s(\tau))^2\,d\tau = A$$

If white noise of one-sided spectral density $\overline{V_n^2}$ is present at the filter input, the mean square output noise voltage will be given by

$$\overline{v_n^2} = \overline{V_n^2} \int_0^{\infty} |H(f)|^2\,df$$

$$= \tfrac{1}{2}\overline{V_n^2} \int_{-\infty}^{\infty} |S(t)|^2\,df$$

$$= \tfrac{1}{2}\overline{V_n^2}$$

Thus, at $t = T_d$ the signal-to-noise ratio is given by

$$\frac{(v(T_d))^2}{\overline{v_n^2}} = \frac{A^2}{\tfrac{1}{2}\overline{V_n^2}}$$

$$= \frac{2 \times \textit{signal energy}}{\textit{noise energy in unit bandwidth}}$$

A.10.1 Pulse compression

The theory of the previous section holds for any signal waveform including the linear frequency modulation on pulse described in Section 8.6 and given by

$$s(t) = \cos(\omega_0 t + \tfrac{1}{2}\mu t^2) \qquad 0 < t < T$$

in which

$$\omega_0 = 2\pi f_0 \qquad \mu = \frac{2\pi B}{T}$$

Ignoring the delay required for practical realization, the impulse response of the matched filter is proportional to $s(-t)$, or

$$h(t) = \cos(\omega_0 t - \tfrac{1}{2}\mu t^2) \qquad -T < t < 0 \tag{A.27}$$

The transfer characteristic of this filter may be calculated from

$$\begin{aligned} H(f) &= \int_{-T}^{0} \cos(\omega_0 t - \tfrac{1}{2}\mu t^2)\exp(-\mathrm{j}\omega t)\,\mathrm{d}t \\ &= \tfrac{1}{2}\int_{-T}^{0} [\exp \mathrm{j}((\omega_0 - \omega)t - \tfrac{1}{2}\mu t^2) \\ &\qquad + \exp(-\mathrm{j}((\omega_0 + \omega)t - \tfrac{1}{2}\mu t^2)]\,\mathrm{d}t \end{aligned}$$

The response to frequencies in the neighbourhood of ω_0 is determined primarily by the first term. Retaining this, and substituting $\omega - \omega_0 = 2\pi\nu B$, we find

$$H(f) = \tfrac{1}{2}\exp(\mathrm{j}\pi BT\nu^2)\int_{-T}^{0} \exp\left(-\mathrm{j}\left(\frac{\pi B}{T}\right)(t + \nu T)^2\right)\mathrm{d}t$$

Change of variable to u defined by

$$\tfrac{1}{2}\pi u^2 = \left(\frac{\pi B}{T}\right)(t + \nu T)^2$$

reduces the integral to the form

$$H(f) = \tfrac{1}{2}\exp(\mathrm{j}\pi BT\nu^2)\left(\frac{T}{2B}\right)^{\frac{1}{2}}\int_{u_1}^{u_2} \exp(-\tfrac{1}{2}\mathrm{j}\pi u^2)\,\mathrm{d}u$$

where

$$u_2 = (1 - \nu)(2BT)^{\frac{1}{2}} \qquad \text{and} \qquad u_1 = -\nu(2BT)^{\frac{1}{2}}$$

In terms of the Fresnel integrals $C(u)$ and $S(u)$ introduced and referenced in Section 9.4.3 we have

$$H(f) = \frac{1}{2}\left(\frac{T}{2B}\right)^{\frac{1}{2}} \exp(\mathrm{j}\pi BT\nu^2)[C(u_2) - C(u_1) + \mathrm{j}(S(u_2) - S(u_1))]$$

The nature of the term in square brackets can be seen from the geometrical interpretation given in Figure A.6 which shows the Cornu spiral, a plot of $C(u)$ against $S(u)$. The line U_1U_2 joining points on the spiral for which $u = u_1, u_2$ respectively has a length equal to the modulus of the term in square brackets, and an angle equal to its argument. If $0 < \nu < 1$ and $BT \gg 1$, then

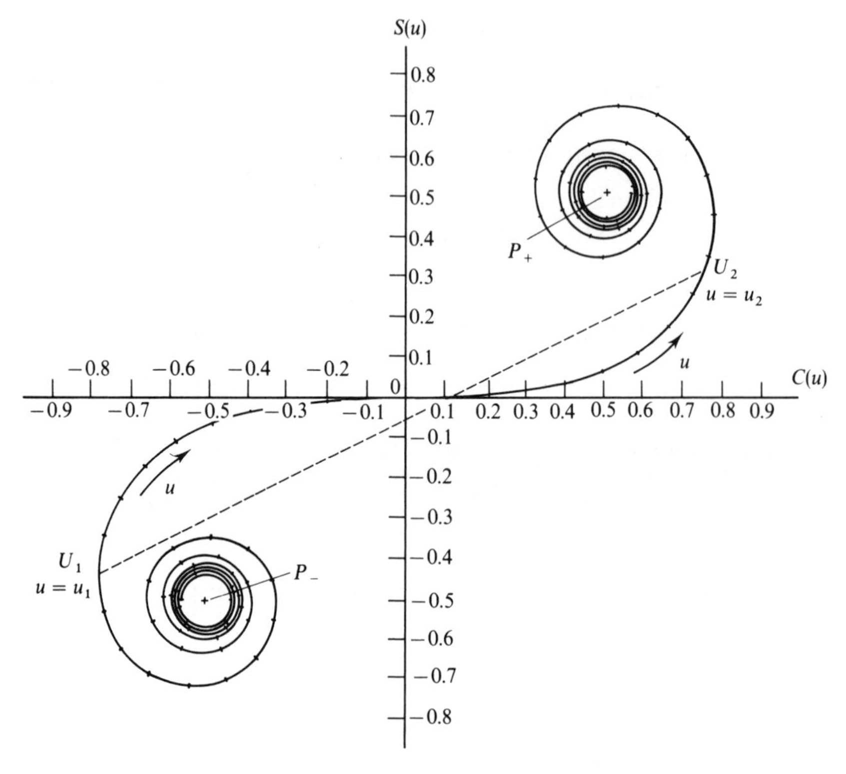

Figure A.6 The Cornu spiral obtained by plotting $S(u)$ against $C(u)$. In the lower left-hand quadrant $u < 0$; in the upper right-hand quadrant $u > 0$. Markers are at intervals of 0.1 in u, with the last points shown $u = \pm 5$. The points U_1 and U_2 correspond to some value of v between 0 and 1, but for clarity the line U_1U_2 represents an unrealistically small value of $(2BT)^{\frac{1}{2}}$.

U_1 will lie in the region of the negative focus P_-, and U_2 will lie near the positive focus P_+, so that the modulus will be in the region of $2^{\frac{1}{2}}$. For $v < 0$ U_1 and U_2 will both be near P_+, for $v > 1$ both will lie near P_-. Thus $|H(f)| \simeq \frac{1}{2}(T/B)^{\frac{1}{2}}$ over the range $f_0 - \frac{1}{2}B < f < f_0 + \frac{1}{2}B$, and falls to zero outside. Depending on the value of BT, exact computation will show oscillations about the constant value, becoming less as BT increases. It becomes convenient to multiply equation (A.27) by $(2B/T)^{\frac{1}{2}}$ in order to have a filter of approximately unity gain. For the case $A = 1$, and ignoring the delay T_d, equation (A.26) becomes

$$v(t) = \left(\frac{2B}{T}\right)^{\frac{1}{2}} \int_{t_1}^{t_2} \cos(\omega_0\tau + \tfrac{1}{2}\mu\tau^2)\cos[\omega_0(t-\tau) - \tfrac{1}{2}\mu(t-\tau)^2]\,d\tau$$

$$= \left(\frac{B}{2T}\right)^{\frac{1}{2}} \int_{t_1}^{t_2} [\cos(\omega_0 t + \tfrac{1}{2}\mu(2t\tau - t^2)) + \cos(\omega_0(2\tau - t)$$

$$+ \tfrac{1}{2}\mu(2\tau^2 - 2t\tau + t^2))]\,d\tau \qquad \text{(A.28)}$$

The limits are determined by the overlap of the functions $s(t)$ and $h(t-\tau)$. These are non-zero in the ranges

$$\begin{array}{ll} s(\tau) & 0 < \tau < T \\ h(t-\tau) & -T < t-\tau < 0 \quad \text{or} \quad t < \tau < T+t \end{array}$$

Thus the limits of the integral are

$$\begin{array}{ll} 0 < t < T & t_1 = t \quad t_2 = T \\ -T < t < 0 & t_1 = 0 \quad t_2 = T - |t| \end{array}$$

For t outside these limits the integrand vanishes. Investigation shows that the lowest frequency components derive from the first term, which may be integrated to yield

$$\left(\frac{B}{2T}\right)^{\frac{1}{2}}\left(\frac{1}{\mu T}\sin\left(\omega_0 t + \tfrac{1}{2}\mu(2t\tau - t^2)\right)\right)_{t_1}^{t_2}$$

Inserting the limits we find for $0 < |t| < T$

$$v(t) = \left(\frac{2B}{T}\right)^{\frac{1}{2}} \frac{1}{\mu|t|}\sin\left(\tfrac{1}{2}\mu|t|(T-|t|)\right)\cos\left(\omega_0 + \tfrac{1}{2}\mu T\right)$$

For $|t| \ll T$, the envelope is of the form

$$\left(\frac{2B}{T}\right)^{\frac{1}{2}} \frac{1}{\mu|t|}\sin\left(\tfrac{1}{2}\mu T|t|\right) = \left(\frac{BT}{2}\right)^{\frac{1}{2}} \frac{\sin \pi B|t|}{\pi B|t|}$$

This is the form stated in Section 8.5 and shown in Figure 8.23.

A.11 A model for fading

Consider the following variation of effective refractive index with height, such as might occur with temperature inversion.

$$\begin{array}{ll} 0 < z < H & n^2 = n_0^2(1 - 2\kappa_1 z) \\ z > H & n^2 = n_0'^2[1 - 2\kappa_2(z - H)] \end{array}$$

in which

$$n_0'^2 = n_0^2(1 - 2\kappa_1 H)$$

It was left as an exercise in question 9.5 to show that the equation of a ray launched at the origin with inclination ε is

$$z = -\tfrac{1}{2}\kappa_1 x^2 + \varepsilon x$$

Such a ray attains its maximum height, $z_m = \varepsilon^2/2\kappa_1$, and returns to ground at $x = 2\varepsilon/\kappa_1$. There is therefore a ray that just reaches the discontinuity in refractive index at $z = H$. The inclination at launch, ε_0, and the point of return to ground, L_0, are given by the expressions

$$\varepsilon_0 = (2\kappa_1 H)^{\frac{1}{2}} \qquad L_0 = 2\left(\frac{2H}{\kappa_1}\right)^{\frac{1}{2}} \tag{A.29}$$

A ray launched at $\varepsilon < \varepsilon_0$ will reach the ground at $x = L < L_0$, given by

$$L = \frac{2\varepsilon}{\kappa_1} = L_0\left(\frac{\varepsilon}{\varepsilon_0}\right) \tag{A.30}$$

Thus a direct ray exists between two points separated by $L < L_0$, not rising above the discontinuity. Consider now rays launched with inclination greater than ε_0. The situation is shown in Figure A.7. Then

$$H = -\tfrac{1}{2}\kappa_1 x_1^2 + \varepsilon x_1$$

whence

$$x_1 = \left(\frac{\varepsilon}{\kappa_1}\right)\left[1 - \left(1 - \frac{2H\kappa_1}{\varepsilon^2}\right)^{\frac{1}{2}}\right] \tag{A.31}$$

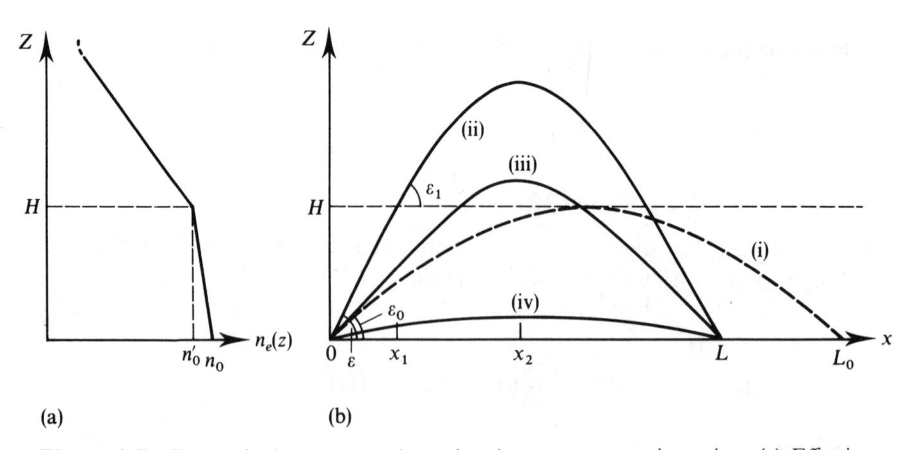

Figure A.7 Ray paths in an atmosphere showing temperature inversion. (a) Effective refractive index. (b) Rays: (i) covering maximum distance (L_0) entirely within lower layer; (ii) and (iii) rays entering the upper layer and covering the same distance $L < L_0$; (iv) direct ray covering distance L.

The inclination at x_1 is given by

$$\varepsilon_1 \simeq \left(\frac{dz}{dx}\right)_1 = \varepsilon - \kappa_1 x_1 = \varepsilon\left(1 - \frac{2H\kappa_1}{\varepsilon^2}\right)^{\frac{1}{2}}$$

We find

$$\varepsilon_1^2 = \varepsilon^2 - \varepsilon_0^2 \tag{A.32}$$

The ray above the discontinuity is launched with inclination ε_1 and reaches its maximum height after a distance ε_1/κ_2. Thus the highest point on the trajectory occurs for $x = x_2$, given by

$$x_2 = x_1 + \frac{\varepsilon_1}{\kappa_2}$$

$$= \left(\frac{\varepsilon}{\kappa_1}\right)\left[1 - \gamma\left(1 - \frac{2H\kappa_1}{\varepsilon^2}\right)^{\frac{1}{2}}\right] \tag{A.33}$$

where $\gamma = 1 - \kappa_1/\kappa_2$. This distance to the point at which the ray returns to

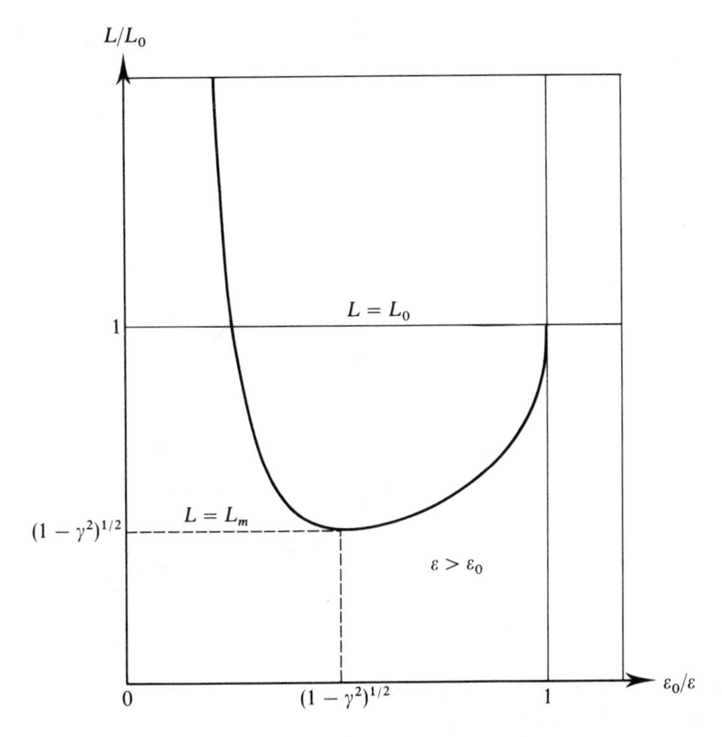

Figure A.8 The variation of L/L_0 with $\varepsilon_0/\varepsilon$, as shown by equation (A.34).

ground, L, is equal to $2x_2$, which may be written in the form

$$L = L_0\left(\frac{\varepsilon}{\varepsilon_0}\right)\left[1 - \gamma\left(1 - \frac{\varepsilon_0^2}{\varepsilon^2}\right)^{\frac{1}{2}}\right] \tag{A.34}$$

This expression is illustrated graphically in Figure A.8. It will be seen that for $L_0 > L > L_m = L_0(1 - \gamma^2)^{\frac{1}{2}}$ there are two possible values of ε, and thus two rays exist. In addition there will be a direct ray, as shown by equation (A.30), making a total of three rays. It is the relative phase delays between these rays that can cause fading.

To calculate the delay time the straight ray model shown in Figure A.9,

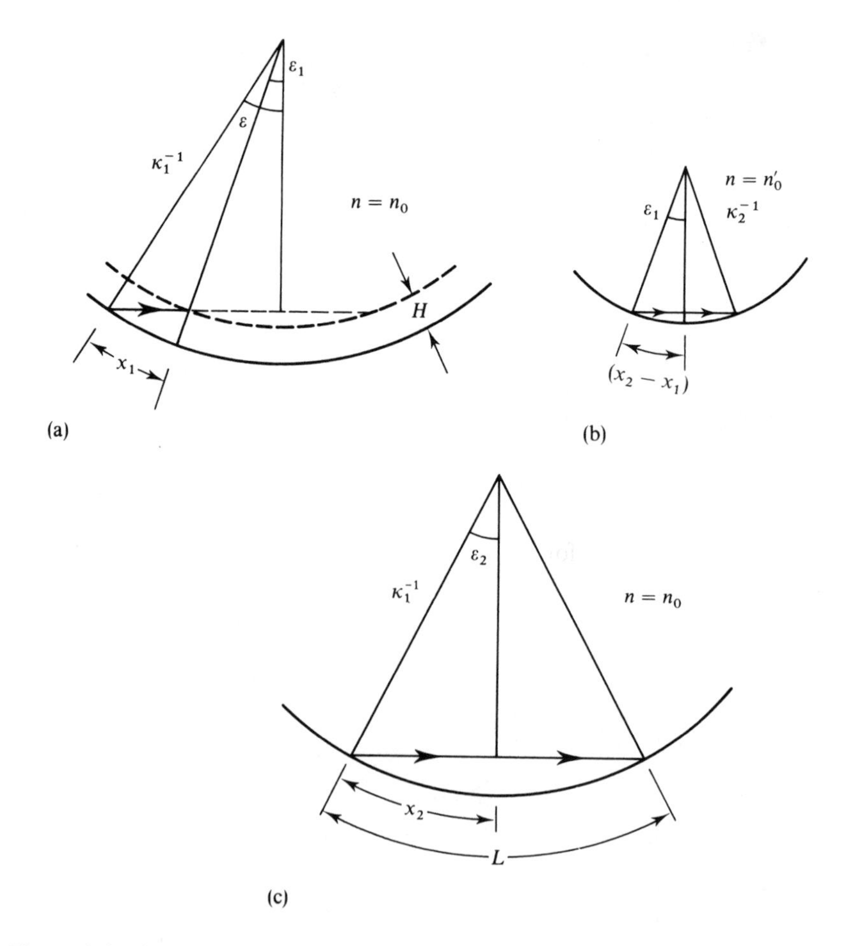

Figure A.9 Straight ray models for delay calculation: (a) region $z < H$, refractive index n_0; (b) region $z > H$, refractive index n_0'; (c) model for direct ray, refractive index n_0.

(a) and (b) is used for the two segments of the ray path. For the total delay

$$\tau = \left(\frac{2n_0}{c}\right)\left(\frac{1}{\kappa_1}\sin\varepsilon - \left(\frac{1}{\kappa_1} - H\right)\sin\varepsilon_1\right) + \left(\frac{2n_0'}{\kappa_2 c}\right)\sin\varepsilon_1$$
$$= \left(\frac{2n_0}{c\kappa_1}\right)[\sin\varepsilon - \gamma(1 - \kappa_1 H)\sin\varepsilon_1]$$

We will compare this expression with that for the delay on the direct ray. This ray is characterized by an angle of launch ε_2 (Figure A.9(c)), where $\varepsilon_2 = \kappa_1 x_2$. Thus using equation (A.33) we find

$$\varepsilon_2 = \varepsilon\left(1 - \gamma\left(1 - \frac{\varepsilon_0^2}{\varepsilon^2}\right)^{\frac{1}{2}}\right) = \varepsilon - \gamma\varepsilon_1 \tag{A.35}$$

The corresponding delay is $(2n_0/c\kappa_1)\sin\varepsilon_2$. The excess delay is therefore given by

$$\Delta\tau = \left(\frac{2n_0}{c\kappa_1}\right)[\sin\varepsilon - \gamma(1 - \kappa_1 H)\sin\varepsilon_1 - \sin\varepsilon_2]$$

Expressing $\kappa_1 H = \varepsilon_0^2/2$ and keeping only terms up to the third order,

$$\Delta\tau = \left(\frac{2n_0}{c\kappa_1}\right)\left(\varepsilon - \frac{\varepsilon^3}{6} - \gamma\left(1 - \frac{\varepsilon_0^2}{2}\right)\left(\varepsilon_1 - \frac{\varepsilon_1^3}{6}\right) - \varepsilon_2 + \frac{\varepsilon_2^3}{6}\right)$$
$$= \left(\frac{2n_0}{c\kappa_1}\right)\left(\varepsilon - \frac{\varepsilon^3}{6} - \gamma\left(\varepsilon_1 - \frac{\varepsilon_1^3}{6} - \frac{\varepsilon_0^2\varepsilon_1}{2}\right) - \varepsilon_2 + \frac{\varepsilon_2^3}{6}\right)$$

When the expression for ε_2 in equation (A.35) is substituted, the first order terms cancel, leaving

$$\Delta\tau = \left(\frac{n_0}{3c\kappa_1}\right)(-\varepsilon^3 + \gamma(\varepsilon_1^3 + 3\varepsilon_0^2\varepsilon_1) + (\varepsilon - \gamma\varepsilon_1)^3)$$
$$= \left(\frac{\gamma n_0}{3c\kappa_1}\right)(\varepsilon_1^3 + 3\varepsilon_0^2\varepsilon_1 - 3\gamma\varepsilon_1\varepsilon^2 + 3\gamma\varepsilon_1^2\varepsilon - \gamma^2\varepsilon_1^3)$$

Writing $\varepsilon^2 = \varepsilon_1^2 + \varepsilon_0^2$ gives

$$\Delta\tau = \left(\frac{\gamma n_0}{3c\kappa_1}\right)\varepsilon_1^2[3\gamma\varepsilon - (2 + \gamma^2)\varepsilon_1] \tag{A.36}$$

The significance of these results is best seen by considering a numerical example. Take

$$H = 27\,\text{m}$$
$$\kappa_1 = 10^{-7}\,\text{m}^{-1}$$
$$\kappa_2 = 10^{-6}\,\text{m}^{-1}$$

for which

$$\varepsilon_0 = 2.32 \times 10^{-3}\,\text{rad}$$
$$L_0 = 46.5\,\text{km}$$
$$\gamma = 0.9$$
$$L_m = 20.3\,\text{km}$$

Consider an intermediate distance $L = 37\,\text{km}$, for which three rays will exist. The direct ray will be found to have an inclination of 1.85×10^{-3} rad. Assuming $n_0 = 1$, the following values may be obtained using equations (A.32), (A.34) and (A.36):

$\varepsilon_0/\varepsilon$	ε (rad)	ε_1(rad)	$\Delta\tau$ (s)
0.9665	2.40×10^{-3}	0.60×10^{-3}	1.79×10^{-11}
0.1362	1.71×10^{-2}	1.69×10^{-2}	-4.12×10^{-9}

The ray corresponding to the lower set of figures penetrates far into the upper region: calculation gives a maximum height of some 150 m. It is clear that at microwave frequencies phase shifts of many periods can occur, leading to reduction of signal strength through cancellation between the several rays. (Further investigation into this problem can be found in Ruthroff, C. L. (1971), *BSTJ*, **70**, 2375.)

SOLUTIONS TO EXERCISES

Chapter 2

2.1 (a) $\eta = 377\,\Omega$, $b/a = 3$; $Z_0 = 66\,\Omega$.
(b) Use the formula for α_c in Table 2.1: $R_s = 8\sqrt{2} = 11.3\,\text{m}\Omega$; attenuation 0.16 dB.
(c) $5.2 \times 10^6\,\text{V m}^{-1}$.
(d) 375 W.
The maximum electric field is very closely $1.2 \times 5.2 = 6.2\,\text{MV m}^{-1}$.

2.2 (a) $a = 2b = 3.75\,\text{cm}$.
(b) Diameter 4.4 cm.

2.3 Use Table 2.5. $\lambda_c/\lambda = 1.5$; $\lambda/\lambda_g = 0.745$; $P = 3.1\,\text{MW}$.

2.4 From Table 2.6, TE_{11} and TM_{01}. Using Table 2.7, $\lambda/\lambda_g = 0.68$, $Z_g = 554\,\Omega$; at $\rho = 0$, $|E| = A/2$; $P = 2\,\text{MW}$, $|E| = 0.96\,\text{MV m}^{-1}$.

2.5 Use equation (2.31); $16\,\text{dB km}^{-1}$.

2.6 Lowest symmetrical mode TM_{01}: $f_c = 2.5/1.5\,\text{GHz}$, $a = 6.9\,\text{cm}$.
At 2.5 GHz, TE_{11}, TM_{01}, TE_{11}.

2.7 $w/h = 100$, $Z_a = 276\,\Omega$; $\varepsilon_{rf} = 5.54$; $Z_0 = 117\,\Omega$, $\lambda/\lambda_f = 2.4$.

2.8 4.5 mm.

2.9 $y = 0.2 + \text{j}0.377$, $\Gamma = 0.703\exp(-42.6°)$, $S = 5.7$. At 0.128λ back from load $y = 1 + \text{j}1.98$. OC stub 0.324λ.

2.10 Taking ports one and two on opposite sides of the common reference plane and matching port two with Z_0, $V_1 = (a_1 + b_1)Z_0^{\frac{1}{2}}$, $V_2 = b_2 Z_0^{\frac{1}{2}}$, $b_1/a_1 = S_{11}$ is the reflection coefficient at port one, $S_{21} = (1 + S_{11})(V_2/V_1)$.
(a) $S_{11} = S_{22} = z/(z + 2)$, $S_{12} = S_{21} = 2/(z + 2)$.
(b) $S_{11} = S_{22} = -y/(y + 2)$, $S_{12} = S_{21} = 2/(y + 2)$.
(c) $S_{11} = -S_{22} = (n^2 - 1)/(n^2 + 1)$, $S_{12} = S_{21} = 2n/(n^2 + 1)$.

2.11 Use 2.10(b): neglecting b^2, $S_{11} \simeq b/2$, VSWR $\simeq 1 + |b|$.
Admittance presented to connector 1 by terminated line + connector $2 \simeq 1 + \text{j}b_2\exp(2\text{j}\beta l)$, i.e. at connector 1, effective shunt is $\text{j}[b_1 + b_2\exp(2\text{j}\beta l)]$, $|S_{11}| \simeq \frac{1}{2}|b_1 + b_2\exp(2\text{j}\beta l)|$ which has extremes $\frac{1}{2}(1 \pm |b_1 - b_2|)$. Result follows.

2.12 Consider the length of nZ_0, with V_1, I_1 at port one, V_2, I_2 at port two, both I_1 and I_2 inward. It is a simple exercise to show that

$$\begin{bmatrix} V_1 \\ I_1 \end{bmatrix} = \begin{bmatrix} \cos\theta & jnZ_0\sin\theta \\ j(Y_0/n)\sin\theta & \cos\theta \end{bmatrix} \begin{bmatrix} V_2 \\ -I_2 \end{bmatrix}$$

In the given configuration

$$V_1 = (a_1 + b_1)Z_0^{\frac{1}{2}} \qquad V_2 = (a_2 + b_2)nZ_0^{\frac{1}{2}}$$
$$I_1 = (a_1 - b_1)Z_0^{-\frac{1}{2}} \qquad I_2 = (a_2 - b_2)Z_0^{-\frac{1}{2}}/n.$$

Expansion of the matrix equation followed by substitution gives two relations between a_1, b_1, a_2 and b_2. As usual, put $a_2 = 0$, $b_1 = S_{11}a_1$, $b_2 = S_{21}a_1$ and find S_{11}, S_{21}. Repeat with $a_1 = 0$. $S_{11} = -S_{22} = A(n - 1/n)\cos\theta$; $S_{12} = S_{21} = 2A$ where $A = ((n + 1/n)\cos\theta + 2j\sin\theta)^{-1}$.

2.13 $P = P_a(1 - |\Gamma|^2) = P_a 4S/(1 + S)^2$. For $P/P_a = 0.9$, $S \simeq 1.9$.

Chapter 3

3.1 $p = \frac{1}{2}EH = \frac{1}{2}E^2/377$; $0.87\,\text{V m}^{-1}$. $P = 4\pi R^2 p$; 314 kW.

3.2 $A = \lambda^2 G/4\pi$; $0.91\,\text{m}^2$.

3.3 EIRP 15.8 kW. 500 m.

3.4 Power 7 dBW, free space loss 137 dB. Received −70 dBW, 0.1 μW. Termination matching, orientation and polarization match.

3.5 $A = 0.063\,\text{m}^2$, $G = 48.5$ dB; 0.63 mW.

3.6 (a) $F_1(\alpha) = E_0\pi a\cos(ka\sin\alpha)/((\pi/2)^2 - (ka\sin\alpha)^2)$ and similarly for $F_2(\beta)$. For pattern in plane $\beta = 0$, beam width between first zeros occurs for $ka\sin\alpha = 3\pi/2$, $\alpha \simeq 3\lambda/4a$; to 3 dB $\alpha \simeq 0.3\lambda/a$; first side lobe is 23 dB below maximum. Compare with results of Section 3.8.2.

(b) $F_1(\alpha) = E_0 w\pi^{\frac{1}{2}}\exp(-(\frac{1}{2}kw\sin\alpha)^2)$.

3.7 Let r_n denote the distance between $(na, 0, 0)$ and the distant observation point. We have $r_n \simeq r_0 - na\sin\theta$ for points in the plane x–z. The signal strength is given by

$$E \propto \sum_{n=0}^{N-1} (I_n/r_n)\exp(-jkr_n)$$

$$\simeq (I_0/r_0)\exp(-jkr_0)\sum_{n=0}^{N-1}\exp(jn(\psi + ka\sin\theta))$$

$$= (I_0/r_0)\exp(-jkr_0)\frac{1 - \exp(jN(\psi + ka\sin\theta))}{1 - \exp(\psi + ka\sin\theta)}$$

Hence

$$|E| \propto \frac{1}{N}\left|\frac{\sin \frac{1}{2}N(\psi + ka \sin \theta)}{\sin \frac{1}{2}(\psi + ka \sin \theta)}\right|$$

Maximum occurs for $\theta = \theta_0$, $\sin \theta_0 = -\psi/\pi$. Zeros at $\theta = \theta_0 + \varepsilon$, $\psi + \pi \sin \theta = 2\pi/N$. Assume $\varepsilon \ll 1$ and result follows.

3.8 Free space loss = 66 dB. Hence gain of horn = 16.5 dB. For a paraboloid $G = 27.5$ dB. $2D^2/\lambda = 67\,\text{m} < 75\,\text{m}$. $4D \simeq 4$ m.

Chapter 4

4.1 $\omega_p^2 = e|\rho_0|/m\varepsilon_0$; $\rho_0 = J_0/u_0 = I_0/Au_0$; $u_0 = (2eV/m)^{\frac{1}{2}}$: $f_p = 586$ MHz. $u_0/f_p = 3.2$ cm. Show $\lambda_p^2 = 8\pi^2\varepsilon_0(e/m)^{\frac{1}{2}}V^{\frac{3}{2}}/J$.

4.2 Extracting the AC term from equation (4.4) when $X \ll 1$, we find $I_1/I_0 \simeq [\omega s V_1/(2u_0 V_0)] \exp[j\omega(t - s/u_0)]$. This agrees with the coefficient in J_1/J_0 in equation (4.13).

4.3 $f_p = 434$ MHz, $\lambda_p = 12.2$ cm. $s = 3.0, 9.1$ cm.

4.4 For growing wave $Re(\gamma_3) = -\beta_0 3^{\frac{1}{2}}\mathscr{C}/2 = -3^{\frac{1}{2}}\pi\mathscr{C}/\lambda_p$. This gives gain for unit length; for N wavelengths, gain is $8.686 \times 3^{\frac{1}{2}}\pi\mathscr{C}N = 47.3\mathscr{C}N$ dB.

4.5 $N = 18$, $\Omega = \omega/9 = 0.908 \times 10^9\,\text{rad}\,\text{s}^{-1}$; $e/m = 1.76 \times 10^{11}$; cut-off 87.4 kV; threshold 31.0 kV.

4.6 From equation (4.9) $\rho_1 = \varepsilon\partial E/\partial z$, $J_1 = u_0\rho_1 + \rho_0 E u_0'$, $\rho_0 = en_0$. Hence result follows. For trial function to be a solution $\gamma(\gamma - \rho_0 u_0'/u_0\varepsilon - j\omega/u_0) = 0$, i.e. $\gamma = 0$ or $\gamma = \rho_0 u_0'/u_0\varepsilon + j\omega/u_0$.

4.7 We find $\bar{\alpha}(E_p) \simeq K(E_p/E_0)^5$ whence $l_a K(E_{p0}/E_0)^5 = 1$ defines E_{p0}; $\bar{\alpha}_0' = 5KE_{p0}^4/E_0^5 = 5/(l_a E_{p0})$; $\tau_a = l_a/v_s = 10^{-11}$ s; $E_{p0} = 3.96 \times 10^7\,\text{V}\,\text{m}^{-1}$, $L = 0.8$ nH, $P = 2$ W ($2 \times 10^8\,\text{W}\,\text{cm}^{-3}$), $C_a = 1.1$ pF, $f = 5.5$ GHz.

Chapter 5

5.1 Expressions for the input impedance can be obtained by repeated use of equation (2.19). Near the design frequency $\tan \beta l \simeq \pi/2 - \varepsilon$ for $\varepsilon \ll 1$. Using this and retaining lowest powers in ε it will be found that for (a) $|\Gamma| \simeq (N^2 - 1)|\varepsilon|/2N$, (b) $|\Gamma| \simeq (N^4 - 1)\varepsilon^2/2N^2$. Thus, as ε alters, the VSWR in (a) shows a cusp at $\varepsilon = 0$, (b) a parabolic maximum. By pursuing the algebra, exact expressions can be obtained.
(a) $|\Gamma|^2 = (N^2 - 1)^2/((N^2 - 1)^2 + 4N^2 \sec^2 \beta l)$.
(b) $|\Gamma|^2 = (N^4 - 1)^2/((N^4 - 1)^2 + 4N^4 \sec^4 \beta l)$.
For VSWR, $|\Gamma| = 1/11 \ll 1$.
(a) $N = 2$, $\cos \beta l \simeq 4K/3$, $1.45 < \beta l < 1.87$.
(b) $N = 2^{\frac{1}{2}}$, $\cos^2 \beta l = 4K/3$, $1.21 < \beta l < 1.93$.

5.2 $Q_u = \omega_0 L/R$, $Q_l = \omega_0 L/(R + Z_0/n^2)$, $Q_e = n^2\omega_0 L/Z_0$.
Note $Q_u/Q_l = 1 + \beta$.

5.3 To input line $\Gamma = (1 - \beta + jx)/(1 + \beta + jx)$, $x = 2Q_u(\Delta\omega/\omega_0)$.
Use Smith Chart to follow change of Γ for $-\infty < x < \infty$.
When $\Delta\omega/\omega_0 = \pm 1/(2Q_l)$, $x = \pm Q_u/Q_l = \pm(1 + \beta)$.

5.4 $\beta > 1$, $\therefore \beta = 1.54$. $S = 6.25$ leads to $|\Gamma|^2 = 0.5244$, and hence occurs for $x = 2Q_u(\Delta\omega/\omega_0) = 2.55$. For this $2\Delta\omega/\omega_0 = 9370/1.2$ whence $Q_u = 1.99 \times 10^4$. $Q_l = Q_u/(1 + \beta) = 7840$.
Note by using $S_0 = 1.54$, the value of S_1 in question 5.3 corresponds to the value 6.25, which leads directly to Q_l.

5.5 The most straightforward approach is to use the circuit transmission matrix, as in Chapter 2, question 2.12. For the circuit of Figure 5.14(e), putting $j\omega L + 1/j\omega C + R = Z$.

$$T_s = \begin{bmatrix} n' & 0 \\ 0 & 1/n' \end{bmatrix}\begin{bmatrix} 1 & Z \\ 0 & 1 \end{bmatrix}\begin{bmatrix} -1/n'' & 0 \\ 0 & -n'' \end{bmatrix} = \begin{bmatrix} -n'/n'' & -n'n''Z \\ 0 & -n''/n' \end{bmatrix}$$

For a $\lambda/4$ length of Z_0-line

$$T = \begin{bmatrix} 0 & jZ_0 \\ jY_0 & 0 \end{bmatrix}$$

Hence new reference planes $\lambda/4$ away correspond to

$$TT_sT = \begin{bmatrix} +n''/n' & 0 \\ Y_0^2 Zn'n'' & n'/n'' \end{bmatrix}$$

Comparison with the transmission matrix for the circuit of Figure 5.14(f) shows agreement with transformer ratios $n':1$ and $1:n''$ and with a shunt admittance $Y_0^2 Z$. The parallel resonant circuit has components $R' = Z_0^2/R$, $L' = Z_0^2 C$, $C' = L/Z_0^2$.

5.6 Transform all components to centre section.

5.7 (a) Terminate two ports in Z_0, then show third port matched. Current into one port divides equally, so $S_{ii} = 0$, $S_{ij} = \frac{1}{2}$, $i \neq j$.
(b)

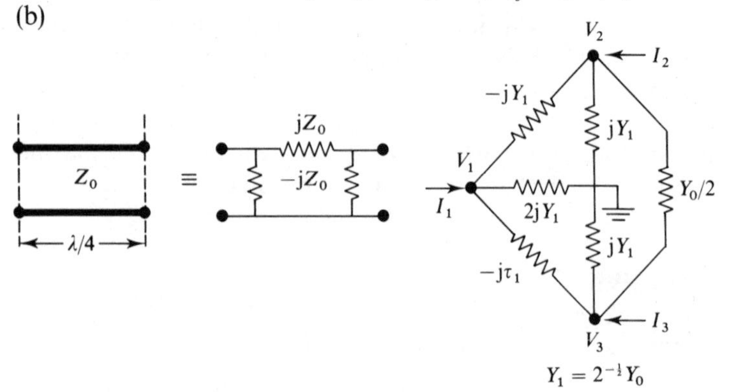

Nodal admittance matrix

$$\begin{bmatrix} I_1 \\ I_2 \\ I_3 \end{bmatrix} = \begin{bmatrix} 0 & jY_1 & jY_1 \\ jY_1 & \frac{1}{2}Y_0 & -\frac{1}{2}Y_0 \\ jY_1 & -\frac{1}{2}Y_0 & \frac{1}{2}Y_0 \end{bmatrix} \begin{bmatrix} V_1 \\ V_2 \\ V_3 \end{bmatrix}$$

By the usual technique of solving with all ports except one matched, the scattering parameters can be found. The result follows. Note that port three is isolated from an input to port two. The same technique can be used off design frequency by use of the equivalent circuit of Figure 5.14(c).

5.9 Scattering matrix must be of form

$$\begin{bmatrix} 0 & \alpha & \beta \\ \alpha & 0 & \gamma \\ \beta & \gamma & 0 \end{bmatrix}$$

Forming $\tilde{S}S^*$, we find that in order to make the off-diagonal terms vanish, two of α, β and γ must vanish. It is then impossible to make the diagonal terms all unity.

5.10 For Figure 5.22(a) we can draw at the design frequency the equivalent circuit.

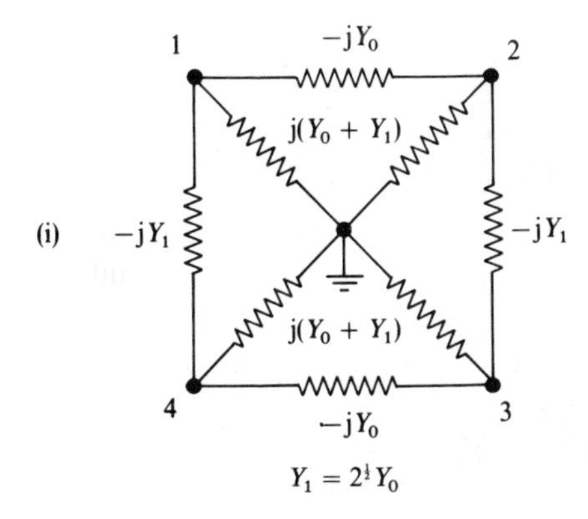

Nodal admittance matrix is

$$\begin{bmatrix} 0 & jY_0 & 0 & jY_1 \\ jY_0 & 0 & jY_1 & 0 \\ 0 & jY_1 & 0 & jY_0 \\ jY_1 & 0 & jY_0 & 0 \end{bmatrix}$$

Proceed as before: terminate ports two, three and four in Z_0 and excite at port one; repeat for excitation at port two.

For 5.22(b), using Figure 5.14(c) for $3\lambda/4$ length between ports one and four, we find

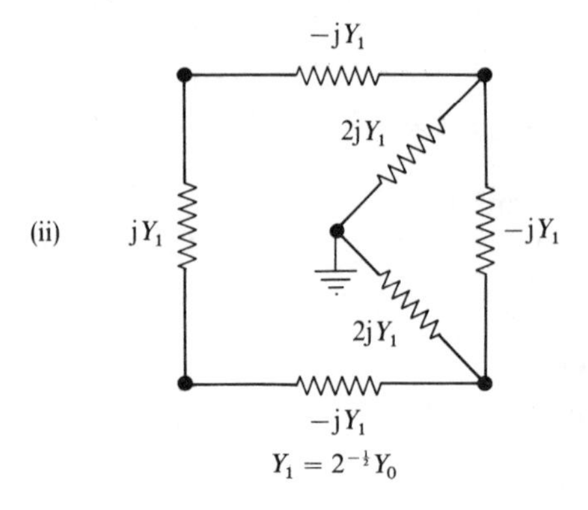

The result is then obtained as before.

5.11 Because of TE_{10} field pattern an input to port one will not excite port four and will split equally in phase to ports two and three.

Presuming reference planes are symmetrically disposed this requires $S_{41} = 0$, $S_{31} = S_{32}$. For the same reason, an input to port four will not couple to port one and will split equally but in antiphase to ports two and three. Hence $S_{14} = 0$, required by reciprocity in any case, and $S_{24} = -S_{23}$. Symmetrical matching components in arms one and four do not upset this argument, so we can make $S_{11} = S_{44} = 0$. Further $S_{22} = S_{33}$. The scattering matrix then reduces to

$$S = \begin{bmatrix} 0 & \alpha & \alpha & 0 \\ \alpha & S_{22} & S_{23} & \beta \\ \alpha & S_{23} & S_{22} & \beta \\ 0 & \beta & -\beta & 0 \end{bmatrix}$$

Application of the unitary conditions then shows $S_{22} = S_{23} = 0$ and $|\alpha|^2 = |\beta|^2 = \frac{1}{2}$.

5.12 Use Table 2.7, $\alpha = j\beta$, $\lambda > \lambda_{11}$. (i) 2.06; (ii) 1.97.

5.13 Obtain expressions for the field components when a forward wave and a backward wave are superimposed, using formulae of Table 2.4. Assuming the ends of the cavity are $z = 0, d$, we must have $E_y = 0$ for $z = 0, z = d$. The quoted expressions result. Since the expressions are in peak phasor form, the average electric and magnetic energy densities are $\varepsilon|E_y|^2/4$, $\mu|H \cdot H^*|/4$. Integration of these expressions shows that $\bar{W}_e = \bar{W}_m = \varepsilon_0 E_0^2 V/16$. The total stored energy is the sum of these, hence result. Average energy density at specimen is $\varepsilon E_0^2/4$. Therefore $\Delta W_e = vE_0^2(\varepsilon_1 - \varepsilon_0)/4$. Using equation (5.28) result follows. In this case $\lambda_g = \lambda_{10} = 2a$, leading to $f = 8.4856$ Hz: $\delta f = 147$ MHz.

Chapter 6

6.1 Denoting the upper amplifier by scattering matrix S' and S'' for lower, and input a_1, we find reflection to input port $\frac{1}{2}(S'_{11} - S''_{11})\ a_1$, to dummy load at input coupler $\frac{1}{2}\mathrm{j}(S'_{11} - S''_{11})a_1$; to output $\frac{1}{2}\mathrm{j}(S'_{21} + S''_{21})a_1$; to dummy load $\frac{1}{2}(S'_{21} - S''_{21})a_1$.

6.2 (a) 9.2 MHz
(b) 8.7 MHz.

6.3 99.6 dB; 6.3 dB; 947 K.

6.4 1.66 dB, 0.47.

6.5 3.2 dB, 1.9 dB.

6.6 97 K, −110 dBm.

6.7 Plotting graphs in the form of Figure 6.15, equation for fundamental output is $P = p + 40$; for intermodulation output $P' = 3p + 60$. Intercept Q is for $P_0 = P'_0$ whence $Q = (-10, 30)$. Gain = 40 dB. For BW = 100 MHz, F = 3 dB, noise power = −91 dBm. This will give an output of −51 dBm. A spurious signal of the same output requires input 37 dBm. Hence dynamic range = 91 − 37 = 54 dB.

6.8 In terms of the coefficients used in Table 6.1 and assuming measurements are at the same impedance we have

$$P_1 = p_1 + 20\log(A_1);\ P_2 = p_2 + 20\log(A_1)$$
$$P_{12} = 2p_1 + p_2 + 20\log(3A_3/4) \quad 2\omega_1 - \omega_2$$
$$P_{21} = p_1 + 2p_2 + 20\log(3A_3/4) \quad 2\omega_2 - \omega_1$$

Using the definition of the intercept point we find

$$p_i = 10\log(4A_1/3A_3) \quad \text{and} \quad P_i = 10\log(4A_1^3/3A_3)$$

It can then be shown $20\log(3A_3/4) = P_i - 3p_i$, and result follows. −80 dBm at 115 MHz, −85 dBm at 130 MHz, plus fundamentals.

6.9 $V = A_1 a_1 - (3|A_3|/4)a_1^3$. 1-dB compression point determined by $A_1 a_1/(A_1 a_1 - 3a_1^3|A_3|/4) = 1$ dB, giving $a_1^2 = 0.21\ (4A_1/3A_3)$. Hence $p_c = p_i - 6.8$ dB.

6.10 Using notation of Figure 6.16, we find

$$b_2 = S_{21}a_1/(1 - \Gamma_l S_{22}) \quad P_0 = \tfrac{1}{2}(|b_2|^2 - |a_2|^2)$$

Result follows.

6.11 Using results of Section 2.11.3 we have

$$a_1 = b_s/(1 - \Gamma_s S_{11})$$

Input power

$$P_{in} = \tfrac{1}{2}|b_s|^2 \frac{1 - |S_{11}|^2}{|1 - \Gamma_s S_{11}|^2}$$

Available power

$$P_a = \tfrac{1}{2}|b_s|^2/(1 - |\Gamma_s|)^2$$

Using the result of question 6.10, expressions follow.

6.12 Regarding Γ_s and Γ_l as variable successive maximization leads to the result. Note $G = G_T$ in this situation.

6.13 Gain 22 dB. Matching can be handled as finding L_1 and L_2 such that termination in S_{11} gives 50 Ω input: $L_1 = 16$ nH, $L_2 = 27$ nH, $L_3 = 9.3$ nH, $C = 0.4$ pF. This output circuit is not very satisfactory and might be better done with transmission line.

6.14 Write $\Gamma = u + jv$, $\Gamma_0 = u_0 + jv_0$ and proceed as in Section 6.12 for Smith Chart.

6.15 $(1 - |\Gamma|^2)/|1 - S_{11}\Gamma|^2 = 3.02$, 4.8 dB. $(1 - |S_{11}|^2)^{-1} = 7.2$ dB. Loss in gain 2.4 dB. 6.4 nH in series with source, then 0.2 pf in shunt.

6.16 $\gamma = |1 + \Gamma_0|^2 |y - y_0|^2 / 2(y + y^*)$. Put $y = (1 - \Gamma)/(1 + \Gamma)$, show $\gamma = |\Gamma_0 - \Gamma|^2(1 - |\Gamma|^2)$, then substitute $\Gamma = u + jv$ and result follows.

6.17

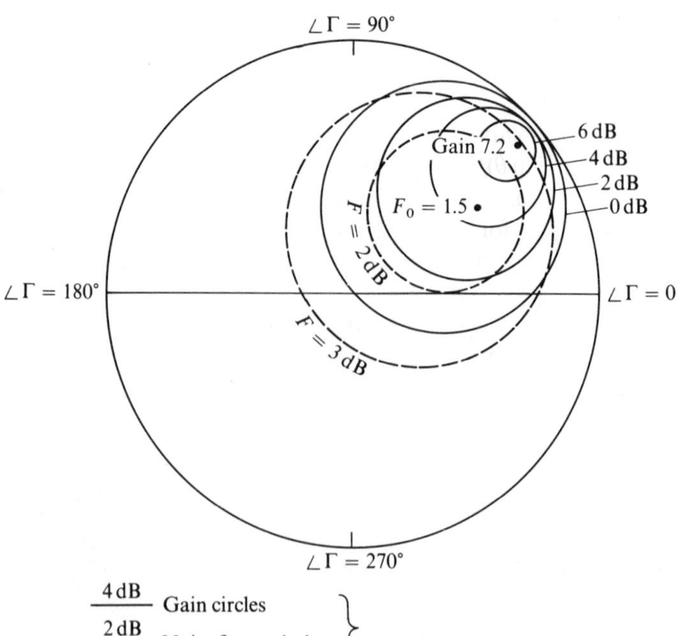

Gain and noise figure circles at 2 GHz

Chapter 7

7.2 With notation given

(a) $A = c_1, a = c_0, B = c_0, b = c_1^2/c_0$.

(b) $A = c_1(c_0 - c_2)/c_0, a = (c_0^2 - c_2^2)/c_0, B = (c_0^2 - c_1^2)/c_0,$
$b = c_1^2(c_0 - c_2)^2/(c_0(c_0^2 - c_1^2))$.

7.3 $c_n = (\alpha E/\pi)\left(\dfrac{\sin(n-1)\pi/2}{n-1} + \dfrac{\sin(n+1)\pi/2}{n+1}\right)$.

$c_0 = 2\alpha E/\pi; \quad c_1 = \alpha E/2; \quad c_2 = 2\alpha E/3\pi.$

$1/G_s = 615\,\Omega, L_m = 5.5\,\text{dB}$.

(a) $1/G_s = 362\,\Omega, L_m = 6.3\,\text{dB}$.

(b) $1/G_s = 517\,\Omega.\quad L_m = 4.1\,\text{dB}$. 0.73 mW.

7.6 An output a_1 to port one produce at port three an output Γa_1, Γ = reflection coefficient of load at port two. Result follows.

7.7 Noise temperatures for combinations: 1140 K;

(a) 416 K.

(b) 676 K.

Required signal levels:

(a) −89 dB m;

(b) −93 dB m;

(c) −91 dB m.

Chapter 8

8.1 215 km, max. unambiguous range 300 km. 400 W.

8.2 $h \simeq (R^2 + 2aR\sin\theta)/2a$, where R = range, a = Earth radius and θ elevation. 13.6 km.

8.3 From Figure 8.2, for the sphere $\sigma \simeq 6.7\,\text{cm}^2$. In this situation σ/r^4 is constant, so for the thrush $\sigma = 400\,\text{cm}^2, r = 140$ miles, 220 km.

8.4 For one drop $\sigma \simeq 3.6 \times 10^{-10}\,\text{m}^2$; Volume of cell $\simeq 5.8 \times 10^6\,\text{m}^3$; 10 drops per cubic metre, σ for cell $= 2.1 \times 10^{-2}\,\text{m}^2$. $P_r = 1.5 \times 10^{-12}\,\text{W}$.

8.5 If $G \propto \operatorname{cosec}^2\theta$, radar return $\propto G^2/r^4 \propto G_0/(r\sin\theta)^4 \propto G_0/h^4$. Echo depends primarily on altitude and not range.

8.6 (a) $|\sin\pi f\tau|/\pi f$

(b) $|\sin\pi f\tau/(2\pi f(1 - f^2\tau^2))|$

(c) $\tau/(1 + (2\pi f\tau)^2)$.

8.7 The amplitude of the incident signal is proportional to $F^{\frac{1}{2}}$, and the antenna gain in reception is also proportional to $F^{\frac{1}{2}}$. Hence signal amplitude $S(t)F(\phi)$, and $\phi = \Omega t, \Omega = 2\pi/T$.

$$F(\Omega t) = \sum_{-\infty}^{\infty} c_n \exp(jn\Omega t) \quad c_n = \frac{1}{2\pi}\int_{-\pi}^{\pi} F(\phi)\exp(-jn\phi)\,d\phi.$$

For the example

$$c_n = \frac{1}{2\pi}\int_{-3\phi_0}^{3\phi_0} \exp((-\phi^2/\phi_0^2) - jn\phi)\,d\phi$$

$$\simeq \frac{1}{2\pi}\int_{-\infty}^{\infty} \exp[(-\phi^2/\phi_0^2) - jn\phi]\,d\phi.$$

The integral was given in question 3.6, hence

$$c_n = (\phi_0/\pi 2^{\frac{1}{2}})\exp(-(n\phi_0/2)^2)$$

We find $\phi_0 = 0.0126$ rad, $n = 240$, $n/T \simeq 16$ Hz.
Low speed targets give this order of Doppler shift.
At 3 cm, 16 Hz corresponds to about 1 km h^{-1}.

8.8 (a) 1.2 kHz.
(b) 14.8 kHz.
If ϕ is the angle between zenith and satellite radius vector, r slant range, $\dot{r} = (aR/r)\sin\phi\dot{\phi} = a\cos\theta\dot{\phi}$, θ = elevation of satellite from observation point. $\dot{\phi} = 2\pi/T$, therefore maximum $\theta = 0$. $f_d = 2\dot{r}/\lambda = (4\pi a/\lambda T)\cos\theta$. Near zenith $\phi \simeq \pi/2 - \theta$, $\dot{r} = a\phi\dot{\phi}$, $\ddot{r} = a\dot{\phi}^2$, $\dot{f}_d = 2\ddot{r}/\lambda$. 44.5 kHz.

8.9 $P_a = B^{-1}/T_a = 1.7 \times 10^{-9}$. From Figure 5.28 $(S/N) \simeq 13$ dB.
Noise $= NkT_0B = 1.2 \times 10^{-14}$ W, signal $= 2.4 \times 10^{-13}$ W.

8.10 Range discrimination $c\tau/2$, $\tau = 0.1\,\mu$s. Bandwidth $1.2/\tau = 12$ MHz. Antenna: in horizontal plane for uniform illumination radiation pattern of form $\sin X/X$, $X = (2\pi a/\lambda)\sin\theta$. At 3 dB $\theta = 0.5°$, $a/\lambda \simeq 25$. To improve side lobes, say $a/\lambda = 40$. For a wavelength of 3 cm, $2a \simeq 2.4$ m. An edge array plus flare would give this, width in vertical direction ~10 cm. A crude estimate of gain: area $= 0.24\,\text{m}^2$, $A \simeq 0.5 \times 0.24 = 0.12\,\text{m}^2$, $G = 4\pi A/\lambda^2 = 32$ dB. $kTB = -133$ dBW, assuming $F = 2$, noise $= -131$ dBW. At 20 km we find, for $S = N$, $P_t = 40$ dBW, 10 kW. This will be too low, but perhaps 25 kW as a beginning estimate. Sea clutter will play a large part in determining target visibility. The repetition frequency is open to choice, say 1 kHz.

Chapter 9

9.1 Effective antenna area $= 4.6\,\text{m}^2$ ∴ at 2 GHz $G = 34$ dB; Antenna noise $= kTB = 3.2 \times 10^{-13} \equiv -125$ dBW; $C_m = -125 + 5 + 23 = -97$ dBW; hence $P_t = 3$ dBW; $L_s = 3 + 97 - 35 + 68 = 133$; $r = 53$ km.

9.2 Fade margin 35 dB, $L = 0.018$. Equation (9.17) gives $T/T_0 = 5.8 \times 10^{-5}$, or 155 s in one month. For 3 months and 10 links, outage = 78 m, or 0.015%.

9.3 2, 4.5, 10 dB. Change of altitude only few tens of metres.

9.4 Let point of least clearance be s_1 from antenna at height h_1, and s_2 from antenna of height h_2. Using equation (9.10) $s_1^2 - s_2^2 = 2a_e(h_1 - h_2)$, $s_1 + s_2 = 40$ km, hence $s_1 = 28.6$ km, $s_2 = 11.4$ km, $z_m = 35$ m, $\psi_1 = s_1/a_e = 0.38°$, $\psi_2 = 0.15°$.

9.5 Use equation (9.10) in the form $z = z_m + (x - x_m)^2/2a_e$. From equation (9.8) $a_e = -1/\kappa$, put in conditions at $x = 0$, $z = 0$. Whence $z_m = \varepsilon^2/2\kappa$, $x_m = \varepsilon/\kappa$, $\varepsilon = \frac{1}{2}\kappa L$, $z_m = \kappa L^2/8$. We have $n\,\mathrm{d}s = n_0(1 - \kappa z)(1 + (\mathrm{d}z/\mathrm{d}x)^2)^{\frac{1}{2}}\,\mathrm{d}x$: both κz and $(\mathrm{d}z/\mathrm{d}x)^2$ are of the order of ε^2, so $n\,\mathrm{d}s \simeq n_0(1 - \kappa z + \frac{1}{2}(\mathrm{d}z/\mathrm{d}x)^2) = n_0(1 + \frac{1}{2}\varepsilon^2 - 2\varepsilon\kappa x + x^2\kappa^2)$. Integrating between $x = 0$ and $x = L$ gives the result.

9.6 Use equation (9.11). Extra path length is $2h_1h_2/d = 2\lambda + (2/15)\lambda$, giving an excess phase of 48°. $|H|$ then gives $+1.7$ dB. In general $1 + R > |H| > 1 - R$, giving $+2.3$ dB to -3.1 dB.

9.7 By calculation using equation (9.8) and a value of u in Section 9.4.3 of 0.8: (a) height of straight line path above a smooth earth, (b) clearance, (c) width of Fresnel surface $u = 0.8$.

	(a)	(b)	(c)
A	245		
C	164	44	24
D	151	61	26
E	179	39	20
B	225		

Hence clearances are satisfactory. Points for discussion: antenna gain, power budget, fading margin.

Chapter 10

10.1 $\lambda = 0.04$, $r = 3.6 \times 10^7$, $L_D = 201$ dB. Received power $39 - 201 + G_R$; noise $10\log(kTB) = -153 + 10\log T$.

Hence

$$39 - 201 + G_R = -153 + 10\log T + 15.$$
$$10\log(G/T) = -153 + 15 - 39 + 201 = 24\ \mathrm{dB\,K^{-1}}(\times 250)$$

Assume a dish diameter d. Approximately

$$G = \frac{4\pi}{\lambda^2}\frac{\pi d^2}{4} \times 0.65 = 4 \times 10^3 d^2 = 250T$$

$d^2 = T/16$; 1 dB noise figure $\equiv 75$ K, allow 50 K for antenna, $d \simeq 2.8$ m. A minimum dish diameter of 3 m is called for.

10.2 $G = G_m \exp(-\alpha\theta^2)$, assuming α large. 20 W $\equiv$ 13 dBW, EIRP 39 dBW. Therefore $G_m = 26$ dB, $\times 400$. Use $\int G\,d\Omega = 4\pi$.

$$\int G\,d\Omega = \int_0^{\pi} G_m \exp(-\alpha\theta^2)\, 2\pi \sin\theta\, d\theta$$

$$\simeq \int_0^{\infty} 2\pi G_m \theta \exp(-\alpha\theta^2)\, d\theta$$

$$= \pi G_m/\alpha = 4\pi,\ G_m = 4\alpha.$$

Therefore $\alpha = 10^2$ and 3 dB for $\theta = 0.083$ rad, 4.8°. Circle approximately 3000 km radius.

10.3 Relevant points:
(a) from beam width estimate gain of satellite antenna;
(b) using given power flux density the satellite transmitter power can then be estimated;
(c) estimate gain of receiving antenna and check power budget;
(d) estimate power flux density at satellite on up-link, and make estimates of signal-to-noise.

INDEX